The Ecology of Tropical
Lakes and Rivers

The Ecology of Tropical Lakes and Rivers

A. I. Payne

Coventry (Lanchester) Polytechnic

JOHN WILEY & SONS

Chichester · New York · Toronto · Brisbane · Singapore

Library of Congress Cataloging in Publication Data:
Payne, A. I.
   The ecology of tropical lakes and rivers.
   Includes index.
   1. Lake ecology—Tropics.   2. Stream ecology—
Tropics.   I. Title.
QH84.5.P38 1986   574.5′2632   85–26368

ISBN 0 471 90524 0

British Library Cataloguing in Publication Data:
Payne, A. I.
   The ecology of tropical lakes and rivers.
   1. Fresh-water ecology—Tropics
   I. Title
   574.5′2632′0913   QH541.5.T7

ISBN 0 471 90524 0

Typeset by Bath Typesetting
Printed and bound in Great Britain

Contents

viii

CHAPTER 1

INTRODUCTION

The idea for this book came when I was teaching courses in aquatic biology to students at a West African university. For such courses, the equipment required can be relatively simple and the implications of the subject are often relevant to the needs of the country. However, one vital ingredient was lacking—a suitable book which dealt with tropical aspects of hydrobiology. There have been many books on temperate freshwater biology at all levels, including a wide selection for the undergraduate, but scarcely any on aquatic biology within the tropics and certainly none which could be used by the undergraduate or the professional person in a related discipline.

The intention has been to produce a self-contained textbook for students from tropical countries, based mainly on material from the tropics and emphasizing those features which make tropical aquatic ecosystems different from temperate ones. The Tropics of Cancer and Capricorn at 23.5° N. and S. have provided approximate geographical limits, although information from subtropical areas has been included where particularly appropriate.

The region covered is consequently very extensive and includes four continents. This, combined with the rather spasmodic development of the subject so far, has meant that sources of information are very scattered and progress in many aspects has lacked continuity. Although the initial intention, therefore, was to provide a basic text, it has also become something of a synthesis, albeit an incomplete one, from very disparate material. Outlines of some basic techniques and biology are also included, which are of more general applicability.

One of the attractions of aquatic ecology in the tropics is the variety and extremes of habitats which are found there. These can range from almost permanently frozen pools high in the Andes to warm saline lakes of the African Rift Valley, and from the inland seas of the East African Great Lakes to the demineralized streams flowing under the deeply shaded canopy of the equatorial rain forests. To document the communities which inhabit these varied environments, to explain how they function and to identify any common principles which govern them is the challenge.

Local people who live or work near water have often developed a considerable understanding of the animals and plants which inhabit lakes and rivers in

1

the vicinity. Unfortunately, however, their information is not widely available or generally, systematic. The method of science is to bring order to observation, and the first people to try to make some order out of tropical aquatic ecosystems were the taxonomists.

During the nineteenth century, explorers and adventurers sent back preserved specimens of aquatic animals and plants to the museums and herbaria of Europe and America. Amongst these collectors was the redoubtable Mary Kingsley, who travelled through West Africa making observations on a wide range of topics, including the animals. One of the fishes she collected, the small climbing perch, *Ctenopoma kingsleyi*, was subsequently named after her. Perhaps one of the most bizarre of these collectors was Emin Pasha who, during his prolonged and arduous retreat from southern Sudan down the Nile and around Lake Victoria following the Mahdist Rebellion, assiduously collected animals, including the swamp worm, *Alma emini*, which now bears his name.

From material gathered in such a haphazard fashion, the taxonomists systematically attempted to sort out the plants and animals into species. Until the early part of this century, taxonomy was one of the major areas of biology, and eventually sufficient information was accumulated to make possible descriptive works such as, *A Catalogue of African Fishes* by Boulanger (1915) which characterized in detail every species of African fish which the British Museum of Natural History had in its collection. This work was so extensive and thorough that it is still a valuable guide to identification today although many more species have naturally been discovered since it was published.

With regard to the position of taxonomy in biology, the pendulum has swung in entirely the opposite direction since the beginning of the century and now it is often considered something of a backwater. However, the species composition of many tropical communities, both aquatic and otherwise, is often poorly documented, very diverse, and may well include previously undescribed species. Giving organisms the correct name, therefore, is particularly important.

Following this early descriptive phase where the animals and plants were being collected, a more sustained scientific interest in tropical waters was initiated by field expeditions working intensively on specific localities for short periods of time. Amongst the most valuable of these was the German Sunda expedition which examined a series of lakes in Java, Sumatra, and Bali in 1928 (Ruttner, 1931, 1952). The Cambridge University expedition to the lakes of East Africa in 1930–31, which was part of a series between 1925 and 1933 (Worthington and Worthington, 1933), laid the foundation for much of the valuable work which has subsequently been carried out there. A most readable autobiographical account of the early days of limnological investigation in East Africa has been produced by Worthington (1984) in *The Ecological Century* which also puts the work into a social and political context. Some analysis of tropical situations was also included in the major work by Hutchinson (1957a, 1967) *A Treatise on Limnology*.

These expeditions produced an illuminating but inevitably limited view of hydrobiology in the tropics, but the interest that they generated led to the creation of a few, more permanent institutions following the Second World War. The Max Planck Institute has for many years made Manaus on the Amazon a centre of research into all aspects of the river and its ecology. In Central America, the Smithsonian Tropical Research Institute, Panama, was established to look at all aspects of tropical biology and it enabled extensive work to be carried out on Gatun Lake and in the small streams and rivers of the nearby rain forests. Another productive research centre, the East African Freshwater Fisheries Research Organization was set up at Jinja on Lake Victoria in the 1950s to provide a base where some important early work on both fundamental and applied aspects of these lake ecosystems was carried out.

A considerable spur to progress of many ecological sciences was provided by the International Biological Programme (IBP) during the 1960s. This brought together scientists from many countries to investigate the functioning of a wide variety of ecosystems. Lake George in Uganda was chosen as a representative tropical lake and a team worked here between 1966 and 1973 (Greenwood and Lund, 1973; Greenwood, 1976). This lake is one of the most thoroughly investigated of all tropical lakes although, since the IBP, teams have also worked on Lake Turkana and Lake Chad (Lévêque and Carmouze, 1983) to provide similar information.

A tropical river was also investigated during the IBP. This was the Sungai Gombak stream in Malaysia but, unlike Lake George, it was not investigated by a team but only a single worker (Bishop, 1973). Nevertheless, our knowledge of this one stream is now so detailed that, like Lake George, it has become the standard against which all similar systems in the tropics must be compared.

The IBP was also responsible for the production of an invaluable series of practical manuals on techniques and methods for various aspects of freshwater investigations including chemical analysis (Golterman et al., 1978), primary production (Vollenweider, 1969), secondary production (Edmondson and Winberg, 1971) and fish production (Bagenal, 1978).

The advent of independence and the pressures of an expanding population mean that most nations in the tropics are aware of the need to make the maximum use of their natural resources. This requires planning, which in turn demands the best information available. Our knowledge of tropical waters and their communities is still relatively slight, and understanding of the way in which they work is only just beginning to emerge. The demand for such commodities as water and fish is great and there is the danger that the means of supply will be irreparably damaged before we accumulate sufficient experience to manage them properly. It is important, therefore, that fundamental research, as well as development, must continue. There cannot be applied science without a basis for the application.

The number of institutions primarily concerned with the rivers and lakes in the tropics is very small and major research teams are infrequent. Individuals or small groups, therefore, still have a great contribution to make as indeed they have in the past.

CHAPTER 2

THE RIVER ENVIRONMENT

2.1 CHARACTERISTICS OF THE WATER

2.1.1 The Concentration of Surface Waters

The total concentration of ions in a river results largely from the interaction of rain-water with the land in its catchment area. Some ions are derived directly from the rain-water itself whilst some are leached from the soil and the underlying parent rock, although the vegetation of the river basin might further modify their availability. Since the water is flowing, there is much less time available for accumulation of dissolved substances than there is in lakes or in the sea. Rivers generally, therefore, have a lower concentration of dissolved salts than lakes in the same vicinity and certainly lower than the sea where evaporation and crystallization have caused salts to build up to a considerable concentration.

With regard to salt concentrations, the sea provides a remarkably constant environment with a total dissolved salt content frequently close to $35 \, \text{g} \, \text{l}^{-1}$. Fresh waters, as well as being much more dilute than the sea, show considerable variability in both concentration and composition from place to place. This is largely a result of being separated into isolated basins and catchment areas, the major chemical features of which are determined by the nature of the underlying rocks, the soil, and the vegetation. The chemically heterogeneous environments presented by inland waters, together with the generally dilute nature of the media, are characteristics which living organisms had to come to terms with as they colonized fresh waters. Moreover, the variability in content and composition has had considerable influence on the production and distribution of plants and animals in these environments.

The total salt concentration in fresh waters can commonly vary between 100 and 1000 times less than that of sea-water although inland waters can be found which are almost completely deionized or which are several times more concentrated than sea-water. Although the detailed chemical composition is not necessarily related to the total ionic concentration, nevertheless the mineral content of a water often does have certain chemical implications. Waters of low

4

ionic content, often regarded as 'soft' waters, have a poor buffering capacity and a neutral to acidic pH reaction, whilst those of high ionic concentration, referred to as 'hard' waters under some circumstances, frequently have a good buffering capacity and a neutral to alkaline pH. Since the total concentration is very variable and also tells us something about the nature of the water itself, a measurement of ionic concentration is often a useful preliminary indication of water type.

2.1.2 The Measurement of Ionic Concentration

The most direct means of measuring dissolved compounds in the water is to evaporate a filtered sample of known volume to dryness and then weigh the residue. This can then be expressed as $mg\,l^{-1}$ total dissolved solids (TDS). A rather less direct but more convenient method is to measure the electrical conductance (EC) or conductivity (K) of the water. The ability of water to conduct electricity is proportional to its ionic concentration. Pure, deionized water is a very bad conductor of electricity but this improves with a progressive increase in ionic concentration. There are a variety of conductivity meters on the market which can measure this property. The older unit of conductivity is the mho which is ohm, the unit of resistance, written backwards since conductivity is the reciprocal of resistance, but the internationally recognized unit is now the siemen, or most commonly the microsiemen (μS) owing to the very dilute concentrations of most fresh waters. Fortunately, these two units are numerically the same so no conversion factor is required. All measurements of conductivity require compensation for temperature since temperature also influences the ability of water to carry an electrical current. Many conductivity meters have an automatic compensating device and it is usual to transform the reading from ambient to a standard temperature which is generally either 20 °C (K_{20}) or 25 °C (K_{25}).

It is possible to derive a factor to convert conductivity units into total dissolved solids, but the magnitude of this factor depends upon the ionic composition of the water. For example, a pure solution of sodium chloride requires a conversion factor of 0.45 times the conductivity whilst a water strongly dominated by calcium carbonate requires a factor of 0.66. Ideally, it is necessary to determine the factor for each type of water by making sample measurements of both conductivity and TDS.

2.1.3 Bicarbonate and Carbonate

The composition of sea-water is remarkably constant, particularly in oceanic regions, and is dominated by sodium and chloride ions unlike fresh waters which, as world averages for rivers show, are frequently dominated by calcium and bicarbonate or carbonate (Table 2.1). Bicarbonates and carbonates have a particularly significant role to play in aquatic systems since they represent the

Table 2.1 Mean chemical composition of river waters from two temperate and two tropical continental areas (adapted from Livingstone 1963 and Goltermann 1975; African averages from Visser 1974 and original data)

	HCO_3^-	SO_4^-	Cl^-	SiO_2-Si	NO_3-N	Ca^{++}	Mg^{++}	Na^+	K^+
				$mg\,l^{-1}$					
North America	67.7	40.3	8.1	4.2	0.23	42.0	10.2	9.0	1.6
South America	31.1	9.6	4.9	5.6	0.16	14.4	3.6	3.9	0.0
Europe	95.2	48.0	6.7	3.5	0.84	62.4	11.4	5.3	1.6
Africa	68.9	9.3	20.2	22.2	0.17	7.9	7.8	21.5	—

carbon reserves for photosynthesis, mainly due to the equilibria that exist between these ions and free carbon dioxide in the water:
photosynthesis:

$$\uparrow CO_2 + H_2O \rightleftharpoons H_2CO_3 \rightleftharpoons H^+ + HCO_3^- \rightleftharpoons 2H^+ + CO_3^{2-}$$
$$> pH\ 8$$

Utilization of carbon dioxide for photosynthesis causes a shift to the left of this system with the effective replenishment of free carbon dioxide. The ratio of carbon dioxide and carbonate to bicarbonate in water is a function of the pH (Fig. 2.1) and carbonate only becomes significant above a pH of 8. The most convenient method of determining bicarbonate and carbonate is titrimetrically against a standard acid; consequently, together they are commonly termed the 'alkalinity' or 'base reserves' of the water. The alkalinity above pH 8 is often

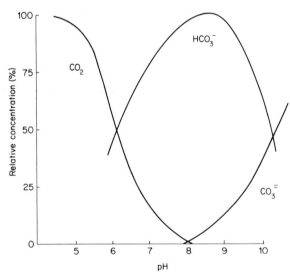

Fig. 2.1 The influence of pH on CO_2, HCO_3^- and CO_3^{2-} concentrations in surface waters (adapted from Fogg, 1972)

assumed to be due to carbonate alone although, in fact, with a knowledge of the total alkalinity and pH the bicarbonate and carbonate contents can be predicted from purely physical considerations (Fig. 2.1). An additional component to the alkalinity determined in this way is hydroxyl (OH^-) ions but concentrations are usually negligible, particularly below pH 9.

2.1.4 Concentration and Composition of Tropical and Temperate Waters

Although bicarbonate and carbonate do not contribute substantially to the relative ionic content of sea-water, their absolute concentrations are rather higher than those of most fresh waters and are sufficient to give sea-water an alkaline pH range of 8.1–8.4. Moreover, amongst fresh waters, there is considerable variation in the concentration of these and other ions from place to place, both on a local and continental scale. A comparison of the average ionic composition of rivers from two temperate continental areas, Europe and North America, with those from two essentially tropical regions, South America and Africa, (Table 2.1), shows marked differences in both concentrations and composition. The tropical river waters tend to have much lower total concentrations than temperate rivers. The dominance of calcium and bicarbonate is much less pronounced in the tropical waters and is partially compensated for by relatively greater sodium and chloride concentrations. In African rivers, for example, Na^+ is most often the dominant cation. Additional contrast is provided by the much higher levels of silicate and iron in the tropical waters. These differences are the product of geological and climatic variation which in turn can reflect some fundamental characteristics of tropical and temperate environments (Section 2.2).

2.1.5 Sources of Major Chemical Components

The principal source of calcium and bicarbonate is provided by weathering and erosion of rocks, particularly those derived from marine sediments from which the ions will dissolve according to

$$CaCO_3 + CO_2 + H_2O \rightarrow Ca^{2+} + 2HCO^-_3$$

Alternatively, silicate-based rocks such as feldspar ($NaAlSi_3O_8$) or anorthite ($CaAl_2Si_2O_8$) can act as a source (Golterman, 1975):

$$2NaAlSi_3O_8 + 2CO_2 + 11H_2O \rightleftharpoons 2HCO^-_3 + 4H_4SiO_4 + kaolinite + 2Na^+$$

By contrast significant quantities of Na^+ and Cl^- can be contributed by rain-water. In tropical rivers, where low ionic concentrations are common, rain-water has been shown to be the major contributor. For example, in the rivers around Kampala, Uganda, Visser (1961) estimated that rain-water could account for all the sodium, ammonium and nitrate, 75% of the chloride and over 70% of the potassium contained in lake and river waters. In the Rio Tefé, a

8

lowland tributary of the Amazon, it was considered that precipitation contributed 81% of the cations in the water compared to 19% from the rocks (Gibbs, 1970). Rain may even supply a significant quantity of the minor water constituents such as phosphorus and nitrogen which are so important to production within the ecosystem. In the Ivory Coast, for example, phosphate concentrations in rain-water varied between 150 and 480 $\mu g\,l^{-1}$, whilst nitrate ranged from 620 to 2480 $\mu g\,l^{-1}$ with the highest concentrations generally being found at the time of lowest rainfall (Lemasson and Pages, 1982). The principal ions of rain-water, sodium and chloride, are mainly derived from the sea spray, although naturally the concentration is generally low and therefore precipitation can only be expected to play a major role when ions from other sources are scarce.

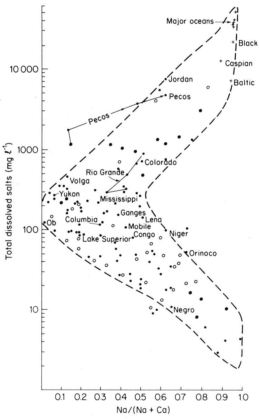

Fig. 2.2 The relationship of the weight ratio of Na/(Na + Ca) to total dissolved salts in world surface waters (●) Rivers, (○) lakes, (+) oceans. (*Reproduced by permission of the American Association for the Advancement of Science from Gibbs, 1970. Copyright 1970 by the AAAS*)

2.1.6 Possible Roles of Rain, Rocks, and Evaporation in Determining the Composition of Tropical River Waters

Gibbs (1970) classified surface waters of the world according to their proportions of sodium to calcium and chloride to bicarbonate in relation to total dissolved salts (Figs. 2.2 and 2.3). The majority of the large rivers draining tropical areas such as the Niger, Zaïre, and Orinoco are grouped together in the lower arm of this < - shaped distribution, indicating common characteristics of low total salt concentrations and relatively high levels of sodium and chloride compared to calcium and bicarbonate. The low total concentration and the general similarity in composition to rain-water led Gibbs to interpret this lower limb of the distribution as being the result of progressive influence of precipitation in determining the ionic content of the rivers. The influence of rain-water is

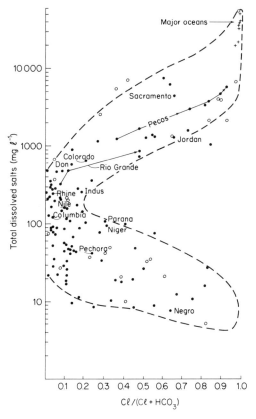

Fig. 2.3 The relationship of the weight ratio of $HCO_3/(HCO_3 + Cl)$ to total dissolved salts in world surface waters.(*Reproduced by permission of the American Association for the Advancement of Science from Gibbs, 1970. Copyright 1970 by the AAAS*)

accentuated since the basins of these tropical rivers are in the equatorial regions, which have very high annual rainfall, often over a restricted period of the year, which consequently gives rise to highly leached conditions in the catchment area. The axis of the distribution in Fig. 2.4 represents waters of relatively high calcium and bicarbonate content where weathering of rocks was considered to be the major influence. The ascending limb contains waters of high salinity with increasing sodium and chloride concentrations and largely includes rivers from hot, arid regions where rates of evaporation are great. The resulting elevated salinities tend to cause the precipitation of calcium carbonate from solution which consequently increases the relative content of sodium and chloride in the water. The progressive changes observed down the course of the Rio Grande and the Pecos River, both in desert regions, are shown by arrows (Figs. 2.3 and 2.4).

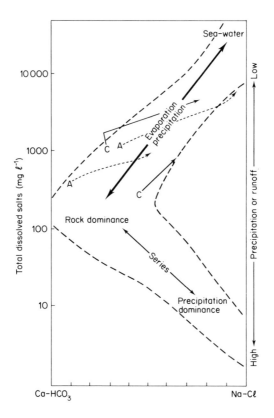

Fig. 2.4 A model of the effects of major determining factors in the composition of world surface waters. (*Reproduced by permission of the American Association for the Advancement of Science from Gibbs, 1970. Copyright 1970 by the AAAS*)

It is possible, therefore, to interpret the primary composition of surface waters in terms of three major processes—atmospheric precipitation, the weathering of rocks, and evaporation/crystallization. Rainfall is of overriding significance in the tropics, not only as a direct source of ions but also because it is the medium for the weathering of rocks and soils. Not all equatorial rivers have relatively high sodium and chloride components, however. The headwaters of the Amazon are rich in calcium carbonate, which can be attributed to the fact that 85% of dissolved salts in the Amazon are derived from the recent rocks of the Andes (Gibbs, 1972). The Purari River in New Guinea has very similar characteristics to the Amazon with regard to ionic ratios and it also drains mountainous regions which are young, geologically (Petr, 1976). In contrast, lowland tributaries of the Amazon such as the Rio Negro (Figs. 2.2 and 2.3) possess low total concentration with a predominance of sodium and chloride common to lowland, equatorial rivers. Slope and altitude proved to be the factors most closely correlated with total salt content and nature of the dissolved and suspended load in the passage of the Amazon from the headwaters in the Andes, where the TDS was $120–140 \, \mu g \, l^{-1}$, to the mouth where a value of $36 \, mg \, l^{-1}$ was found (Gibbs, 1967a,b). The progressive dilution is due to the influence of tributaries of low salt concentration derived from humid, lowland regions of the Amazon basin. Steep slopes, therefore, tend to enhance the weathering effects of rainfall on rocks whilst, in low relief tropical areas, weathering is impeded by a relatively uneroded soil cover and its vegetation. These lowland soils are often free draining, sometimes down to 20 m, highly leached, owing to the intense flushing effect of the rain, and have little available soluble material to add to the water.

2.2   WATER AND LAND IN A RIVER BASIN

2.2.1 Soil and Vegetation

The interaction of rainfall, rock, soil, and vegetation has a considerable bearing upon some of the minor constituents of the water, several of which are of great biological significance as essential plant nutrients. Silicate levels, for example, are consistently high in tropical rivers owing to the interaction of soil pH and the fact that silica, like iron and aluminium, has a solubility which is pH dependent (Viner, 1975a). Iron and aluminium produce amphoteric hydroxides which are least soluble at pH 7 and increasingly soluble under more acid or alkaline conditions. The solubility of silica is unaffected up to a pH of 9, beyond which it increases. Tropical soils tend to have a higher pH range, around 5.8–7.5, compared to temperate soils where a range of 4.0–5.8 is more normal owing to the generally higher organic content. Under the more acidic conditions, larger quantities of aluminium and iron are dissolved whilst under neutral conditions, more silica than aluminium and iron passes into solution. This, with the consistently higher temperatures which also enhance the solubility of silica, contributes to the selective weathering of silica and consequently to the high values observed in tropical rivers. The iron left behind in the soils results in the red coloration so characteristic of tropical lateritic soils. Although the solubility of iron is low at the normal soil pH ranges in the tropics, relatively high concentrations find their way into the water because it is so plentiful.

Phosphate is initially derived from the weathering of rocks, most commonly igneous types such as apatite, and volcanic rocks are frequently a particularly rich source. Even in volcanic areas, however, the availability of phosphate can be modified by secondary factors such as vegetation. For example, in Uganda the soils around Mount Elgon and also in the Karamoja show considerable volcanic influence and the streams in the regions have high phosphate levels. However, streams of the Karamoja have a rather lower total ionic concentration (mean conductivity $81.5\,\mu S\,cm^{-1}$) than those of Mount Elgon (mean conductivity $156\,\mu S\,cm^{-1}$) but the relative phosphate content is three times higher. The soils of the Karamoja steppes are very thinly vegetated compared to those of Mount Elgon which support wooded savanna and forest vegetation and it is the sparser vegetation which allows much greater erosion and leaching of phosphorus (Viner, 1975b).

The modifying effects of soil and vegetation are particularly important with respect to nitrogen concentrations in runoff waters, although several features of the interrelationships of these with rainfall are common to most dissolved substances. Combinations of high temperatures, prolonged and intense sunlight, and high rainfall, at least over part of the year, produce extremely weathered ferrulite soils, which are low in available ions. The vegetation cover helps to damp out changes in ground-water flow which otherwise would be highly seasonal and lead to severe erosion, as indicated in the example of the Karamoja soils mentioned above. The heavy leaching of tropical soils means that the amounts of essential elements, such as phosphorus and nitrogen, accumulated in the vegetation itself are extremely important. Tropical rain forests (Fig. 2.5), in particular, are thought to minimize loss of nutrients by having a close physical relationship with decomposers, largely in the form of mycorrhizal associations with fungi, which reduces the length of time that free ions are available in the soil for leaching. Consequently, it might be anticipated that runoff water from rain-forest areas would be particularly low in elements such as nitrogen. From a detailed analysis of the influence of vegetation types on surface waters in Uganda, the rivers from the Ituri and Ruwenzori forests showed very low NO_3-N concentrations, whilst the highest were found in moist savanna areas (Viner, 1975b) suggesting that the relative loss from forest areas was smaller. However, since the rainfall over the forest areas is considerably greater, the total quantities of nitrogen passing down the rivers annually from the forests must be high.

The principal sources of nitrogen are the rain-water itself, which in some cases may account for all nitrogen in surface waters (Visser, 1974), the organic material in the soil and also the fixation of atmospheric nitrogen by soil organisms and plants. There is often a high percentage of leguminous trees in tropical woodlands and forests, which must contribute a substantial proportion of

Fig. 2.5 A forest river in West Africa, the River Taia, Sierra Leone

nitrogen to the system and there are also other non-leguminous plants also capable of nitrogen fixation. Nevertheless, the concentration of inorganic nitrogen in tropical rivers is considerably less than that of temperate rivers (Table 2.1) although the total amounts available are also a function of river discharge.

2.2.2 Swamps, Floodplains, and Phytoplankton

Vegetation or plant growth can further influence the composition of river waters during passage of water through a swamp or lake. Swamps are generally characterized by abundant plant growth, often associated with quantities of decaying organic material under conditions of low flow. Essential plant nutrients such as nitrogen and phosphorus can be reduced to very low levels as water passes through papyrus swamps, owing to their utilization by the plants themselves (Carter, 1955; Viner, 1975a). The extensive decomposition which takes place produces conditions of low oxygen and acidic pH, both of which favour the release of iron from organic complexes in the sediments. Increased concentrations of iron and the chemically similar manganese are, therefore, frequently found in the water flowing from both large and small swamp areas. In Ugandan papyrus swamps, iron concentrations average some $1970 \, \mu g \, l^{-1}$, about ten to twenty times more than other sites (Viner, 1975a). Similar conditions exist also in the extensive papyrus swamp of the Sudd on the White Nile (Talling, 1957). The Sudd, however, does show some distinguishing features which appear to be a factor of its large size. For example, inorganic nitrogen levels increase through the swamp although largely in the reduced form of ammonia. Phosphate levels also rise, probably due to the tendency of phosphate to be released from the mud/water interface under deoxygenated conditions.

Floodplains on rivers can have similar effects upon the ionic composition of the water as do swamps owing to the large amount of decomposition which occurs when the surrounding terrestrial areas are flooded. Thus, the iron concentration of the Kilombero River in Tanzania was up to $1800\,\mu g\,l^{-1}$ as it passed through its floodplain, whilst 30 km below the floodplain the concentration was $450\,\mu g\,l^{-1}$. Greater solubility of iron is induced by deoxygenation and low pH conditions close to the soil surface and, moreover, a distinctive 'iron pan' may be found in soil profiles of the floodplain area as the lowered pH has increased the solubility of iron and it becomes selectively leached down through the soil.

Just as the passage of water through a region of vigorous plant growth such as a swamp can remove substantial proportions of essential plant nutrients, so the occurrence of a lake, with its phytoplankton community, can have a similar effect. The process has been observed in action as the rising Amazon, which is deficient in phytoplankton but relatively nutrient rich, begins to flood lakes which occur along its floodplain. The lakes are normally poor in nutrients owing to the demand from the phytoplankton and the low nutrient quality of runoff in their immediate catchment. As the Amazon water invades a lake such as the Lago Janauaca, the phosphate concentration has been observed to decline rapidly from 29.9 to $4.3\,\mu g\,l^{-1}$ and the combined nitrate and nitrite concentration is reduced from 168 to $1.4\,\mu g\,l^{-1}$ as large phytoplankton blooms develop under the stimulus of the nutrient influx (Fisher and Parsley, 1979). For similar reasons, the Semliki River, which connects Lake Edward to Lake Albert in Uganda, has proportionally very low concentrations of PO_4–P $(127\,\mu g\,l^{-1})$, NO_3–N $(57.7\,\mu g\,l^{-1})$ and silicate $(12.9\,\mu g\,l^{-1})$ compared to other rivers in Uganda, owing to the removal of these by phytoplankton in Lake Edward (Viner, 1975b). The reduction in silicate is largely due to the activities of diatoms which require this for their silica-based frustules.

2.2.3 The Influence of Animals

Herbivorous and fish-eating animals will also have an influence on ionic composition and nutrient availability, either by removal of material from the river as food or addition as faecal waste and excretory products. Little information is available on the relative importance of animals on rivers in this respect, but, in Africa, one of the most potentially important must be the hippopotamus. Although aquatic during the day, these animals feed on land at night and can cover quite large distances in their foraging. A proportion of the food intake —around 50% is common for a herbivore—is voided as faecal material, much of it during the day when the animals are in the water. The hippos, therefore, represent a means of transporting both organic and inorganic materials from the terrestrial to the aquatic system. Many hippo populations have been much

Table 2.2 The chemical contribution of the hippopotamus to the Rufiji River

	Total (kg day^{-1} animal^{-1}) (Viner, 1975a)	Dissolved and suspended (kg day^{-1} animal^{-1}) (Viner, 1975a)	Total (tons yr^{-1} total population^{-1})	Dissolved and suspended (tons yr^{-1} total population^{-1})	Rufiji (tons yr^{-1} × 10^2)	Hippo input dissolved and suspended as % of Rufiji
Carbon	1.6 − 2.2	0.46 − 0.63	7762–10 596	2225 − 3046	35 351	0.063–0.086
Nitrogen (NH$_4^+$ + NO$_3^-$ + NO$_2^-$ − N)	0.04 − 0.056	0.05 − 0.07	246– 336	201 − 262	10.75[a]	1.9 − 2.5
Phosphorus (PO$_4$ − P)	0.144–0.2	0.01 − 0.014	548– 954	38.7 − 67.5	74.4[b]	0.52 − 0.91
Sodium[c]	—	0.043–0.059	—	207 − 286	2 206	0.09 − 0.13
Calcium[c]	—	0.017–0.023	—	82 − 111	848	0.1 − 0.13
Silica (SiO$_2$–Si)[c]	—	0.028–0.038	—	135 − 183	2 344	0.06 − 0.08

[a] As dissolved NO$_3$ only.
[b] As dissolved orthophosphate only.
[c] Only occur as soluble form.

Note: The hippo population above Stiegler's Gorge is taken to be 13 248 (from Rodgers, 1980).

reduced by hunting but on the Rufiji River, as it runs through the Selous Game Reserve in Tanzania, there remains probably the largest undisturbed population in Africa. Combining population counts on the river with estimates of faecal and urinary outputs per animal (from Viner, 1975b) and comparing these to the total loads of the various ionic components in the river (Table 2.2), shows the actual contribution of hippos to be relatively small, the most significant being nitrogen and phosphorus. However, the form of the contribution may make it more readily available to the organisms of the river community. For example, much of the carbon must be in the form of vegetation partly processed by passage through the digestive tract of the hippopotamus.

Predators may also play an important role in mineral recycling. Since excessive hunting has caused the disappearance of the caiman (*Melanosuchus niger*), a type of crocodile from the Amazon, the rate of fish production seems to have declined. This is possibly due to the fish populations accumulating nutrients such as nitrogen which had previously been recycled as a large proportion of the fishes were consumed, digested, and remineralized by these voracious carnivores (Fittkau, 1970). In this way, a rapid recycling of nutrients would contribute to the general fertility of the water.

2.3 SUSPENDED MATERIALS

As well as dissolved compounds, rivers also transport suspended material which can either be inorganic, resulting from the erosion of rocks or soil, or organic. The organic material can be derived from outside the river, mainly from the terrestrial system (allochthonous) or from production within the river (autochthonous). Suspended solids are often a prominent feature of tropical rivers although their role is poorly understood (Fig. 2.6).

It has been suggested that whilst the amount of dissolved solids is less than suspended solids, the proportion can vary from 10% in dry environments to 50% in humid climates (Langbeim and Dawdy, 1964). Available data from tropical rivers suggest, however, that within these limits it is difficult to draw conclusions with regard to particular environments; thus an average for the Amazon is 32% (Gibbs, 1967a,b), for the Purari in New Guinea, 22% (Petr, 1976), for the Rio de la Plata, 35% (Bonetto, 1972), and for the savanna rivers of West Africa—the Niger and Benue—35% and 17% respectively (Grove, 1972). The volume of suspended solids can be influenced by a number of factors such as vegetation cover, which tends to inhibit erosion, as well as agriculture and forest clearance, which tend to promote it. The weight of suspended material can be quite considerable. For example, in the Rangoon River, values of 3.2–$7.4 \, g \, l^{-1}$ have been observed whilst the Nile in flood can carry $2.95 \, g \, l^{-1}$ (Ramadan, 1972) although levels of less than $300 \, mg \, l^{-1}$ are probably more normal. The average for the lower Amazon is $90 \, mg \, l^{-1}$ and for the lower Zaïre, $32 \, mg \, l^{-1}$. The relative proportion of organic to inorganic material in the suspended fraction is more or less unknown in most cases, yet this may be an

Fig. 2.6 The confluence of the Rufiji and the Great Ruaha, Tanzania, showing the heavy load of pale silt in the Rufiji contrasting with the relative transparency of the Ruaha

important feature in river productivity. The inorganic component, particularly silt, can have considerable ion-exchange capacity which enables it to transport other chemical compounds. Phosphates, for example, are readily absorbed on to silt particles so that the Kiasi, which runs into Lake Edward, had 200 mg l^{-1} suspended material and 0.8 mg l^{-1} particulate phosphorus, whilst other rivers in the area contained 0.08 mg l^{-1} phosphorus including 0.04 mg l^{-1} in the colloidal form (Golterman, 1973). Moreover, it is the smallest particles with diameters less than 4 μm which appear most important in this respect (Viner, 1982).

2.4 VARIABILITY OF TROPICAL RIVERS

2.4.1 Geographical

Tropical river waters do have some common features, as demonstrated in the previous section; they tend to have low salt contents, more sodium than calcium, high silica and iron contents, and a pH within the range 7.5–4.3. The last feature is partly a result of the low buffering capacity of the waters which in turn is a function of their low total ion content. However, the factors such as rainfall, rock, soil, and vegetation, which determines these characteristics, do themselves vary from place to place and can, in consequence, cause significant differences even within rivers of the same system. The Amazon and its tributaries, for example, have been classified into white, clear, and black waters each with their own characteristics. White-water rivers generally have their sources in geologically recent rocks of the Andes and, therefore, contain large quantities of inorganic sediment, which gives them their colour and also a low transparency of only 0.1–0.5 m. Since they originate in mountainous, volcanic areas, white waters also tend to have the highest ionic contents with conductivities of 44.8–83.8 $\mu S\ cm^{-1}$ in the middle Amazon (Sioli, 1975) and even higher in the Andean headwaters themselves. The pH ranges from neutral to slightly acid.

The clear waters tend to originate either from the old, weathered massifs of central Brazil and the Guianas or from the forested regions of the central Amazon where latosols occur. Apart from their lack of sediment, which gives a high transparency, they generally have lower ionic contents, particularly with respect to calcium, sulphate, and bicarbonate, than white waters, with conductivities of 0.57–5.3 $\mu S\ cm^{-1}$. The pH range is also generally lower, 4.4–6.6 (Sioli, 1975).

The black waters are products of rather singular circumstances. They are characterized by reduced transparency compared to that of clear waters but contain very little suspended inorganic material. The reduced transparency comes from the brown, humic compounds derived from the breakdown products of leaves from the forests through which the black-water rivers run. Most natural waters contain a yellow or brown colouring agent sometimes called *Gelbstoff* although it is not always apparent to the naked eye; in black waters, however, this effect is greatly enhanced. In the Amazon basin, black waters mainly occur in forests which grow on very poor soils of bleached white sand principally found on the upper Rio Negro. These rivers are very low in inorganic ions and have conductivities of 6.8–10.4 $\mu S\ cm^{-1}$ but the pH tends to be extremely acidic, 4.6–5.2, due to the accumulated organic acids and the poor buffering capacity of the water. If waters are extremely demineralized, as many forest streams and rivers are, then even small amounts of dissolved carbon dioxide are sufficient to give the water an acidic reaction, as can be observed with distilled water which can have a pH of 6.4–6.6 due to carbon dioxide dissolved from the atmosphere, but when distinctly acidic products result from decomposition, as they do in the black-water areas, then the effect becomes pronounced.

This classification of the rivers of the Amazon basin finds some parallels in other systems. For example, Welcomme (1979) identified black-water rivers in the Zaïre basin and also in Malaysia, both essentially in forested areas on poor, podsolitic soils. The black waters found in the Tasek Bera swamp in Malaysia appear to have a somewhat different origin to those of the Amazon basin such as the Rio Negro (Sato *et al.*, 1982). Whilst the colour in the Rio Negro is picked up as water passes through the forested podsols of the catchment area, in Tasek Bera, where there are no such podsols, it has been shown to originate from the decomposition of litter from swamp forest and the reed *Lepironia*. Very little colour is produced by the mud itself. The colouring agent in this instance has been characterized as humic rather than fulvic acids. These organic acids play an important role in the recycling of iron in the water through the formation of ferric humate complexes which can be reduced by sunlight. In Tasek Bera, the water colour becomes more dilute in the rains in contrast to that of the Rio Negro, which becomes more pronounced during the rains as the organic compounds are flushed from the soil (Sioli, 1964, 1968).

The Purari River from the New Guinea highlands resembles white-water composition in the Amazon and there are also some similarities between this and the Blue Nile which comes down off the Ethiopian plateau from which it gains a high ionic content (140–390 $\mu S\,cm^{-1}$), a predominance of calcium over sodium, and a suspended silt level of several grams per litre (Talling and Rzoska, 1967). Essentially, however, the Amazon classification reflects the difference between chemically rich, silt-laden rivers coming down from high elevations with rocks of recent origin, and those of older low relief, rain-forest areas, which are nutrient poor and normally possess a low suspended load. Smaller streams arising in rain-forest areas also often have low dissolved salt concentrations. The Gombak stream in Malaysia had a conductivity of 29–41 $\mu S\,cm^{-1}$ and a pH of 6.6–7.0 (Bishop, 1973), and a forest stream on the Freetown peninsula, West Africa, showed a conductivity of 18–39 $\mu S\,cm^{-1}$ and a pH of 5.6–6.6 (Payne, 1975).

Rivers of savanna, grassland, or open woodland may have different characteristics to any of those found in Amazonia. Their pH tends to range from neutral to slightly alkaline, which could reflect the lower quantities of organic material recycled through the soils of the catchment area compared to those from lowland forests. Their waters generally have an appreciable suspended load which may reflect a greater erosion from savanna than from forested areas.

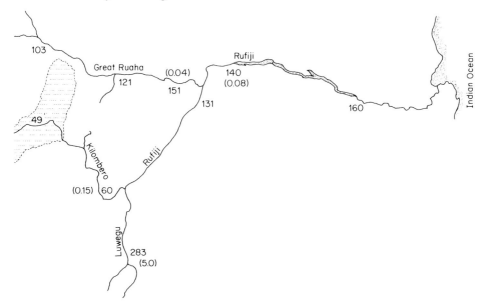

Fig. 2.7 Changes in the total ionic concentration as indicated by conductivity and calcium/sodium ratio down the Rufiji River system, Tanzania. Unbracketed values give conductivity (μS) whilst those in brackets denote Ca/Na ratio

The mean suspended load for the Niger is 152 mg l^{-1} compared to 32 mg l^{-1} for the Zaïre which has a largely forested basin (Eisma *et al.*, 1978). The Paraguay and Parana Rivers at their confluence have a mean suspended load of 160 mg l^{-1} (from Bonetto, 1972) compared to the Amazon which has 90 g l^{-1} (Gibbs, 1967a).

Ions tend to accumulate down a river system rather than be diluted, as occurs in the main channel of the Amazon in its passage from the Andes to the Atlantic. A more typical pattern is shown by the rivers of the Rufiji basin, Tanzania, where increase in ions occurs both along the tributaries, such as the Ruaha, and down the main body of the Rufiji River itself (Fig. 2.7). This example also shows how different tributaries within a single basin can be. The Kilombero is a river of low elevation which meanders through an extensive floodplain and consequently has a lower ionic concentration than the Great Ruaha which arises in a mountainous area. The Luwegu, by contrast, does not drain a high-relief area yet has a very ionic concentration, of which calcium and not sodium is the dominant cation. This indicates some major geological difference in the rocks of the Luwegu catchment area. The lower Rufiji, therefore, reflects the net effect of all of these diverse influences and exemplifies the general tendency of ions to accumulate down through a river system.

2.4.2 Seasonal

Most tropical rivers have an annual cycle which, like so many features of the river, is dictated by the pattern of rainfall. Since many tropical areas have distinct rainy seasons when most of the annual precipitation occurs, there tends to be a seasonal cycle in the discharge rate of the rivers. Headwater streams respond most rapidly and can change from quiet, trickling streams to torrents in an hour or two following the storms which often occur at the advent of the rains. Further down a river system, particularly if it has an extensive series of tributaries, these rapid responses become rather damped down. Nevertheless, as the rains build up to a peak, so the mean water levels in all channels begin to rise and the flow rate increases as the river discharge rate reflects the influence of the rains. If the channel is unable to contain the rising river then the water reaches the 'bankful' level after which it spills out, often fairly slowly, in the fashion known as 'creeping flow'. In many rivers such flooding is rarely very extensive but in some instances, where there is some impediment to drainage, immense areas of floodplain can be inundated—up to 53 000 km^2 in the case of the Mekong, for example, or 80 000–100 000 km^2 for the Gran Pantanal of the Paraguay River.

Particularly long rivers can have more than one flood per year. The Niger has two—the 'black floods' and the 'white floods', so called because of the colour caused by the associated sediments. The black floods are caused by the local rains in December–February whilst the rather more extensive white floods appear in the dry season, during July–September as a result of rainfall in the headwaters in the Guinea highlands which has taken some time to reach the savanna areas of the middle and lower Niger.

Some features, such as swamps and forests, can damp down flood oscillations, and a similar effect is also shown by large river systems which have tributaries from a variety of latitudes. Even the Amazon, however, shows a marked period of high water, generally in May, although the rains occur at different times from place to place in its catchment area.

During the dry season river discharge can be considerably reduced, in some cases to the point where the river ceases to flow and becomes a series of isolated pools. This is commoner in the drier, savanna areas such as the Sokoto River in northern Nigeria (Holden, 1963) and in the upper reaches of the Esequibo in Guyana (Lowe-McConnell, 1964). Thus, in a savanna river such as the Senegal, the average maximum flow is 132 times greater than the minimum, whilst for the Amazon, which receives water from such a wide variety of geographical zones, the factor is only 2.3 (Welcomme, 1979).

Such fluctuations in the discharge rate can have marked effects on the water chemistry and the suspended material. Usually the concentration of ions in the water is at a maximum during the dry season when the influence of ground water is greatest. The rains, because of the low ionic content of rain-water, tend to have a diluting effect. However, during the dry season, mineralization of organic matter in the soils of the basin proceeds, so that when the first rains arrive there is a flushing effect and a spike of ions can sometimes be detected, at least in the smaller streams. This effect may be particularly significant with respect to nitrate since nitrate liberation is enhanced by alternate wetting and drying of the soil. Nitrates accumulate in tropical soils during the dry season; they show a brief increase in the earliest rains, which declines sharply as the rains flush out the ions. There generally exists an inverse relationship between flow rate and ion concentration, which manifests itself as a general reduction in ionic concentrations as the rains reach their peak.

The change in the suspended load caused by the rains is the reverse of that of the dissolved component. During the dry season, the amount of the suspended material is lowest owing to the lack of material being washed in, but also to the increased possibility of sedimentation caused by the flow and lack of turbulence. The increase in sediment closely follows the increase in flow caused by the rains so that, for example, the amount of suspended material in the Bermejo River, Argentina, is forty times higher at the peak of the discharge cycle than at the lowest point (Golterman, 1975).

When a river floods its banks, however, there is a transition from high flow rate in the central channel to rather slow rates as the water advances by 'creeping flow' over the floodplain. In this process the heaviest particles are quickly deposited, whilst the lightest particles remain in suspension for longer and are deposited over much greater distances from the river. Since a considerable proportion of the lighter fraction is made up of organic material, this can be widely distributed over the floodplain and therefore considerably enhances the fertility of the soil. The most striking example of this is the Nile Valley where the fertile

Fig. 2.8 The profound influence a river can have on the surrounding land is demonstrated by the sharp change from cultivation to desert in the Nile valley

area is very sharply defined by the extent of the river flood (Fig. 2.8) although, since the completion of the Aswan High Dam, this is now accomplished by irrigation rather than by flood and there is less sediment in the water because much of it settles out in Lake Nasser.

GENERAL READING

Golterman, H. L. 1975. *Chemistry In River Ecology*. Whitton, B. A. (ed), pp. 39–80. Oxford, Blackwell.

Mynes, H. B. N. 1970. *The Biology of Running Waters*. Liverpool, University Press.

CHAPTER 3

THE LAKE ENVIRONMENT

3.1 CHARACTERISTICS OF LAKES

3.1.1 Size and Formation

Standing inland waters are perhaps the most diverse of aquatic systems. They vary greatly in size, shape, and the dynamics of their water exchange. They also show a particular wide range of chemical environments. In their natural form, they can vary from muddy cattle wallows to virtual inland seas such as Lake Victoria (Fig. 3.1) which, at 69 000 km², has the second largest surface area of any lake in the world and the largest in the tropics. It is the bigger lakes which invite comparisons with the marine system, but lakes vary greatly in their physical and chemical composition whereas the sea is remarkably constant. Lakes are bodies of water usually isolated from each other, a fact which is the basis of this variability. There are, however, certain regional similarities and tropical lakes have a number of characteristics in common.

Lakes can be brought about by the formation of basins from geological upheavals or faulting or can be a result of volcanic activities. The sites of extinct volcanoes themselves can accumulate water to give the type of crater or caldera lakes often found in areas of recent volcanic activity such as East Africa, the Andes, Papua New Guinea, and Hawaii. Occasionally, lava flows can dam previously existing watercourses. This appears to have been the case in the formation of Lake Kivu which lies close to the Virunga volcanic field in Uganda.

3.1.2 Water Loss and Gain

Most lakes gain water either, directly, from rain falling on the surface of the lake or, less directly, from rainfall within the catchment area. Water from this source can appear rapidly in the lake as surface runoff over the land or there may be a time-lag when water percolates down through the soil to become part of the ground water which then eventually finds its way into the lake by seepage.

23

Fig. 3.1 The southern end of Lake Victoria, the largest tropical lake

Water from runoff or seepage, however, most commonly collects first in the discrete watercourses or inflowing rivers and streams.

The loss of water from lakes can be due to evaporation, seepage to ground water, or by means of an outflowing river or stream. The existence of an outflow means that the lake has an open drainage basin, but some lakes have a closed basin with no outflow. In the tropics, where the rate of water loss through evaporation can be very high, salts can become extremely concentrated in lakes with a closed basin which can lead to the development of salt or soda lakes. The extreme example of lakes with open basins are those which might be called 'river lakes' where some constriction in the course of a river has led to the accumulation of water behind the constriction. This can occur naturally, as in the case of Stanley Pool on the lower Zaïre River, but many man-made lakes of this kind have been created recently by the damming of major rivers, such as Lake Kariba on the Zambezi, Lake Nasser/Nubia on the Nile, and, the largest of these, Lake Itaipu on the Parana in Paraguay. Man-made lakes or reservoirs, both large and small, are now a prominent feature of tropical areas.

There is an additional type of lake associated particularly with river floodplains. These lakes rely upon the seasonal flood to bring about their annual water exchange although they may, in fact, be isolated for most of the year. Lakes of this type are very common in the Amazon basin where, on the alluvial plain of the lower and middle Amazon Rivers, there are a large number of shallow *varzea* lakes which can measure up to 1000 km^2 (Sioli, 1975). These may have permanent channels of several kilometres in length connecting them to the main river although, again, the principal period for water exchange occurs when the river rises.

3.1.3 High-altitude Lakes

Most tropical aquatic environments would normally experience high ambient temperatures, but this is not true of lakes occurring at high altitudes. In these, particularly at altitudes greater than 4000 m, temperatures close to zero are common. Perhaps the most well-known of these lakes is Lake Titicaca in the Andes between Peru and Bolivia, but the 'altiplano' of the Andes contains a large number of smaller lakes, often with rather singular chemical characteristics including very high salt content. Some of the smaller high-altitude lakes may be frozen over for part of the year, such as Lake Waiau at 3969 m on Hawaii, where an ice cover is common during the winter months (Maciolek, 1969). Similarly, Curling Pond close to the summit of Mount Kenya, East Africa, is frozen over most of the time (Löffler, 1964). Snow melt may also be important in the water budget of these ponds.

3.1.4 General Distribution

Lakes of one type or another occur throughout the tropical region, although the largest concentration forms the Great Lakes of East Africa which include several of the most extensive, such as Lake Victoria and Lake Malawi as well as Lake Tanganyika, which is the deepest at 1470 m. Their formation was associated with the recent geological changes which produced the two rift valleys of eastern Africa. These lakes have attracted much scientific attention and Lake George, in Uganda, was selected for particularly intensive study during the IBP of the 1960s, as being representative of shallow, equatorial lakes. In contrast, other areas of the tropics are dominated by river systems, including the Amazon basin and the rain-forest areas of South-east Asia and have very few natural, true lakes. Throughout the tropics, however, man-made reservoirs for drinking water, irrigation, or hydroelectric power have a long history and are now ubiquitous.

3.2 HYDROLOGY AND WATER BUDGETS

3.2.1 Routes of Gain and Loss

Lakes tend to gain water from inflowing streams and rivers, by seepage and by direct rainfall on the surface, whilst water is lost via outflows, seepage, or through evaporation. If gain and loss are equal then the lake level and volume will remain the same; the simplest system is exemplified by a lake such as Lake Lanao in the Philippines where the mean annual lake level has not varied by more than 0.8 m in 27 years of records. The discharge of the outflowing River Agus has varied from $233\,m^{-3}\,sec^{-1}$ to less than $12.8\,m^{-3}\,sec^{-1}$, which has effectively achieved a balance with the inflowing rivers (Frey, 1969). In this case, rainfall over the catchment area and loss through the outflow are the main routes of flux, with evaporation accounting for only some 31% of the total loss (Table 3.1). Any tendency for the lake level to rise quickly results in an increase in the discharge rate.

In many lakes the inflow provides the major water input, particularly in a 'river lake' such as Kariba, where the annual inflow is equivalent to a quarter of the lake volume, therefore giving a complete exchange of water every four years. In smaller reservoirs and man-made lakes the annual inflow considerably exceeds the volume, leading to a replacement time of less than a year. A similar situation may be found in natural lakes with copious inflows (Table 3.1). Lake George, for example, has the Ruwenzori Mountains within its catchment area, which provides sufficient water for complete water exchange 2.8 times per year (Viner and Smith, 1973).

Table 3.1 Proportional contributions of the major routes of water gain and loss in Lake Lanao (from Frey, 1969), Lake George (from Viner and Smith, 1973) and Lake Victoria (from Talling, 1966)

Lake	Surface area (km^2)	Mean depth (m)	Gain Combined ($m^3 \times 10^{12}$)	Inflow (%)	Rainfall (%)	Loss Combined ($m^3 \times 10^{12}$)	Outflow (%)	Evaporation (%)
Lanao	357	60.3	4.827	78.8	21.2	4.816	68.4	31.6
George	250	2.4	2.153	90.5	9.5	2.153	78.8	21.2
Victoria	69 000	40	120.06	17	83	120.06	17	83

Not all lakes have this kind of water budget, however. In Lake Victoria, for example, inflowing rivers supply only about 0.6% of the volume each year, equivalent to the outflow down the Nile. Direct rainfall on to the lake surface accounts for some 83% of the total water input and a similar amount is lost through evaporation (Table 3.1). Moreover, the total water turnover within the lake is very restricted with complete replacement taking at least 100 years. In other lakes with a similar type of cycle, replacement can take even longer so that, for example, the replacement time for Lake Tanganyika may be as long as 1700 years.

Rather different water budgets are found in lakes with closed basins since, as there is no outlet, loss is entirely due to evaporation. Where such lakes occur in arid areas, as with Lake Turkana in the East African Sahel region, direct rainfall on the lakes is low and evaporation may exceed inflow over the long term to create saline conditions. However, in several cases, including Lake Turkana and Lake Chad in West Africa, there is evidence of seepage from ground-water sources since the waters are less saline than they should be if only inflow and evaporation were taken into account.

Many small ponds and crater lakes have neither inflow nor outflow. Here, water is gained from seepage or by rainfall either over the surface or by runoff from a very restricted catchment area. Loss can again be by seepage or by evaporation. Lakes situated in volcanic craters naturally have a catchment area sharply defined by the rim of the crater. The steep inner slope gives a very limited catchment. For example, in Lake Sonachi, Kenya, it is only 21% of the lake surface area (Melack, 1981).

In lakes with open basins it is very difficult to determine if significant amounts of seepage occur directly from the ground water into the water body of the lake. However, detection was possible in Subang Lake, a reservoir in Malaysia, since the very low ionic concentration of the ground water was detectable at the bottom of the water column (Arumagam and Furtado, 1980). Subang had, in fact, been constructed to take advantage of underground seepage water flow, as with many reservoirs.

3.2.2 Periodic Changes

Any discrepancy between gain and loss of water over an annual cycle will result in a change in lake-water volume and consequently in lake level. Some lakes are consistently stable, such as Lake Lanao (Section 2.2.1) which has a mean annual variation of 0.8 m (range 0.3–1.5 m) whilst the mean depth is 60.3 m. Lakes can, however, show year-to-year changes and often these reflect long-term variations in the hydrological regime. Throughout this century, Lake Victoria has experienced periodic fluctuations in level usually of $\pm$ 1 m, but during its history more dramatic changes have occurred. One such period has been recorded from 1961–1964 when, generally following changes in rainfall pattern, the water rose by almost 3 m, causing extensive flooding of marginal areas. The lake is currently in the declining phase (Beadle, 1981). Even more remarkable were the declining levels in Lake Chad on the south-western edge of the Sahara, during the 1960s, culminating in the Sahel drought, 1972–73. This drought reduced the flood of the main inflow, the Chari River, to insignificant proportions with the result that the larger north basin of the lake dried up leaving only the south basin. The vegetation which grew on the natural barrier between the basins stabilized this so that when the floods from the Chari returned to more typical proportions the vegetated barrier acted as a dam which prevented water from refilling the north basin. Lake Chad, therefore, has become very much curtailed in size (Chouret, 1978). Changes in water level of this nature obviously have considerable effects upon the resources available to the lacustrine communities and also upon their degree of isolation. Consequently, there can be substantial ecological and evolutionary implications to such changes, which are dealt with in Chapter 7.

3.2.3 Seasonal Changes

Superimposed upon long-term annual changes in lake level are seasonal changes due to peaks in gain and loss occurring out of phase, at different times of year.

This tends to be characteristic of the tropics where wet and dry seasons give periods alternately dominated by precipitation and evaporation. In Lake George, annual evaporation amounts to three-quarters of the lake volume. The actual rate of evaporation from this equatorial lake shows little variation with season, with an average of around 150 mm month^{-1}, but during the dry season, when actual loss amounts to 1.25×10^6 m^3 day^{-1}, it greatly exceeds gain from runoff or precipitation with the result that during the 3 months between rainy seasons the level of this lake, which has a mean depth of only 2.5 m, can drop by 0.2 m (Viner and Smith, 1973). In fact, if all water input were to cease, the lake would dry up within 500 days. As it is, however, the amount of water which comes down from the Ruwenzori Mountains during the rainy periods is several times that which is lost annually by evaporation; thus, the net loss in the dry season can be made up during the rains. An additional point demonstrated by Lake George, concerning lakes where water gain is dominated by runoff from extensive or distant catchment areas, is the time-lag between rainfall over the area and maximum observable effect on refilling the lake. The lag in the case of Lake George is unusually long—1.4–2.5 months. A heavily vegetated catchment area can retard flow by 1–4 weeks but the situation in Lake George is additionally complicated by the unusual water-exchange pattern between this lake and Lake Edward which is connected to Lake George via the Kazinga channel, the main outlet of Lake George. Thus, the seasonal pattern of water level and volume changes depend upon the interaction of water exchange with the vegetation and morphology of the area in the vicinity of the lake.

In reservoirs and man-made lakes, seasonal changes can be considerable because the main requirement for water from these impoundments, whether for drinking, irrigation, or hydroelectric power, is during the dry season. Man-made lakes are created from rivers, usually with the intention of storing all or some of the floodwaters and releasing this water during the dry season. This often means that large areas of the lake bed are exposed during the dry season as the lake level goes down. The extent to which the level subsides is called the 'drawdown' of the reservoir. In some of the larger lakes, such as Lake Kariba, the drawdown may only be 1 or 2 m, but in others it can be considerable; for example, in Lake Kainji, Nigeria, the drawdown is 16 m and a substantial proportion of the lake bed can be exposed. In hydroelectric schemes the aim is usually to obtain a steady flow rate through the generating turbines. Ideally, therefore, the floodwaters of the rivers are caught behind the dam wall in the rains with any excess going over the spillway, the level of which determines the maximum lake level. The same rate of discharge through the dry season is then maintained by controlled discharge of the impounded water through the turbine outlet tunnels. The volume used is then replaced by the next floods. One effect of this is that it considerably reduces the fluctuations in river discharge rates and in river levels below the dam, which is good for flood control in areas either side of the river, but can have potentially damaging effects on biological communities and fisheries below the dam which previously relied upon seasonal changes in the river for food and information.

3.3 CHEMICAL ENVIRONMENT

3.3.1 Ionic Concentration

Lakes tend to be supplied with dissolved substances largely through their inflowing rivers and, to a lesser extent, from rainwater. Frequently, the concentration of lake water is higher than that of the major inflows as ions tend to accumulate within lakes owing to evaporation, biological turnover, and interaction with the sediments. Even in recently formed lakes this can occur. For example, the average conductivity of Lake Kariba, formed in 1958, is $70\,\mu S\,cm^{-1}$, whilst that of the inflowing Zambezi is $40\,\mu S\,cm^{-1}$. However, lakes with a rapid turnover time tend to reflect both the total concentration and the composition of the inflow very closely. The situation may appear more complicated when the lake has several inflows. In Lake Lanao in the Philippines, for example, which has an average conductivity of $120\,\mu S\,cm^{-1}$, three of the inflowing rivers have concentrations less than this whilst the fourth, probably as a result of Tertiary coral limestones in its watershed, has a higher conductivity. The lake concentration itself is the resultant of these (Frey, 1969). In some cases, exceptional circumstances can influence the ionic concentration. In Lake Kivu, East Africa, the lower layers of water are extremely saline, in excess of $4000\,\mu S\,cm^{-1}$, due to emergence of warm saline waters of volcanic origin through the lake bed. Lakes which are close to the sea such as Lake Izabal in Guatemala, are often influenced by it. In this particular lake, which has a normal conductivity range of 175–$200\,\mu S\,cm^{-1}$ (Brinson and Nordlie, 1975), the basin is largely below sea-level so that, although some 42 km from the sea, occasionally incursions of brackish water up the connecting Rio Dulce in the dry season can elevate the water of the seaward end to $5000\,\mu S\,cm^{-1}$ (the conductivity of sea-water would be some $46\,000\,\mu S\,cm^{-1}$).

The varzea lakes of the Amazon floodplains are virtually dependent upon the river as a source of dissolved salts. As the river rises, a fresh influx of water reaches the lakes along connecting channels or *furos* which may be several kilometres long and, consequently, during the early flood period, the ionic concentration of the lakes is close to that of the river which could be in excess of $50\,\mu S\,cm^{-1}$. However, precipitation is also heavy and, as the rains reach a peak, the lakes become progressively diluted. This dilution continues even after the lake levels start to fall, since there is always some rain even in the dry season so that, eventually, the lowest level of the lake coincides with the lowest concentrations—generally less than $20\,\mu S\,cm^{-1}$—immediately prior to the recommencement of the cycle (Schmidt, 1973a,b). A close link between a lake and a river will often lead to seasonal fluctuations in the ionic concentrations of the lake, but usually, in drier parts of the tropics, the cycle in the lake reflects that of the river with concentrations which are lower in the rains and higher in the dry season.

The influence of inflowing rivers on the ionic content of lakes with open basins must be considerable, although the relationship becomes more indirect as precipitation and evaporation play larger parts. In lakes with closed basins there are considerable possibilities for evaporation to increase the concentration of the water since there is no outflow. There are examples of lakes with no outflow remaining relatively fresh, such as Lake Chad in West Africa and Lake Naivasha in Kenya, but these are usually suspected of having subterranean outflows. In arid areas, lakes with close basins almost invariably develop into lakes with a high salt content. The largest of these is probably Lake Turkana in the East African Sahel region, although whether this can truly be regarded as saline depends upon the definition. Beadle (1981) suggests three criteria for 'saline' waters:

(a) salt can just be detected by human taste;
(b) water just too saline to act as a permanent source of drinking water for men and other mammals;
(c) water is too saline to support typical freshwater fauna and flora.

Lake Turkana does have a freshwater fauna and flora and it can be drunk by those habituated to it, but it does have a salty taste. The actual concentration is between 2860 and 3300 $\mu S\,cm^{-1}$, but some of the smaller East African lakes considerably exceed this, for example: Lake Elmentitia, 43 750 $\mu S\,cm^{-1}$; Lake Manyara, 94 000 $\mu S\,cm^{-1}$; and Lake Nakuru, 162 500 $\mu S\,cm^{-1}$ (Talling and Talling, 1965). The conductivity of sea water, around 46 000 $\mu S\,cm^{-1}$, indicates just how concentrated these lakes can become.

In some lakes the concentration can be observed to rise measurably with time. Lake Qarun in the Fayum depression of the Nile Valley has increased from 12 $g\,l^{-1}$ in 1922 up to 30 $g\,l^{-1}$ at present. Such a situation can result in the whole lake drying up to a salt pan. This was the fate of the Great Saharan Lakes which covered much of what is now desert, from the Niger to the Nile, in Pleistocene times up to perhaps 5000 years ago. The present conditions mean that, although considerable amounts of fresh water are to be found in aquifers below the desert, wherever the water table reaches the surface the rate of evaporation is considerable and the concentration readily becomes elevated.

In Lake Valencia, Venezeula, deforestation and development of the catchment area may have initiated or certainly enhanced its desiccation over the last 250 years (Lewis and Weibezahn, 1976). Abstraction for urban and agricultural use, particularly, has increased the rate at which the lake is drying up.

Hot, dry regions are not the only places where saline lakes occur; lakes at extreme altitudes in the Andes, above 3500 m, where it is dry but cold, can also have very high salinities. The best known of these is Lake Titicaca, but there are also a large number of closed basin saline lakes in the altiplano desert close to the Andean summits of southern Peru, Bolivia, and northern Chile. Altiplano lakes can actually be divided into two categories—those of the dry desert high-altitude zones of the Andes, and those of wetter, equatorial, or seaward-facing ranges, which are found among the higher mountains of East Africa, New Guinea, and in the wetter parts of the Andes. These zones have been categorized as 'Puna' and 'Paramos' respectively by Löffler (1964) and each has its own chemical characteristics. In the Paramos, the high rainfall is generally related to very low salt concentrations whilst those of the dry Puna have remarkably high salinities, although all intermediate types exist. Some indication of the range can be seen from a survey along the Peruvian Andes (Hegewald *et al.*, 1976). Here, the ionic concentrations ranged from 78 μS cm^{-1} in Laguna Sausacocha at the wetter northern section of the chain, to 1028 μS cm^{-1} in Lake Titicaca to the south, with a number of lakes showing intermediate values which are influenced by geological composition of the rocks. All of these, however, contrasted with a small lake, Laguna Quistococha, occurring in the lowland, forested edge of the Amazon basin in eastern Peru. Here, a conductivity of 36 μS cm^{-1} was found, which is typical of the very soft waters characteristic of lowland tropical conditions.

A similar pattern emerged from analyses of lakes in Ecuador with the high, wet Paramos lakes having low ionic concentrations and a low pH whilst those of the high, dry Puna possessed elevated salt and pH values. The lowland forest waters showed very low nutrient concentrations and a low pH owing to the weathered, leached nature of the surrounding larosal soils and the low buffering capacity of the demineralized waters (Steinitz-Kannan *et al.*, 1983).

3.3.2 Ionic Composition

In the case of rivers, we have seen how their composition depends upon factors such as the nature of the rocks and soil in the catchment area, together with the type of vegetation and the composition and leaching effects of rain-water. Moreover, just as the ionic concentration of a lake is the net effect of the concentration of its inflows, including direct rainfall, so the composition is the resultant of these. In some cases it has been suggested that selective elimination of some major ions by biological processes must have occurred to explain the differences between an inflow and the lake. However, when such cases have been examined closely with respect to the relative contribution of all the inflows it has not been necessary to invoke any other explanation (Talling and Talling, 1965). In closed basin lakes the pattern can be more complicated as evaporation and progressive concentration cause crystallization and precipitation of some elements. Minor constituents, such as nitrate, phosphate, and to a lesser extent silicate, are extremely variable in most lake waters because they are involved in biological cycles and can, therefore, be incorporated into organic or structural compounds within living organisms.

It was concluded that, in tropical rivers, most frequently the major ions were Na^+ and HCO_3^-, although Ca^{2+} can be the dominant cation when sedimentary rock is present within the catchment area (Chapter 2). Therefore, as the sodium/calcium balance does vary to some degree and since bicarbonate dominance is shared by many temperate waters, probably the most characteristic features of tropical lakes are their high silicate and low nitrogen (Talling and Talling, 1965) derived from the rivers by the processes explained in Chapter 2.

The wide range of total ionic concentration found in lakes has been emphasized in the previous section and this concentration itself bears some relationship to composition of the water (Talling and Talling, 1965). There is a very strong correlation between total ionic concentration and alkalinity ($HCO_3^- + CO_3^{2-}$, determined by titration against a standard acid; Section 2.1.3). In fact, if the alkalinity expressed in $meq\,l^{-1}$ is multiplied by 100 this gives a reasonable approximation to the conductivity at 20 °C (Talling and Talling, 1965). Many of the major physical and chemical characteristics of waters are so closely linked that perhaps some consideration of units is necessary. Alkalinity is related to conductivity because HCO_3^-, together with CO_3^{2-}, above pH 8, is the major anion and the conductivity is a measure of the ionic concentration. However, a more direct way of measuring the ionic concentration is by weighing the salt residues following evaporation of a sample, which then provides the weight of TDS in $mg\,l^{-1}$ (Section 2.1.2). An alternative method of approximating this is to determine individually and to add up the concentrations of the major constituents, Na^+, K^+, Ca^{2+}, Mg^{3+} and HCO_3^-, CO_3^{2-}, Cl^-, and SO_4^{2-}. Conductivity is correspondingly strongly correlated with TDS. However, the precise relationship varies with the nature of the ions involved (Section 2.1.2). Lakes in the middle Amazon in which calcium and bicarbonate were dominant had factors ranging from 0.78 to 0.94 with an average of 0.83 (Schmidt, 1973a,b). Amongst African lake waters, where sodium and bicarbonate tend to be the principal ions, Talling and Talling (1965) considered the most reliable interconversion to be given by a factor of $85\,\mu S\,cm^{-1}$ per $1\,meq\,l^{-1}$ total ionic content. However, since conductivity is most frequently used as a measure of total salt concentration, it is not often necessary to convert to TDS, but the implications of the relationship need to be borne in mind.

One further point in considering the general interrelationships amongst the ions is that water should be electrically neutral, which means that the sum of the major positive ions in milliequivalents per litre (1 meq = molecular weight × equivalent) should equal the sum of the major negative ions. Bicarbonate and carbonate concentrations as alkalinity determinations are most frequently given in $meq\,l^{-1}$ but, since concentration = meq × molecular weight, alkalinities expressed in this way can be converted to HCO_3 or CO_3 concentrations by being multiplied by 61 or 60 respectively. An unsatisfactory measure of alkalinity, which has been used, is to convert alkalinities to mg $CaCO_3\,l^{-1}$, which means multiplying the value in meq by 100 but, since in tropical waters carbonate is often not associated with calcium, this is particularly inappropriate.

Amongst othe major constituents of the water, sodium, like bicarbonate, increases positively with concentration, even in the highest salinities. Chloride can also show parallel increases although it becomes proportionally higher at the greater end of the salinity range. At lower salinity ranges, calcium increases with total concentration but, like magnesium, tends to precipitate out at moderate concentrations. Silicate, which often exceeds $10\,mg\,l^{-1}$, tends to increase progressively with total salt content and can become proportionally very high in soda lakes since these often have a high pH, which increases the solubility of silicate and can lead to concentrations greater than $100\,mg\,l^{-1}$ (Talling and Talling, 1965).

The pH of tropical lakes can be extremely variable, from 4 in some lakes of the soft, black-water areas of the Amazon and Zaïre basins, to 11 in some of the closed-basin soda lakes of East Africa. There is not, however, a marked correlation between conductivity and pH owing to the considerable biological influence upon this factor. High rates of photosynthesis can elevate pH from less than 8 up to 10 within a few hours in a way that is not directly related to the chemical nature of the water. The pH, therefore, is rather labile in tropical waters where high sunlight can promote rapid photosynthesis.

In the same way some of the minor constituents, particularly phosphate and nitrate, are also variable owing to incorporation into aquatic communities. For this reason total phosphorus and total nitrogen, both estimated from complete acid digests of all material present in the water, are better guides as to the actual nutrient status of the water. The nitrogen content is generally low, with that portion due to NO_3 often less than $30\,\mu g\,l^{-1}$, which is lower than most temperate waters. Phosphorus levels can sometimes be quite high, particularly in areas of recent volcanic activity such as the Andes and parts of East Africa. In Lake Bogoria, a crater lake in Kenya, for example, a phosphorus concentration of $3300\,\mu g\,l^{-1}$ was found which is extremely high when compared to continental averages (see Table 2.1).

3.4 CLASSIFICATION OF LAKES

The previous sections have shown how different tropical lakes can be, ranging from waters which are almost deionized to those with salinities several times that of sea-water. Their ionic composition can also vary widely. Moreover, lake basins are more or less isolated from each other, unlike marine basins which are interconnected thus giving rise to gross similarities between oceans both in concentration and composition. Lakes are probably the most varied aquatic environments and, therefore, it is important to obtain a systematic view of tropical lakes as biological environments.

The most comprehensive attempt has been made by Talling and Talling (1965) based on a review of all African lake waters. The basis of the classification which they adopted was conductivity since this and its close correlate, alkalinity, have a profound influence on biological systems (see Chapter 5). Lakes with a conductivity of less than $600\,\mu S\,cm^{-1}$ were regarded as Class I. These include many equatorial lakes in forest regions, 'River' lakes, lakes of open basins fed by rivers of moderate salt content and also lakes fed largely by rain-water, such as small volcanic lakes. In these Class I lakes, the dominant ions are often sodium and bicarbonate, although the calcium content may be higher than sodium where underlying rock formations provide a soluble source of this element. The alkalinity extends to $8\,meq\,l^{-1}$. Ionic concentrations are often particularly low where the inflow has to enter the lake through a swamp area since many ions are removed through absorption and accumulation by the vegetation. The pH of Class I lakes, whilst changeable, generally lies between 6.5 and 8.7, although it can be considerably lower in demineralized situations where buffering capacity is low and there is much organic debris. Phosphate levels are frequently less than $150\,\mu g\,l^{-1}$.

Class II lakes have conductivities within the range $600-6000\,\mu S\,cm^{-1}$. These are often lakes with closed basins, such as Lake Turkana ($3300\,\mu S\,cm^{-1}$) where evaporation plays a significant role although it does also include some open-basin lakes such as Kivu ($1240-4000\,\mu S\,cm^{-1}$) or Tanganyika ($620\,\mu S\,cm^{-1}$) where there are inflows from alkaline volcanic areas. Class II lakes are largely alkaline, with a pH of perhaps 8.8–9.5, and through this series a decline in calcium is found, being lost by precipitation at the higher ionic concentrations and higher pH—almost all, therefore, are dominated by sodium and bicarbonate. Chloride concentration may also be high and, in some situations, is far more important than bicarbonate. This is the case in Saharan waters and lakes of the dry altiplano in the Andes whilst the high sodium chloride concentrations are often accompanied by relatively high sulphate. These lakes have much lower alkalinities and lower pH values (7.5–8.5) than those previously described amongst the Class I series.

Class III lakes are categorized by conductivities of $6000-160\,000\,\mu S\,cm^{-1}$. These are usually shallow closed-basin lakes where evaporation processes are well advanced and crystallization of various salts may have begun to occur, often in the form of deposits around the lake. These deposits can be *trona* ($Na_2CO_3.NaHCO_3.2H_2O$) or sodium chloride depending upon the predominant ions. Alkalinity and pH are usually high with the latter commonly between 9.5 and 10.5. Silicate concentrations are often considerable due to increased solubility at the high pH. Some of these lakes, such as Lake Magadi, Tanzania, are actually derived from springs, often warm, rising over salt deposits to produce very saline waters and large areas of salt deposits (Talling and Talling, 1965).

Amongst tropical lakes those of Class I are in the majority and, although they share some fundamental characteristics, they do show considerable variation. Rai and Hill (1980) classified lakes in the Amazon basin by their chemical characteristics, which in turn were derived from the type of water received from the river. Those receiving water from the River Solimoes (middle Amazon) were termed 'white-water' lakes and had conductivities of $31-80\,\mu S\,cm^{-1}$ whilst those receiving water from the Rio Negro, a 'black-water' tributary, had conductivities of $5.1-50\,\mu S\,cm^{-1}$ with a mean at 16.7. Lakes receiving both types constituted a third 'mixed-water' category. The white waters are supplied from the Andes which explains their higher salt content and also why calcium, from the rocks of the mountains, exceeds sodium (see Chapter 2). The black-water lakes have a low pH range of 4.0–7.0 which is partly due to the low buffering capacity and partly due to the organic acids resulting from the incomplete decomposition of organic material. They may not be perpetually dark in colouring; for example the Lago Crystallino which is fed by the Rio Negro, is deeply coloured at high water, but as the river recedes the humic substances begin to precipitate out leaving the lake clear during the low-water period (Tundisi *et al.*, 1984). The phosphate levels in the black-water lakes were found to be relatively stable between 61.5 and 74.2 μg l^{-1}, but this only reflected the lack of biological phosphate utilization. In the white-water lakes, with a pH range of 6.6–8.4, the phosphate range was 0–115.7 μg l^{-1} which indicates considerable biological utilization of phosphate. Accumulation in the planktonic community was demonstrated by total phosphorus which averaged 246.5 μg l^{-1} in the white-water lakes compared to 8.4 μg l^{-1} in the black-water types, showing the lack of accumulated organic phosphorus in the latter case. The lack of phytoplankton and suspended organic material also gives the black-water lakes a higher trans-parency. In fact, these are extremely unproductive (Section 5.3). Lakes of this type can be created when black-water rivers are impounded to form reservoirs. The tea-coloured water of the Guri reservoir in Venezuela is derived from the humic substances in the black-water Caroni River. This drains wet forest areas and supplies the main inflow into the reservoir. The water has a very low mineral content, as indicated by the conductivity of 9 μS cm^{-1} and an acidic pH of 6.4 (Lewis and Weibezahn, 1976).

This scheme of Rai and Hill is not applicable to all lakes in the Amazon basin or other lowland tropical areas because there are more water types than just 'white' and 'black'. Most often, however, lakes in the forest regions do have low conductivity, low pH, and a dominance of sodium over calcium owing to the predominant effects of abundant rainfall and the weathered nature of rocks and soils.

In all of these categories the biological characterisitics are a product of the varied chemical features of the water. The biological process can further act to modify aspects of the chemistry through photosynthesis, respiration, the pres-ence of swamps or even, in some instances, large numbers of animals. In Lake George, for example, the input of nitrogen and phosporus from the hippopotamus population accounts for 2% and 10% respectively of the annual loss of these elements (Viner, 1975b). Whilst the pelican population of Lake Nakuru was responsible for the annual export equivalent of 10% of the total phosphate from the Lake (Vareschi, 1979). The primary features of rainfall and leaching from rocks and soil play a dominant role, particularly in Class I lakes, in establishing the major chemical features of lakes, as they do with rivers (Chapter 2), either through the river inflows themselves or by more direct means, depending upon the nature of the lake. In Class II and Class III lakes, evaporation crystallization become progressively more important in just the same way as Gibbs (1970) suggested for rivers (Chapter 2).

3.5 STRATIFICATION AND ITS EFFECTS

3.5.1 General Features

Tropical lakes receive a greater degree of insolation than do their temperate counterparts and this may be spread over longer periods of the year. Whilst the

36

sun provides light energy, it is also a source of heat which will have its maximum effect on the surface of standing waters. Insolation does vary with season in the tropics and this is reflected in the temperature changes in the upper-water layers of lakes, although wind patterns, which exert a cooling effect, can modify this. A survey down the African continent (Fig. 3.2) shows the shift from the cool season in January–February in northern latitudes to July–August in the south.

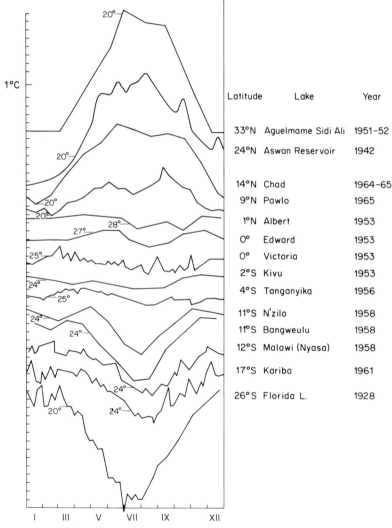

Latitude	Lake	Year
33°N	Aguelmame Sidi Ali	1951–52
24°N	Aswan Reservoir	1942
14°N	Chad	1964–65
9°N	Pawlo	1965
1°N	Albert	1953
0°	Edward	1953
0°	Victoria	1953
2°S	Kivu	1953
4°S	Tanganyika	1956
11°S	N'zilo	1958
11°S	Bangweulu	1958
12°S	Malawi (Nyasa)	1958
17°S	Kariba	1961
26°S	Florida L.	1928

Fig. 3.2 The annual variation of surface temperature in a series of fourteen African lakes, arranged by latitude. Successive curves are displaced downwards and the common temperature scale (left) refers to differences only. Absolute values can be located using the single temperature marked on each curve. (*Reproduced by permission of Schweizerbart'sche Verlagsbuchhandlung from Talling, 1969*)

The amplitude of the oscillations is considerably reduced towards the equator and it can also be seen that lakes just north of the equator possess maxima and minima at periods corresponding to those in the southerly lakes owing to the wide influence of the trade winds (Talling, 1969).

The warming of the upper layer of water is not transferred equally down through the whole water column. As the temperature of water is increased beyond 4 °C, the point of maximum density, there is a progressive decrease in its density. Therefore, as the surface layer is warmed, it becomes light and effectively floats on the cooler, denser water below. These two layers can become quite distinct and are detectable by a significant change in temperature from the warm upper water to the cooler lower depths over a very restricted depth of water. In a temperature profile down through the water column this appears as a thermal discontinuity (Fig. 3.3), and the transitional zone between the warm lighter water and the cooler denser water is called the thermocline. In the presence of a thermocline the water is said to be stratified. The water body above the thermocline is referred to as epilimnion and that below it as hypolimnion, whilst that in the discontinuity itself is the metalimnion. The sig-nificance of the thermocline is that it acts as a physicochemical barrier to diffusion and consequently can isolate the epilimnion from the hypolimnion with respect to the circulation of dissolved substances.

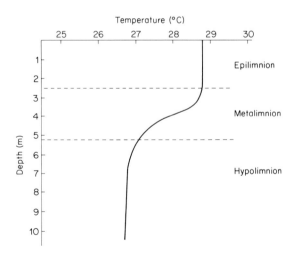

Fig. 3.3 Diagrammatic temperature profile of a stratified water column showing thermocline structure

The intensity and duration of sunlight in the tropics means that the thermocline is a significant feature of many tropical lakes. However, unlike temperate lakes where the formation of the thermocline in the summer is marked by a change of several degrees centigrade across the discontinuity, in the tropics the difference in temperature may be quite small. In Lake Malawi, for example, there is a stable, enduring thermocline at between 200 and 250 m depth, but the difference in temperature is 1 °C (Eccles, 1974). Such a small temperature difference constituting a thermocline is not confined to large lakes, but can also be found in small ones such as Subang Lake in Malaysia where temperature discontinuities of the order of 0.5 °C at 3–4 m were sufficient to establish thermocline effects (Arumagam and Furtado, 1980). The reason why such small differences can create a thermocline in the tropics is due to the fact that the density of water decreases disproportionately with increase in temperature so that, for example, an increase from 26.5 to 27.5 °C gives a density decrease equivalent to a change from 4 to 10 °C at the lower end of the scale. In short, small changes in the tropics, where water temperatures tend to be between 20 and 30 °C, can give quite a large change in water density, certainly sufficient to create a stable thermocline with its resulting barrier effect.

Once a thermocline is formed, organic material can sink through, but salts or gases in solution cannot diffuse across it. Typically this means that, since the productive zone where light is sufficient to promote photosynthesis comes within or includes the epilimnion, dying phytoplankton and detritus sink through the thermocline into the hypolimnion where decomposition takes place. Accumulation of organic material in the hypolimnion results in oxygen utilization during decomposition which cannot be replaced owing to the inability of oxygen to diffuse across the thermocline from the oxygenated epilimnion. This can lead to severe deoxygenation of the hypolimnion. The sinking of the algae and other organisms through the thermocline also constitutes a drain of nutrients from the productive zone of the epilimnion, since they incorporate phosphorus and nitrogen which are released as inorganic compounds by decomposition in the hypolimnion. The inorganic phosphorus and nitrogen are potential nutrients to sustain primary production, but they are lost to the epilimnion because they cannot diffuse back across the thermocline. The presence of a thermocline can therefore lead to progressive impoverishment of the surface waters.

3.5.2 Effect on Physical and Chemical Factors

We can obtain a more complete picture of these and other implications of thermocline formation from a series of observations made on the Lago do Castanho, one of the floodplain varzea lakes of the middle Amazon (Schmidt, 1973a,b). Following the Amazon floods, this lake fills up and rapidly becomes stratified with the principal thermocline forming usually between 3 and 5 m down. In Fig. 3.4 a significant change in temperature can be seen between 1 and 3 m which, although only amounting to slightly more than 1.5 °C, is sufficient to establish a functional thermocline. The most immediately apparent effect of this is the great decline in oxygen between 2 and 3 m in depth and its complete elimination below 5 m (Fig. 3.5). The hypolimnion has become completely anaerobic owing to the use of oxygen for bacterial decomposition of organic material sinking from the surface layers. The oxygen content is highest close to the surface,

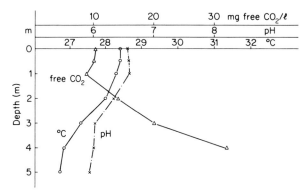

Fig. 3.4 Concentrations of free carbon dioxide and pH in the presence of a thermocline in the Lago do Castanho, a varzea lake of the Amazon. (*Reproduced with permission from Schmidt, 1973a*)

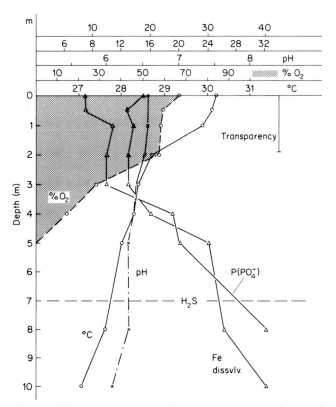

Fig. 3.5 Effects of the thermocline on oxygen, phosphate and dissolved iron in the Lago do Castanho. (*Reproduced with permission from Schmidt, 1973a.*) The depth of transparency measured by secchi disc and depth at which hydrogen sulphide is present are also indicated

partly due to uptake from the atmosphere and partly due to oxygen produced by the phytoplankton during photosynthesis. The concentration diminishes with depth as respiratory demand from bacteria, zooplankton, and from the algae themselves exceeds oxygen gain. Conversely, free carbon dioxide is low in the epilimnion where it is used for photosynthesis, but reaches very high levels in the hypolimnion where carbon dioxide is produced from bacterial respiration (Fig. 3.4). The highest concentrations are found close to the lake bed since it is here that the greatest amount of decomposition takes place. These processes are also reflected in the pH which is high above the thermocline and lower beneath it. We have already noted that rapid photosynthesis elevates the pH, hence the higher value in the epilimnion. In the hypolimnion the production of carbonic acid from carbon dioxide dissolving in the water, together with organic acids from the incomplete anaerobic degradation of organic material, combine to give the reduced pH in this region.

One further effect, mentioned previously, of decomposition taking place beneath the physicochemical barrier of the thermocline was the mineralization of the organic material to produce inorganic compounds, including those of phosphorus and nitrogen which are essential for plant production. The discontinuity in the distribution of these inorganic compounds within the water columns can be referred to as the chemocline. In the Lago do Castanho, phosphates do accumulate appreciably in the hypolimniom whilst remaining in very low concentrations, generally between 0 and $0.46 \, \mu g \, l^{-1}$, in the epilimnion where utilization and turnover are rapid owing to high rates of photosynthesis. It is evident, however, that particularly high phosphate concentrations are found at the bottom of the lake (Fig. 3.5) and this introduces a new element—interrelationship of the water with the mud surface. At a pH of 7–8 and under aerobic conditions, the mud surface readily adsorbs many ions including phosphate, but under acidic and/or anaerobic conditions the situation is reversed and ions are released. It has been suggested that the mechanism is a ferric hydroxide/organic humus complex with amphoteric properties and that the complex breaks down under the latter set of conditions (Mortimer, 1941, 1942). The anaerobic conditions and low pH are present in the hypolimnion and this has caused additional release of phosphate from the mud as well as that being produced directly from decomposition. It can also be seen that iron levels increase towards the sediment, which suggests some iron release. This can often take the form of colloidal iron which gives a rather pearly appearance to the water. In addition, the reducing conditions in this particular instance may lead to the production of FeS which would precipitate out and would also tend to increase the iron concentration close to the lake bed.

Phosphate, therefore, appears in elevated concentrations below the thermocline as a result of decomposition, but the same is not true of nitrates. The concentration of nitrate in the surface waters is never high. In the Lago do Castanho it is frequently less than $20 \, \mu g \, NO_3–N \, l^{-1}$. There may be a slight increase at the immediate surface, possibly due to nitrate in rain-water or due to suppression of algal growth and, therefore, demand by ultraviolet (UV) light.

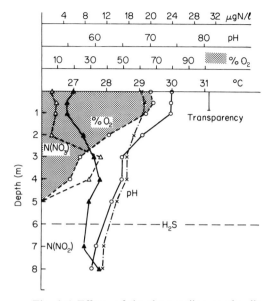

Fig. 3.6 Effects of the thermocline on the distribution of nitrate and nitrite nitrogen in the Lago do Castanho. (*Reproduced with permission from Schmidt, 1973a*)

There is often also a discernible elevation at the thermocline (Fig. 3.6) which suggests a sharply defined zone of intensive nitrification and denitrification of the type found in eutrophic stratified lakes elsewhere (Schmidt, 1973a,b); there is, however, no nitrate to be found in the hypolimnion below 5 m. This is because, although nitrogen is being liberated by decomposition, it will be in its reduced form—ammonia—under these anaerobic conditions. Ammonia was not measured in the particular study under consideration, but in a similar situation on the stratified Lake Subang in Malaysia, the ammonia concentration varied from $0.1 \, mg \, NH_3–N \, l^{-1}$ at the surface to $7 \, mg \, NH_3–N \, l^{-1}$ at the bottom at a time when the hypolimnion was completely anaerobic (Arumugam and Furtado, 1980). One effect of the reducing conditions below the thermocline is the reduction of sulphate to hydrogen sulphide. The level at which this could just be detected by smell was included in the records from the Lago do Castanho (Fig. 3.5).

Many of these characteristics are observable in most stratified tropical lakes, although they may vary by degree. In lakes where the thermocline persists for a few days or weeks, anaerobic regions tend to develop. Lake Carioca, for example, a small forest lake in Brazil, shows a marked thermocline between 5 and 7 m depth mainly representing the limits of mixing due to nocturnal cooling and it is completely anaerobic between 6 m and the bottom at 9 m (Barbosa and Tundisi, 1980; Reynolds *et al.*, 1983). Lake Valencia, Venezuela, has a primary thermocline at 30 m and is anaerobic below this although the discontinuity lies only 6 m above the lake bed (Lewis and Weibezahn, 1976). The greatest part of the lake is, therefore, epilimnetic.

The depth and stability of the thermocline depends upon the duration of the calm, warming period and the frequency and magnitude of mixing events. In Lake Victoria where the thermocline most often occurs at 30–40 m depth, complete mixing of this enormous water body occurs once a year and partial mixing occurs at other times. Totally deoxygenated conditions and the presence of ammonia, therefore, are only found for a few weeks in February close to the bed of the lake at 60 m.

Even in high-altitude lakes where temperatures are generally low, some degree of stratification can occur. Lake Mucubaji, a Paramos lake of the Andes, showed a weak stratification at 8–9 m, below which a significant reduction in oxygen concentration could be detected (Lewis and Weibezahn, 1976). Some of the East African high-altitude lakes also possess small thermal discontinuities (Löffler, 1964), whilst the small Hawaiian lake, Lake Waian, showed an inverse thermocline in December (Maciolek, 1969) with cold, light water approaching 0 °C floating on top of water close to 4 °C, the temperature at which water is most dense.

In some cases thermal stratification can be reinforced. In the Lagartijo Reservoir in Venezuela, for example, the cold inflowing rivers keep the hypolimnion at a lower temperature than the surface waters even during the cool season. Destratification is, therefore, prevented (Lewis and Weibezahn, 1976).

3.5.3 Multiple Thermoclines

The stratified situation may become complicated by the formation of more than one thermocline.

In Lake Lanao the primary epilimnion, which may lie at depths of 40–60 m, can split into two or more layers during hot, calm periods, as additional thermoclines form closer to the surface (Lewis, 1974). Production only takes place in the uppermost layer which is also subject to the usual nutrient drain across the thermocline. At the same time, decomposition of senescent plankton in the lower epilimnetic layers gives rise to the characteristic chemical differences normally found across a thermocline, as described in the previous section, including a reduced oxygen concentration.

This effect can be clearly seen in a profile taken from Lake Valencia, Venezuela (Fig. 3.7). The most superficial discontinuity at 5 m probably represents the limits of daily mixing caused by nocturnal cooling under relatively calm conditions. It is accompanied by a reduction in oxygen content from 7.5 to 6.0 μg l^{-1}. There is a further discontinuity, although less sharply defined, between 10 and 15 m, again accompanied by a reduction in oxygen concentration, which could coincide with mixing on moderately windy days. Considerable winds or cooling would, however, be required to mix the whole epilimnion down to the primary thermocline at 28 m depth. This is almost certainly an infrequent event and it is only below this that completely anaerobic conditions occur (Lewis and Weibezahn, 1976).

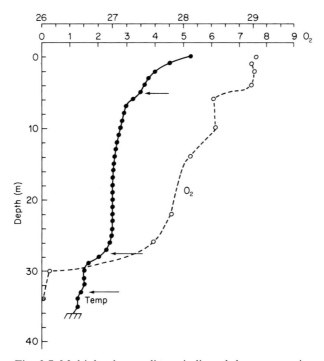

Fig. 3.7 Multiple thermoclines, indicated by arrows, in Lake Valencia, Venezuela. (*Reproduced by permission of Schweizerbart'sche Verlagsbuchhandlung from Lewis and Weibezahn, 1976*)

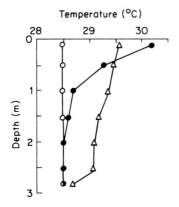

Fig. 3.8 Vertical temperature profiles in Lago Calado at different times of day: 6.10 (○), 12.15 (●), 15.30 (△). (*Reproduced by permission of Schweizerbart'sche Verlagsbuch-handlung from Melack and Fisher, 1983*)

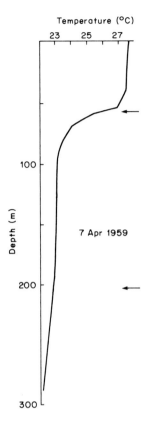

Fig. 3.9 Multiple thermoclines in a deep, permanently stratified lake, Lake Malawi. (*Reproduced by permission of the American Society of Limnology and Oceanography Inc. from Eccles, 1974*)

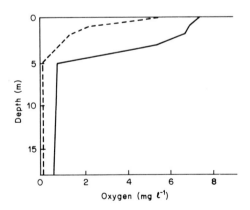

Fig. 3.10 Difference in depth of oxycline at the northern (———) and southern (– – – –) ends of Ranu Lamongan, Indonesia, when the wind was blowing from the south. (*Reproduced by permission of the Zoological Society of London from Green* et al., *1976*)

Even in relatively shallow lakes such as Lake Carioca, Brazil, the increase in temperature of the surface waters during the day can give a marked stratification on top of the persistent thermocline lower down. Nocturnal cooling, however, causes this to break down and disappear during the night (Barbosa and Tundisi, 1980). The formation of temporary daily thermoclines affecting the first few metres of the water column is a common feature of tropical lakes as in the shallow Lago Calado (Fig. 3.8).

In deeper lakes the most stable thermocline may be at a considerable depth, as has been mentioned for Lake Malawi and Lake Victoria, but above this, more marked and less stable discontinuities can develop. In Lake Malawi there is always a discontinuity of about 0.5 °C at 250 m, but a marked seasonal thermocline develops in the upper 100 m which can amount to a difference of 2–3 °C in temperature (Fig. 3.9). On top of this there may be diurnal or more short-term effects closer to the surface still. The precise depth of a thermocline may vary since it can become deeper as the water body warms up or be driven down by turbulence caused by sustained winds. Persistent winds can also cause piling up of surface waters at the windward end of a lake with the consequent tilting of the thermocline downwards at that end. When a south wind was blowing the epilimnion was some 2 m lower at the northern end of Ranu Lamongan in Indonesia (Fig. 3.10) than at the southern end as surface water evidently accumulated near the downwind shore (Green et al., 1976). In this case the position of the thermocline was detected by the sharp drop in oxygen, the oxycline, across the thermal discontinuity. Surface turbulence caused by the wind can also cause oscillations or internal waves called seiches between the two layers separated by the thermocline. This effect could also be found under windy conditions in Ranu Lamongan as the thermocline became re-established, resulting in a varying depth of the epilimnion at one particular station during the day (Fig. 3.11).

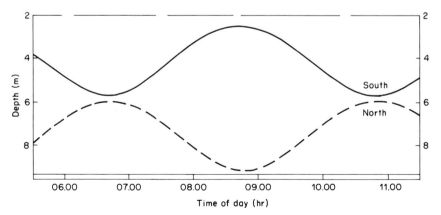

Fig. 3.11 Calculated oscillations in the oxyclines at southern and northern stations of Ranu Lamongan during a northerly wind. (*Reproduced by permission of the Zoological Society of London from Green* et al., *1976*)

3.5.4 The Importance of Mixing

If thermoclines are permanent then the continual drain on nutrients from the epilimnion to the hypolimnion causes productivity to become very limited, even though light energy levels are high. This is widely the case in tropical oceans with the deep, clear water indicating a lack of phytoplankton through a scarcity of nutrients. In some deep lakes, such as Lake Tanganyika and Lake Malawi, this is also the case and a permanent thermocline is present but, in many, some degree of mixing does occur causing enrichment of surface waters with nitrates and phosphates from deeper areas.

Mixing is generally brought about by cooling at the surface, which itself may be enhanced by wind or by physical turbulence resulting from wind action. Cooling causes the water to become denser and therefore to sink down through the water column. In the simplest situation the annual temperature fluctuation is great enough for the surface water to cool sufficiently to cause this sinking. Thus in Lake Lanao (8° N) in the Philippines, at the end of December, the water body becomes isothermal at 24.4 °C through surface cooling, and nutrients such as nitrate become more evenly distributed through the water column. During May and June the water warms to almost 28 °C, although below the thermocline the water was found to remain at 24.3 whilst the range of mean monthly air temperatures was 21–23.8 °C (Frey, 1969). Lakes which show a single seasonal mixing are termed monomictic.

The predictability of such seasonal events in tropical lakes can be quite high although long time series of data are few. Observations on Lake Valencia in Venezuela showed that the lake was stratified from April to November and that seasonal mixing always began in November or December. Moreover, variation between years in dates of overturn, maximum stability of stratification, and the duration of stratification was very small (Lewis, 1984a,b).

Considerably smaller lakes may also be affected in the same way. Lake Carioca in Brazil has an isothermal period in the cool season during July and August, which appears to be solely the result of surface cooling since the dense forests surrounding the lake protect it from wind action and there are no cool inflows (Barbosa and Tundisi, 1980). Surface cooling alone, therefore, can cause complete mixing of a lake in the tropics.

In many lakes, however, surface cooling is augmented by wind action. Air movement both speeds up the rate of cooling and also causes turbulence in the water which promotes mixing. This may be particularly important in more equatorial lakes where seasonal changes in temperature may be insufficient alone to produce complete mixing. In Lake Victoria (0–2° S), for example, the extreme surface-water temperatures only range from 23.8 to 25.8 °C over a year, but the lake still becomes completely mixed for a brief period in July. It seems most likely that this is due to the cooling of water in shallow areas. These shallows are usually 0.5 °C less than the deeper areas and, at the cooler time of the year, appear to lose heat more rapidly. The cooler, heavier water sinks down from the shallows into the deeper regions of the basin and promotes mixing in this way (Talling, 1966). A principal factor in this cooling is the effect of the south-east trade winds which blow from June to August, and quite frequently in tropical lakes it is the stormy windy period at the advent of the rainy season which is responsible for mixing.

In Lake Lanao, turbulence caused by wind action assists in the breakdown of the thermocline during the major period of circulation in January and February. Episodes of strong winds can also cause some mixing, however, by breaking down subsidiary stratification within the primary epilimnion and by depressing the thermocline when stratification exists (Lewis, 1974).

Even in lakes with a permanent thermocline winds can have an effect. In Lake Malawi the strong south-easterly winds are channelled along the north/south axis of the lake and give a tilting of the thermocline. This is driven deeper at the northern end and approaches the surface at the south (Fig. 3.6) in a similar fashion to that observed in the small Indonesian lake, Ranu Lamongan, mentioned in the previous section. As it happens, the southern area of the Lake Malawi is shallow, and nutrient-rich water from the hypolimnion emerges into this arm by the process of upwelling from beneath the intact thermocline. As the winds drop so the thermocline returns to its normal position, although with some oscillation caused by seiches as it settles back (Eccles, 1974). This tilting of the thermocline through prolonged unidirectional wind action is a common feature of stratified lakes and it may result in upwelling. In Lake Malawi this results in a rich fishery in the southern area where nutrient replenishment for the phytoplankton takes place.

These events can cause a short-lived resurgence in surface nutrient levels, as those trapped within the secondary layers are mixed throughout the primary epilimnion. This process has been termed atelomixis (Lewis, 1974) and generally means that storms are followed by a brief bloom in the phytoplankton which diminishes as the more superficial stratification becomes re-established. The cooling effect of wind can have quite dramatic results, particularly when the hypolimnion has become anoxic. During the high-water period in the Amazon, in May and September, a very cold wind, the *friagem*, may blow for up to four days when the air temperature can drop 15 °C. This gives a rapid cooling of the varzea lakes with the result that the cold, heavy surface waters sink rapidly giving a complete overturn of the water body with the consequent distribution of anaerobic water rich in hydrogen sulphide which causes massive fish-kills in the lakes (Schmidt, 1973a,b; Tundisi *et al.*, 1984).

Extensive fish-kills have also been observed in the Indonesian crater of Ranu Lamongan. This lake normally has a very stable thermocline, but seasonal cooling can cause a complete overturn with anaerobic water being brought to the surface resulting in substantial fish mortalities (Green *et al.*, 1976).

3.5.5 Shallow Lakes

Stable stratification does not occur in shallow lakes where the body can be heated right through and is more readily mixed. Lake George, which is right on the equator, has a very shallow water column of about 2.4 m, but whereas it can show sharp daytime stratification with, in extreme cases, the surface waters reaching 36 °C by 17.00 hours whilst the bottom remains at 25 °C, this breaks down every night and mixing is generally promoted both by nocturnal cooling and a fairly persistent wind regime. This, together with the relative constancy of the climate, means that the lake is dominated by a 24-hour cycle rather than an annual one (Viner and Smith, 1973; Ganf and Horne, 1975).

In another East African lake, Lake Nakuru, which has a mean depth of some 4 m, the water column tended to be strongly stratified for 3 hours, weakly stratified for 6 hours, and more or less unstratified for 15 hours over a 24-hour period (Vareschi, 1982).

Similar effects can be seen even in shallow lakes some distance from the equator such as the shallow man-made Lake Moondarra in Queensland, Australia (25° S). Owing to the cool, dry season and warm, wet season in this part of the tropics, the thermal stratification breaks down at night in the cool season and intense rainstorms prevent persistent stratification in the warmer months (Finlayson et al., 1980).

Lakes in which there is frequent mixing, sometimes at irregular intervals, are termed polymictic, as opposed to the infrequent mixing or oligomictic condition of deeper lakes. In these shallow lakes without stable stratification there is no drain of nutrients and anaerobic conditions do not occur, which has considerable implications for their productivity. In deeper lakes stratification may not be found in the shallower areas, for example in Lake Victoria bays of less than 10 m depth do not become stratified except on a daily basis (Talling, 1965).

When the varzea lakes of the Amazon basin are at their shallowest, during the low-water period, they also show stratification during the day (Fig. 3.8) with complete mixing at night (Schmidt, 1973a,b; Tundisi et al., 1984). They are usually less than 4 m in depth at this stage and because of the mixing also remain well oxygenated throughout the 24-hour cycle, although the concentrations tended to remain higher and the diel fluctuations much less in the unproductive black-water Lago Cristalino than in the white-water, heavily vegetated Lago Jacaretinga (Tundisi et al., 1984). There can also be something of an accumulation of nutrients below these daily thermoclines, but these become mixed at night. Resuspension of sediments from the bottom of these shallow lakes also contributes to the pulse of nutrients caused by nocturnal mixing, and benthic algae are also mixed into the water column, as observed in Lago Jacaretinga (Tundisi et al., 1984).

During the high-water season the varzea lakes become much deeper and consequently, like the Lago Cristalino which increases to 8 m depth (Tundisi et al., 1984), change their mixing pattern from the polymictic type typical of shallow waters to an oligomictic one. However, the Lago Jacaretinga at high water only increases to 5 m and daily mixing can be observed throughout the year. Nevertheless, the oxygen concentrations in the water became very low as the extent of mixing in the bottom layers is insufficient to give complete reoxygenation. The effect is cumulative and the bottom water can decline in oxygen to

$0.5\,\mathrm{mg\,l^{-1}}$ and, because of nocturnal mixing, water of low oxygenation can be mixed throughout the water column in the early morning (Tundisi *et al.*, 1984). In this case the lake is just at the limit of the depth which is influenced by nocturnal cooling. In general it appears that lakes shallow enough to experience complete mixing over a 24-hour period are less than 4–5 m deep.

GENERAL READING

Beadle, L. C. 1981. *The Inland Waters of Tropical Africa*, London, Longman, Chs. 5 and 6.

Serruya, C. and Pollingher, U. 1983. *Lakes of the Warm Belt*. Cambridge, CUP.

CHAPTER **4**

COMMUNITY STRUCTURE

4.1 BASIC ORGANIZATION

We have seen in previous chapters the range of abiotic physical and chemical factors that provide the environments for living organisms of lakes and rivers in the tropics. Within these environments live mutually dependent assemblages of plants and animals, including mircro-organisms, forming more or less identifiable communities. The underlying structure of these is common to most others. The principal transformers of light to chemical energy are the autotrophs, commonly plants, which are, therefore, the primary producers of chemical energy for the community. Feeding upon the plants, since they are unable to synthesize their own chemical energy, are the heterotrophic animals which could commonly be termed herbivores although they are more properly called first-stage consumers or secondary producers. At the next level are animals acting as predators which obtain their energy from the first-stage consumers whilst they, themselves, may act as prey for other predators. In this way a tiered structure becomes apparent which was first formalized into the concept of trophic levels by the American ecologist, Lindeman, in 1942. Each trophic level can be seen as an accumulation of chemical energy in the form of organic compounds incorporated into tissues of the organisms. There is a continual supply of energy to each trophic level—light in the case of the first level occupied by plants—and also a continual loss due to consumption and mortality. The actual amount of energy accumulated in one trophic level, therefore, represents an equilibrium between energy input and removal. The maximum number of levels in the trophic system between the plants and the last predator in the chain has been found to be five, although this is often not achieved. The reason why the number of levels is limited is because within each trophic level a certain amount of incoming energy is used by the plants or animals in that level and is lost through respiration, excretion, or defecation; consequently there is less energy available for use by organisms of the next trophic level. In other words there is a net loss of energy as it passes from one level to the next and after the fifth transfer it has all been dissipated.

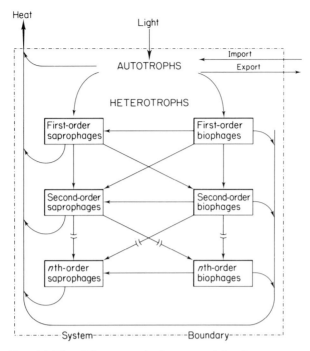

Fig. 4.1 The Wiegert and Owen model of ecosystem structure and function showing parallel grazer (biophage) and decomposer (saprophage) trophic levels. (*Reproduced by permission of Academic Press Inc. (London) Ltd. from Wiegert and Owen, 1971*)

This sequence of events, as energy is transferred from plant to herbivore to carnivore, also has its parallel amongst those organisms, the decomposers, which depend upon non-living material. These also depend upon the primary producers but in this case they use dead material commonly referred to as detritus. In a community, therefore, there are two parallel sequences of energy transfer, one relying ultimately upon living plant material, which can conveniently be termed the 'grazing chain' and the other upon dead material which can be called the 'detrital chain'. Like the grazing chain the detrital chain has a trophic structure with dead plant material being acted upon by a first level largely composed of the micro-flora, bacteria, and fungi, which in turn provide an energy source for a range of invertebrate animals. These largely constitute the second level, although the distinction between levels can be uncertain where it is not clear if the animal can actually feed directly upon detritus and is therefore operating at the first level, or whether it is only utilizing the microflora associated with the detritus. This can be very difficult to ascertain in practice, but where it has been possible the animals appear to feed largely upon the microflora alone. In addition to plant material these two trophic levels also have access to the dead bodies from animals of any trophic level in the grazing chain together with faecal debris and excretory products from this chain. The detrital chain also has its associated predators, giving a third and also possibly a fourth trophic level. There are, therefore, two complementary pathways for energy utilization within the community, one based upon live plant material, the other upon detritus; each has an identifiable structure and, whilst they operate in parallel (Fig. 4.1), they are also ultimately linked (Wiegert and Owen, 1971). In organic systems, however, the distinction between the chains becomes blurred where animals which feed principally upon small particles, that is, microphagous feeders, are common. The fine particles may consist of detritus or bacteria or microscopic plants, largely algae or most often a combination of these organisms from different trophic levels and, theoretically, different chains. A system such as that shown in Fig. 4.1, therefore, provides only a framework or starting-point for consideration of the structure and function of ecosystems.

In terms of the animals and plants which make up the community it is often difficult to confine species to specific trophic levels because many animals have a varied diet and therefore may obtain food items from more than one level. It is this variability in diet which results in the web-like feeding relationships amongst organisms than a rather simple linear chain-like transfer of energy. Such a situation can be particularly complex in the tropics where communities are often characterized by the large number of species they contain. Nevertheless, trophic levels do remain distinct stages in the processing of energy once this has been incorporated into organic compounds by the primary producers. It may, however, be necessary to partition the contribution of individual species between more than one of these to determine the dimensions of energy and material input, loss, and accumulation in relation to each level.

4.2 TYPES AND DISTRIBUTION OF ORGANISMS IN AQUATIC COMMUNITIES

As well as having a functional structure, as outlined above, aquatic communities also have a well defined three-dimensional spatial structure owing to the support water can impart to organisms. These two aspects, however, are often closely related.

In aquatic environments a considerable proportion of the organisms live suspended in the water column. These types, which are often microscopic and have only limited powers of locomotion, constitute the plankton. Owing to their very restricted mobility, they have very little power to influence their own distribution which is largely determined by movements of the water body. Amongst these small suspended drifting organisms the primary producers are the phytoplankton which are mainly single-celled algae although quite commonly they can be aggregated together into colonies or chains. Associated with the phytoplankton are the small animals which feed upon them or their detrital remains, the zooplankton. All suspended material, living or dead, is called the seston.

Naturally, since the phytoplankton and zooplankton are largely at the mercy of water currents, they tend to be more important in lakes than in rivers where there is always a continual drift downstream to the sea.

As the phytoplankton are primary producers and carry out photosynthesis, light is of considerable importance to them. Light is absorbed much more rapidly by water than by air, and this capacity is greatly increased if there is also suspended material in the water. Light is decreased exponentially rather than in a linear fashion with depth (Fig. 4.2) and therefore the intensity diminishes rapidly below the surface. This means that energy for powering photosynthesis declines rapidly with depth and it is generally regarded that when light diminishes to 1% of the incident surface intensity that there is insufficient for useful photosynthesis to take place. The layer of water above this depth, where there is sufficient light for photosynthesis to occur, is called the euphotic or trophogenic zone. The euphotic zone, therefore, defines where effective photosynthesis is possible and its depth can vary considerably from 50 m or more in deep, clear lakes such as Lake Tanganyika to around 1 cm in turbid rivers. Its depth can vary from day to day, depending upon cloud cover, or seasonally, depending upon the pattern of insolation or seasonal occurrence of suspended material which in turn may be related to the production cycle.

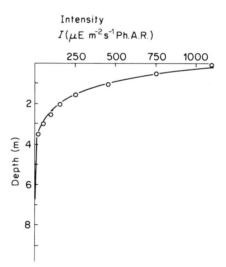

Fig. 4.2 Penetration of photosynthetically active light (Ph.A.R.) in the Lago Carioca, Brazil, showing the exponential attenuation of light with depth. (*Reproduced by permission of Schweizerbart'sche Verlagsbuchhandlung from Reynolds* et al., *1983*)

Below the euphotic zone the light intensity becomes progressively dimmer although it may still be sufficient to provide some visual information. A simple way of obtaining an indication of relative light penetration or water transparency is by lowering a weighted disc of 25 cm diameter which has been painted alternate black and white quarters until it just disappears from view in the water and then noting the depth. This device is called a Secchi disc and experience shows that the disc disappears at a depth where light in the water column is about 15% of that at the surface although this can vary, for example, with the visual accuity of the operator. The depth at which the Secchi disc disappears also tends to be between 0.3 and 0.5 of that of the euphotic zone and so can be used to provide an approximation of this. If the water is extremely turbid or the water column is very deep then all light may be extinguished. Some of the deepest lakes such as Lake Malawi or Lake Tanganyika have zones, at several hundred metres depth, corresponding to those in the deeper oceans, where the light has become extremely attenuated or absent.

In addition to the plankton the water is also occupied by free-swimming types such as fish, turtles and amphibians, which have sufficiently well-developed powers of locomotion to determine their own distribution within the lake or river. As a group within the ecosystem these can be referred to as the nekton (Figs. 4.3, 4.4).

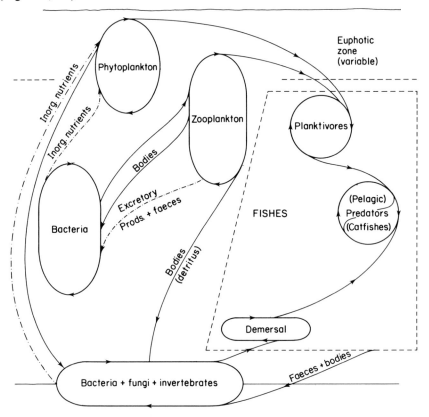

Fig. 4.3 Spatial community structure in a tropical lake ecosystem

55

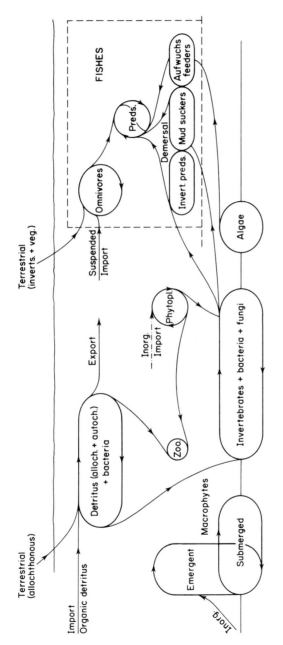

Fig. 4.4 Spatial community structure in a tropical river ecosystem

The organisms which live in the open waters, the phytoplankton, zooplankton, and fishes tend to contribute to the grazing chain, whilst those that live on the bed of the lake or river, the benthos, are more likely to be associated with the detrital chain (Fig. 4.3) if for no other reason that dead material sinks naturally to the bottom thus emphasizing the spatial and functional links within the community. The distinction is far from complete, however, since there can also be a rich bacterial flora suspended in the water amongst the plankton and some of their effects have been previously mentioned in respect to decomposition below the thermocline (Section 3.3). Nevertheless, the sediments or rocks at the bottom of the water column can have a rich microflora of bacteria and fungi associated with invertebrates such as oligochaete worms, crustacea and insect larvae, which play a major role in breaking down detritus and recycling minerals. Once again, however, this is not the sole role of the benthic organisms; the bed of a lake or river can also provide a place for primary producers although, owing to the limited light penetration through the overlying water, the extent to which this is possible depends upon the depth of the water. For benthic primary producers, which could be either algae or higher plants, to be able to photosynthesize effectively, the euphotic zone must extend down to the bottom; they therefore tend to be confined to shallower or less turbid parts of rivers or lakes and this explains why the bottoms of the majority of lakes are not covered with water plants. The benthic producers themselves are divided between microscopic single-celled types and the large multicellular plants, the macrophytes. Many algal groups have representatives that live in or on the sediment or on hard surfaces. The latter types are to be found coating the surfaces of rocks or sunken tree trunks along with other micro-organisms and small invertebrates. These assemblages on underwater surfaces have become known by the German word *aufwuchs* and they provide an important source of food for grazers in all shallow waters but particularly in rivers where the constant water flow makes life difficult for suspended micro-organisms. The benthic region provides a rooting medium for the macrophytes which generally grow up through the water column and either approach the surface or actually emerge, depending upon the nature of the species. The macrophytes themselves are mainly vascular plants which are not often directly grazed by aquatic animals but rather contribute their dead remains which can become a major souce of detritus in conditions where they become abundant, such as swamps (Fig. 4.4). Because they are rooted, macrophytes are often confined to the shallowest regions along the margins of lakes and rivers, although where considerable areas of shallow standing water accumulate they can become the dominant form of vegetation. The submerged parts of these plants, like other hard surfaces, provide an ideal site for the attachment of microscopic algae, and often a layer of these can be formed, becoming evident when the scrapings from a stem surface are examined under a microscope. These algae and also the bacteria, which may form a thin slimy coat on submerged plant stems, are also included in the aufwuchs or are sometimes alternatively named the periphyton.

4.3   STRUCTURE IN LAKES AND RIVERS

The main elements and pathways in the communities of tropical lakes and rivers are seen marked in Figs. 4.3 and 4.4. They are not very different patterns from those of temperate zones since major distinctions tend to be due to the number of species involved, the relative magnitude of the niches, and the timing of events rather than to different basic components. There are some significant differences particularly in the form of phytoplankton-feeding fishes in the tropics which are commonly absent in temperate aquatic environments where zooplankton are the major herbivores. There are significant differences between lakes and rivers, however, even though the basic structure may be the same. The lake, through being confined to a discrete basin, provides suitable, relatively stable conditions for suspended organisms, therefore the system is largely dependent upon phytoplankton and primary production.

Rivers, by contrast, provide a system of continuous movement where any suspended particle will, often within a few days, be carried to the sea. Such a situation is far from ideal for plankton and this, combined with the fact that the water movement is inclined to promote turbidity and thereby reduce light penetration, means that the phytoplankton is a relatively minor element. Instead, much of the material available is in the form of detritus either washed into the river from the bed within its catchment area or from dead macrophytes within the river itself. The aufwuchs or periphyton also have increased importance in rivers compared to lakes under such conditions.

Out of the mainstream of rivers, however, other sorts of habitats are to be found. There are backwaters and semi-isolated pools where the current is slight or absent and the suspended material in the water can sediment out, leaving conditions more suitable for animals and plants which prefer still waters. Then there are the deep still 'mouthbays' found at the confluence of major tributaries of the Amazon, or the extensive flooded areas produced by many rivers during the rains or swamps which can be found alongside many rivers. Therefore, although rivers are essentially flowing water systems they provide a great variety of habitats for living organisms.

4.4   QUALITATIVE AND QUANTITATIVE CONSIDERATIONS

Up to this point the communities of organisms have been separated into groups according to what they do and where they are found but, for further understanding, the detailed composition of these groups needs attention along with a consideration of the more important environmental requirements. In addition, another aspect of the structure of these freshwater communities is how much or how many of the organisms there are within each of the various components. It has already been mentioned that at each trophic level—although there is a constant flux of energy and materials through it—a certain quantity accumulates. This is equally true for each constituent plant or animal population.

As a measure of this accumulation the total weight of organisms from a specified area or volume, at one point in time, gives an estimate of the standing crop or biomass. Weight or biomass, therefore, has greater ecological significance than numbers of organisms but numbers, as frequency or abundance, do add another dimension to community structure. Moreover, the species composition within communities is often a function of the particular environmental conditions encountered by that community.

4.5 THE PHYTOPLANKTON

4.5.1 Algae Large and Small

Most groups of algae have representatives in the phytoplankton and their cells range in size from 1 µm to colonial types such as *Volvox* which can be seen by the unaided eye. However, the majority of phytoplankton species fall within the range 2–200 µm. They are generally subdivided by size, originally due to the shortcomings of sampling methods. In the early days of limnology the main method for sampling phytoplankton was with a plankton net which is a conical net ending in a container. The bag of the net is constructed of fine fibre, originally silk, woven to produce a very small pore size through which water can be filtered, often by towing the net behind a boat. There is, however, a limit to how small the pore size can be made so that even the finest mesh nets tend to select the largest cell sizes. If, however, filtration or centrifugation is used to concentrate the algae then a wide variety of smaller cells are also found in the sample. The larger cells, generally between 20 and 200 µm can be collectively grouped together and termed the netplankton, for obvious reasons, whilst the smaller types within the 2–20 µm range are called the nannoplankton. Although individually very small, the nannoplankton together can form a significant, and even on occasions the majority, of phytoplankton biomass and have corresponding effects in productivity. Whilst all major algae groups have evolved some small species the nannoplankton most commonly consists of flagellates such as *Chlorella* together with coccoid green algae and a few types of centric diatoms such as *Stephanodiscus*.

4.5.2 Composition

Some common examples of phytoplankton types are shown in Fig. 4.5, and these cells fall into four major phyla each with quite distinctive characteristics although there are also a number of smaller families.

(a) Bacillariophyta

This group is commonly called the diatoms. They occur in a wide variety of shapes and sizes from the single-celled, circular concentric types to those which occur in chains, like *Melosira*, and other patterns. They do, however, have one

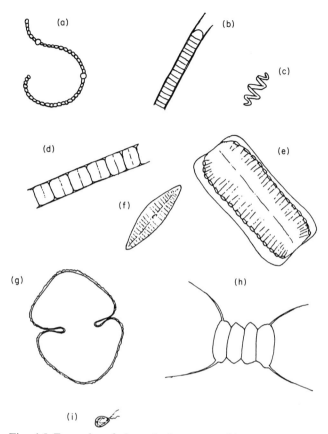

Fig. 4.5 Examples of phytoplankton types: blue-green bacteria/algae (Cyanobacteria)—(a) *Anabaena*, showing larger heterocysts in chain of cells, (b) filamentous *Lyngbia*, (c) unicellular *Spirulina*; diatoms (Bacillariophyta)—(d) *Melosira*, (e) *Surirella*, (f) pennate, benthic *Navicula*; green algae (Chlorophyta)—(g) a desmid, *Cosmarium* (now often placed in the Charaphyta), (h) *Scenedesmus*, (i) a small, nannoplanktonic flagellate, *Chlamydomonas*

feature in common—they all secrete around themselves an outer shell composed of silica. This shell is transparent and occurs in two halves, called frustules, which fit together rather like the two halves of a petri dish. The solid nature of the silica frustules allows the profusion of shapes found in diatom cells. It also means that the frustules, which often have intricate patterns etched on their surfaces, are resistant to destruction and are generally egested intact with the faeces after being consumed by herbivores, even though their contents have been entirely digested. Empty diatom frustules are, therefore, a common feature of lake sediments. The requirement for silica to construct the frustules makes it an essential nutrient for diatoms and therefore a potential limiting factor in some waters.

Truly planktonic, suspended diatoms are completely without a means of swimming or locomotion, but there are those which live on or in the sediments which have the power to glide over surfaces. These, termed the 'pennate' diatoms, are much more uniform in shape than the planktonic types being generally an elongated boat shape, such as *Navicula*, or with slight variations on this including a suggestion of an S-shape as in *Gyrosigma*. Being mobile, these pennate types can creep over the surface of the mud and even migrate up or down from the surface over the top centimetre or two. These movements in response to light intensity are responsible for the contribution of pennate diatoms to the green algal patches which can appear on damp areas of sandbanks during the daytime. More generally they are found associated with sand, mud, or any hard surfaces in the bed of lakes or rivers and if they appear in phytoplankton samples this is often a sign that the sample has been taken from shallow waters where these algae have been temporarily suspended by local turbulence.

(b) Chlorophyta

These are the green algae which, as their common name suggests, tend to possess bright green chloroplasts as opposed to some of the diatoms, for example, where the pigment may have a slight brownish or golden tinge. The green algae are a very diverse group which can be single-celled or colonial, flagellate or non-flagellate. Those with flagella, such as *Chlamydomonas*, have a certain, limited, mobility—usually to adjust their vertical position in the water—but the scale of movement remains very small. There is a tremendous size variation amongst the green algae from some of the largest types found in the phytoplankton such as the colonial *Volvox* to many of the genera amongst the diverse order of Chlorococcales, such as *Chlorella*, which are to be found in the nannoplankton. Another group of green algae common particularly in rather acidic waters, are those commonly known as the desmids which often have a constriction in the centre of the cell. In benthic regions are to be found the filamentous green algae which can sometimes blanket substantial areas of the bottom.

(c) Dinophyta

The dinoflagellates are almost entirely planktonic and have a very distinctive appearance. The cell is entirely covered in plates made of cellulose which are drawn out into three characteristic prongs. Around the centre of the alga is a groove or girdle around which is a flagellum. This gives the cell a limited mobility. They are generally unicellular and sufficiently large as to be part of the netplankton.

(d) Cyanobacteria (Cyanophyta)

The blue-green algae can more properly be considered as bacteria since the nucleus is not bounded by a well-defined membrane, a distinguishing feature of prokaryotes. Moreover, the photosynthetic pigment is not confined to chloro-plasts and appears to be dispersed generally through the cytoplasm although it is actually associated with the membranes of the endoplasmic reticulum within the cell. This does, however, provide a distinguishing feature between the blue-greens and algal cells generally since their cells just look homogeneously green with no discernible chloroplasts. This is most evident when comparing filamentous green algae with filamentous blue-greens, since the chloroplasts are clearly defined in the filamentous green cell and also these cells are generally elongated compared to those of blue-green cells which tend to be rather shorter than broad.

The cyanobacteria can be planktonic or benthic, single-celled or colonial. Many of the colonial, filamentous types, such as *Oscillatoria* and *Anabaena* are planktonic and several of these are able to fix dissolved nitrogen in the water to produce ammonia. This property has been widely associated with those genera, such as *Anabaena* and *Aphanizomenon*, which form the enlarged empty cells termed heterocysts in their filaments, and it has been demonstrated that 90% of nitrogenase (a key enzyme in the conversion of N_2 to NH_3) activity lies within the heterocysts. However, detectable nitrogenase activity has been found in a few other genera, such as *Pleotonema* and *Gloeocapsa*, which do not contain heterocysts although, so far, this remains the exception. In tropical waters, where inorganic nitrogen for plant production tends to be very low, nitrogen fixation in this way could be very important and in the only tropical freshwater system which has been examined in this way, Lake George, Uganda, the annual quantity of nitrogen fixed amounted to 88% of that contributed by all other sources including inflowing rivers (Ganf and Horne, 1975).

(e) Others

There are a number of other algal phyla which occur in the phytoplankton of tropical fresh waters and which may be important in some circumstances. Notable amongst these are the Euglenophyta and the Cryptophyta. Both are flagellate groups the former typified by *Euglena* and the latter by *Cryptomanas*. The cells are generally small, naked, and unadorned; the euglenoids can either swim using their single flagellum or creep over hard surfaces whilst cryptoma-nads are confined to swimming with their two rather unequal flagella.

Although the vast majority of primary producers in the plankton are algae, bearing in mind the rather unique position of the blue-greens here, there are also a few types of bacteria which possess photosynthetic pigments. These, however, use reduced compounds such as H_2S as the hydrogen donor for photosynthesis rather than H_2O in the more conventional pathway, and therefore they only tend to become significant under anaerobic conditions such as might occur when a deoxygenated hypolimnion extends into the euphotic zone. By far the greatest proportion of bacteria in the plankton are saprobic and are involved in decompositon leading to the mineralization of organic compounds which is of vital importance for the community as a whole (Fig. 4.4). The bacteria are generally smaller than the phytoplankton although the largest are similar in size to the smallest algae. They are often to be found adhering to particles of organic matter in the water.

4.5.3 Weight and Numbers

(a) Sampling

Phytoplankton are extremely small and therefore very difficult to weigh directly; consequently indirect methods must be used to obtain estimates of biomass. Cells from phytoplankton samples can be counted, although in natural circumstances they may be highly dispersed and therefore need to be concentrated prior to subsampling. For qualitative purposes plankton can be concentrated in a net towed behind a boat, but this can be a difficult technique to adopt for quantitative purposes due to the progressive effects of clogging of the net pores. For quantitative purposes the phytoplankton are extracted from a known volume of water which can be taken from a given depth or from an integrated sample of all depths in the euphotic zone. The integrated sample can readily be obtained by lowering a weighted, open polythene tube of around 1.5 cm diameter vertically down through the water, thus effectively isolating a column of the water. The upper end can then be stoppered at the surface whilst the lower, weighted end is then drawn up by the operator and the sampled water column released into a container, once the tube has been taken out of the water, by removal of the cork. In this way an indication of average algal density and species composition down through the whole water column can be obtained. Samples from particular depths can be obtained by using one of the specialized water samples such as the Ruttner bottle or by attaching a small hand pump to a polythene tube which can then be lowered to a particular depth and a sample pumped up. For surface samples, of course, only a bucket is required.

From samples of known volume the plankton can be most easily concentrated by sedimentation which uses the fact that the cells do tend to sink. A few drops of a preservative such as Lugol's Iodine (a mixture of iodine, potassium iodine, and glacial acetic acid in water) or dilute formalin enhances sedimentation and allows storage. Much of the original water can then be removed leaving a concentrated layer of algae on the bottom of the container which is then subsampled and counted under a microscope. Some form of chamber of calibrated volume such as the Lund cell can be used to estimate numbers in cells per ml, but the numbers in a standard drop from a vertically held pasteur pipette can provide good approximations (Edmondson, 1969). Counting filamentous algae presents a problem, particularly blue-greens since these are often composed of long chains of tiny cells. Usually, although there are obvious inconsistencies, owing to the varying lengths of the chains, each filament is scored as one although in some of the spiral types such as *Anabaena* the number of complete spirals can give a more consistent measure.

Counting cells from a known volume of water in this way provides an estimate of total numbers of cells per ml and can also be used to assess species abundance. It is possible to convert these numbers into weights by obtaining an estimate of the average weight of each type of cell, bearing in mind that just as algal cells vary in size so they vary in weight. This can be done by measuring the dimensions of each kind of cell under the microscope and then obtaining the volume by using the formula for the geometric shape (e.g. a sphere, a cylinder, a cone), which most closely approximates the shape of the algal species. By assuming a density equivalent to water, since the specific gravities are very close, mean volume/cell can be converted directly to mean weight/cell which can then be multiplied by the cell totals to give an estimate of biomass.

This method will obviously only give an approximation of biomass, although in many circumstances an order of magnitude can be more important than a precise estimate. However, an alternative approach is to consider the phytoplankton basically as primary producers and to estimate the concentration of photosynthetic pigment in the algae as a measure of their abundance. This is perhaps a better indication of the photosynthetic capacity of the phytoplankton but, even so, an analysis of world-wide records shows that there is a high degree of correlation between phytoplankton chlorophyll a content and biomass (Brylinsky and Mann, 1978). Pigment extraction is accomplished from phytoplankton filtered from a known volume of water using an organic solvent such as acetone or methanol (Marker and Jinks, 1982). With acetone extraction from algae on the filter paper is carried out by leaving in a refrigerator for 24 hours whilst, if boiling methanol is used, the extraction is complete within 30 minutes. Although there are a number of pigments extracted by these methods, chlorophyll a is generally considered the most important and therefore the light absorption of the extract of a wavelength 665 nm, the peak for chlorophyll a, is determined.

Some techniques try to correct the estimate for contributions from other pigments with overlapping spectra and this led to the development of the 'trichromatic method' where the absorbances at 630, 645, and 665 nm were determined and then correction factors incorporated to allow for contributions from the pigments with peaks at 630 and 645 nm. Other methods have doubted that these corrections are worth while and with the methanol extraction technique Talling and Driver (1963) arrived at the following relationships for chlorophyll a (μg ml^{-1} solvent):

$$\text{Chl. } a = 13.9 D_{665}$$

where D_{665} is the optical density at 665 nm. Since the chlorophyll has been extracted from a known volume of water then the results can be expressed in μg l^{-1} or mg m^{-3}. However, if an attempt has been made to sample the phytoplankton right down through the water column then the values can be put as mg m^{-2}, in other words the total amount of chlorophyll a (or biomass), through the whole water column, below 1 m^2 of surface.

(b) How much Phytoplankton in tropical waters?

The phytoplankton biomass or pigment density in tropical lakes and rivers covers an enormous range, although it seems that the tropics themselves can be considered together since within this zone no effect of latitude among chlorophyll *a* concentrations or production can be detected (Melack, 1981). In shallow soda lakes chlorophyll *a* concentrations can exceed 500 mg m^{-3} (Melack, 1981) whilst the waters of many tropical lakes contain some 20–200 mg m^{-3}. In some of the larger lakes the chlorophyll *a* content of the water appears to be rather low, for example 1.2–5.5 mg m^{-3} in Lake Victoria (Talling, 1966) or 0.2–1.6 mg m^{-3} in Lake Tanganyika (Melack, 1980). However, the depth of these lakes means that the algae are suspended in a water column of some considerable depth and therefore, although the pigment may be well dispersed, the actual amount per square metre can be quite high. In Lake Victoria this can amount to 44 mg m^{-2}.

The expression of chlorophyll per unit area is a particularly useful indicator of the total abundance of photosynthetic organisms in the water as the concentration per unit volume varies with depth since the phytoplankton are rarely homogeneously distributed (see Sections 4.5.4 and 5.2.3).

There are few estimates of the actual biomass of phytoplankton in the tropics. In Lake George where the mean chlorophyll *a* concentration approximates 500 mg m^{-2}, the average algal biomass was 46 g dry wt. m^{-2} (Ganf and Viner, 1973; Greenwood, 1976). Dry weight of the productive soda lake, Lake Nakuru, estimated on a volume rather than an area basis, showed biomass to be between 58 and 194 g m^{-3} when equivalent chlorophyll *a* concentrations were 155–1160 mg m^{-3} (Vareschi, 1982). Since the water column is something over 2 m the biomass per square metre must be more than twice these volumes. Both of these shallow lakes, however, carry very dense standing crops of phytoplankton. Chlorophyll values of 2 g m^{-2} which is approached in Lake Nakuru, are comparable to the higher range shown by terrestrial communities and algae in hypertrophic conditions (Aruga and Monsi, 1963).

Flowing waters tend to have low phytoplankton densities which are reflected in the low chlorophyll values obtained in the few cases where these have been determined. In the Rio Solimoes (upper Amazon), for example, the river itself had chlorophyll values ranging from 0.1 to 9.4 mg m^{-3} (Fisher and Parsley, 1979). In the associated white-water varzea lakes, which receive the floodwaters, the average content is 28 mg m^{-3}, compared to less hospitable black-water lakes with their high humic contents and low pH, where the average can be 5.2 mg m^{-3} (Rai and Hill, 1980).

Due to their better light climate compared to the white- or black-water systems, clear-water tributaries of the Amazon can develop a substantial phytoplankton; the Rio Tapajoz, for example, possessed chlorophyll *a* concentrations of 61–162 mg m^{-2} (Schmidt, 1982).

As the White and Blue Nile flow away from their sources, Lake Victoria and Lake Tana respectively, there is a big reduction in the number of phytoplankton cells. However, by the time the two unite at Khartoum large seasonal blooms of algae can be found although this seems largely to be due to the construction of large reservoirs upstream (Talling, 1976). Between Aswan and Cairo the chlorophyll a content of the river was measured as 2–13 mg m^{-3} although this was downstream of Aswan High Dam and has consequently no doubt been greatly influenced by this.

Phytoplankton is only found in any significant quantities in slow-flowing rivers or calm backwaters where the rate of cell division of the algal cells more than compensates for the loss downstream. Such conditions are enhanced when a river becomes impounded as in the Nile at Aswan. In headwater or high-altitude rivers and streams the waters are too fast flowing to support a functional phytoplankton and in these situations the periphyton, anchored to hard substances, take on the role of the phytoplankton when the waters are shallow and very clear. In Gombak stream, Malaysia, the biomass of the benthic algae was determined by measuring the material and phytosynthetic pigments taken off asbestos cement tiles left for some time in the river as artificial substrates. It was evident that a number of factors influenced biomass in the river. For example in shady areas chlorophyll a values of 1.3–4.1 mg m^{-2} were found, whilst in open stretches 35 mg m^{-2} were attained (Bishop, 1973). Further downstream, below a source of pollution which enhanced the nutrient content of the water, the pigment density fell to 3.8 mg m^{-2}. This was because although the water had a higher nutrient status it was also very turbid. Moreover, other non-photosynthetic organisms of the aufwuchs such as bacteria do profit from the slightly polluted conditions, as shown by the estimates of total biomass of 3.5 mg m^{-2} at the slightly polluted site compared to 3.9 mg m^{-2} at the open site where the algae are more plentiful.

The chlorophyll values in the Gombak stream suggest that biomass of the epiphytic algae is quite low. In open water the density may attain 35 mg chl. a m^{-2} which is comparable to the 44 mg chl. a m^{-2} found in the euphotic zone of Lake Victoria. In the shaded areas of the forest stream, however, concentrations are much lower.

Artificial substrates have also been used to estimate the biomass of aufwuchs in the black-water swamp of Tasek Bera in Malaysia. Values of 2–35 g m^{-2} were obtained (Watanabe et al., 1982) which are substantially higher than those reported from the flowing-water Gombak stream. The standing crop of periphyton attached to submerged weed stems was also determined in Tasek Bera where values up to 40 g m^{-2} were found (Watanabe et al., 1982). These values demonstrate how dense the mats of algae which form the periphyton can be and suggest that they make a significant contribution to primary production in shallow waters.

4.5.4 Maintaining a Position

Members of the phytoplankton face one overriding difficulty; since they require light for photosynthesis they must remain within the euphotic zone yet their cells are denser than water and tend to sink. Many algae, therefore, have developed anti-sink mechanisms which, even if they do not make the cells positively buoyant, at least slow down the rate of sinking. The simplest means of accomplishing this is to have as large a surface area to volume ratio as possible as this will retard sinking. A flattened object will sink more slowly than a sphere of the same weight, hence many types are flattened or disc shaped. A similar effect is obtained by the cells being joined into ribbons or chains or, alternatively, the development of spines, hairs, or other projections will also increase the resistance and drag. The three prongs of the dinoflagellates (Fig. 4.5) may have this effect.

There are also physiological mechanisms including the use of oils as storage compounds by several groups such as diatoms and dinoflagellates. Lipids are generally less dense than water and will, therefore, enhance the buoyancy. Rather less obvious than this is the apparent ionic regulation of the cell sap noted in a few species where heavier, divalent ions such as Mg^{2+} and SO_4^{2-} have been selectively eliminated and replaced with lighter monovalent ions like Na^+ and Cl^-. The extent to which this mechanism is used by algae is not clear since it is difficult to measure the density of the cell contents.

Blue-green algae are able to produce gas vacuoles in their cells which can actually give positive buoyancy. Under certain circumstances blue-green algae can accumulate as scum on the surface of water which may actually have deleterious effects on the community as a whole. They appear to be able to regulate their buoyancy through control of the proportion of the cell volume occupied by the gas vacuole, and this may be mediated by light intensity and the rate of photosynthesis. Regulation of buoyancy enables these algae to maintain themselves at an optimum depth.

Sinking need not necessarily be a bad thing under all circumstances. When the upper layers of water become depleted of nutrients then, as we have seen in the previous chapter, the availability of nutrients can increase with depth and it could be advantageous for the algae to sink into this zone. The full profit from this advantage will only be felt if the cell then stands a reasonable chance of being carried up again towards the surface and the light, in other words when the upper layers of water are turbulent through wind action or for other reasons. It is curious that diatoms, which are generally non-mobile, accumulate such a heavy substance as silica, but this may be seen as an advantage for these types which thrive in well-circulated waters (Walsby and Reynolds, 1980).

The interplay of all the factors controlling buoyancy and sinking results in a non-random distribution of phytoplankton with depth. In stable situations the phytoplankton do tend to accumulate within the euphotic zone and often also above the thermocline. There may also be some accumulation of algal cells close to the bottom through the sedimentation of cells which have sunk and cannot, under stable conditions, be resuspended (Fig. 4.4). Often the highest concentration of phytoplankton is not directly at the surface but a metre or so below. This may be due to surface turbulence forcing the algae down but functionally, it is advantageous since the photosynthesis of many algae is inhibited by high light intensities such as occur at the surface of tropical waters. The peak in algal densities tends to coincide more with light levels closer to the optimum for the species. Some stratification of the phytoplankton by species has in fact been observed, for example in Sebang Lake, Malaysia, where *Anabaena* was prevalent at 0–0.5 m, *Crucigera* at 0–1.5 m, *Staurastrum* 1.0–2.0 m, *Mallomonas* at 1.0–2.5 m, *Peridinium* 1.0 m, euglenoids 1.0–2.5 m, and *Merismopedia* at 3.0–6.0 m (Arumugam and Furtado, 1980). The colonial blue-green, *Merismopedia*, did, in fact, occur principally below the thermocline where light intensity was low. In a similar situation another blue-green type, *Lyngbia*, was found to be strongly stratified within the metalimnion of Lake Carioca in Brazil, well below the limits of the euphotic zone (Reynolds *et all.*, 1983). In these cases it seems possible that there is a certain amount of heterotrophic uptake or photo-assimilation of small organic molecules since there can be insufficient photosynthesis to sustain the organisms at this depth. This could be advantageous since at these times the epilimnion is depleted of nutrients, particularly nitrogen, whilst the hypolimnion is much more nutrient rich. However, it could also be that the cells at this time are just physiologically quiescent and are waiting for more favourable environmental conditions to arrive (Reynolds *et al.*, 1983).

In shallow lakes, such as Lake George which is never more than 2.4 m deep, the whole water column is susceptible to turbulence and mixing caused by the wind. Such lakes tend to be polymictic (Section 3.4) and prone to stratification during the day which breaks down with nocturnal cooling. During the mixing at night the phytoplankton are more evenly distributed through the water column with perhaps a slightly greater concentration at the surface of chlorophyll *a* of around 200 mg m^{-3}, compared with 140 mg m^{-3} at the bottom. As the surface waters heat up during the day and the water becomes stratified the algal densities at the surface fall whilst those at the bottom increase. By 16.00 hours the surface chlorophyll *a* concentration can be 100 mg m^{-3} whilst that at the bottom has increased to 400 mg m^{-3} due to sinking algae. The onset of night again causes resuspension of the algae due to cooling and wind action. One of the reasons for these changes in algal position could be that many of the algae in Lake George are cyanophytes such as *Microcystis* and *Anabaenopsis* which may control their position through their gas vacuoles. However, it seems more likely that the daily descent is due partly to increased rate of sinking as the density of water in the epilimnion decreases with rising temperature during the day, and partly to the suspended algae being transferred from the more turbulent upper layers to the calmer lower layers due to the higher wind action characteristic of Lake George in the day compared to the stronger night winds. The most important function of the gas vacuoles could be to lift the algae off the bottom mud and thereby increase their chances of being carried up towards the surface once more (Ganf and Horne, 1975). It seems, therefore, that the distribution of phytoplankton with depth can change quite markedly over the daily cycle.

4.6 ZOOPLANKTON

4.6.1 Composition and Characteristics

The commonest constituents of freshwater zooplankton belong to two crustacean groups, the Copepoda and Cladocera, and the Rotifera, a phylum containing the smallest multicellular animals. Of these, the rotifers are generally the smallest in size for although some can be a millimetre or so in length the majority remain microscopic and similar in size to ciliate protozoa at around $110-140\,\mu m$. They are characterized by having a ciliatory organ called the corona at the anterior of the animal, around the mouth, and the beating of the cilia is associated with gathering of the food and also with swimming. The main part of the body is composed of a rather sac-like trunk although the cuticle surrounding this frequently becomes much thickened to form a conspicuous case, the lorica, which may be divided into plates and extended into spines at the anterior and posterior borders (Fig. 4.6). There may also be a narrow extensible appendage at the terminal portion of the body which in crawling or sessile rotifers can be used for attachment.

The Cladocera are considerably larger than the rotifers; they may be up to 1–2 mm in length and can therefore be seen with the naked eye upon close inspection of a water sample. Commonly known as water-fleas, these animals tend to have a rather upright attitude to the water (Fig. 4.6). The head with its eye and rather beak-like extension gives the animal the appearance of a plump bird, and the carapace around the body encloses both the trunk and its appendages which are involved in feeding. On the head are a very much reduced pair of first antennae and a tremendously well-developed pair of second antennae which provide the propulsion for swimming.

The direction of movement is generally vertical with the downstroke of the second antennae propelling the animal sharply upwards. It does tend to sink between strokes although this is slowed owing to the resistance of the antennae acting rather like a parachute. This rapid upward movement and slow descent gives the whole movement a rather jerky appearance when the animal is observed in the water.

Copepods, like cladocerans, are visible without a microscope. They have a fairly uniform appearance of a large elongated thoracic region with a much narrower abdominal section which terminates in two caudal appendages. The group can, however, be divided into two types, the calanoid and the slightly smaller cyclopoid copepods. The principal distinction is the rather more tapering thorax of the cyclopoids and the relatively shorter antennae (Fig. 4.6). The paired limbs along the ventral side of the thorax provide forward impetus through oar-like contractions. During such a contraction the antennae are laid back along the body to reduce resistance but, between contractions, the antennae are extended to reduce sinking, leading to a rather jerky overall motion.

There are other types of animals in the zooplankton including the protozoa which can be quite common both in the water and in the sediments although they have, as yet, received little attention. There are also the small clam-like crustaceans, the ostracods, with a bivalve carapace. One group of particularly specialized habitat are the brine shrimps, such as *Artemia*, of the Anostraca which can be several millimetres long, have no carapace, and swim on their backs. They are able to tolerate salinities up to several times that of sea-water and therefore are to be found in salt lakes.

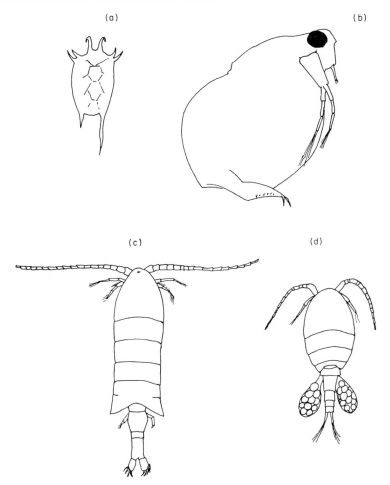

Fig. 4.6 Examples of zooplankton types: rotifers (Rotifera)—(a) the lorica of *Keratella* Crustacea, (b) a cladoceran, *Moina*, (c) a calanoid copepod *Lovenula*, (d) a cyclopoid copepod, *Mesocyclops*, with egg sacs.

Although, like the phytoplankton, the zooplankton are suspended in the water, the problems they face are rather different. There is no particular requirement for the zooplankton to stay in the euphotic zone although they do need to visit the areas of phytoplankton abundance for purposes of grazing. Although planktonic, the members of the zooplankton show quite well-developed powers of locomotion (as mentioned above) but this is usually employed in vertical movement. In fact, diel migration with concentrations of zooplankton at the surface at dawn and dusk and sinking to lower depths during the day is a very widespread phenomenon in all aquatic environments including tropical waters (Section 4.6.5).

For the small animals which spend a good deal of time in the upper layers of water, predation can be a major problem and there is nowhere to hide. Consequently, most zooplankton are transparent which at least makes them more difficult to see.

4.6.2 Feeding

Most of the zooplankton feed upon suspended organic material, either phytoplankton or detritus, but because of their own different sizes and the differences in their mechanisms there is some variation in the size and nature of the particles that are utilized. The rotifers are suspension feeders, using the currents generated by the ciliary whorls of the corona to draw fine particulate matter towards the mouth. Being essentially rather small animals themselves they tend to take rather small particles within the range 1–20 μm which includes nannoplankton and the finer detrital material.

Cladocera are filter feeders with the trunk appendages furnished with very fine hair-like processes to form long comb-like structures through which water is filtered by beating of the limbs. Although they can be relatively large, 0.3–3.0 mm, the filtering mechanism of the Cladocera is very effective in retaining a wide range of particles from 1 to 50 μm and some species have been shown to be able to survive using bacterial suspensions alone, although naturally they would take phytoplankton and detritus as well.

Many of the Copepoda, particularly the calanoid type, are, like the Cladocera, filter feeders. The appendages around the mouth beat vigorously at between 600 and 2640 times a minute, creating two swirls of water passing backwards on either side of the body. These currents are then redirected into the thoracic gill chamber which the combined currents leave anteriorly, passing through a dense screen of hair-like setae on the second maxilla. This acts as a filter and accumulated food particles are transferred to the mouth.

The range of food particles taken is rather wider than that in the other groups at some 5–100 μm, which therefore excludes the smaller algae.

The cyclopoid copepods are raptorial rather than filter feeders in that they seize their food particles, which in some cases may be other small members of the zooplankton. *Thermocyclops hyalinus*, the dominant constituent of the Lake George zooplankton, is able to bite pieces out of large colonial algae, such as the blue-green *Microcystis* or detritus, and to this extent is less constrained by particle size.

Whilst there is selection by particle size generally amongst the zooplankton there appears to be less selectivity as to the nature of the food. Both living and dead material are accepted, and therefore the zooplankton are involved in both the grazing and detrital chain, particularly since some may also take bacteria (Figs. 4.3, 4.4). This becomes further complicated by considering how much of the phytoplankton and detritus is actually absorbed or assimilated in the passage through the gut and therefore available to the zooplankton, compared to how much passes through the gut to be egested as faeces for recirculation within the detritus component. Nevertheless, the zooplankton are major consumers of phytoplankton and therefore must be major elements in the grazing chain.

4.6.3 Reproduction

The generation time of most zooplankton is quite short, perhaps a matter of days or weeks. There are, however, considerable differences in their reproductive strategies. In rotifers reproduction can be by parthenogenesis with a population of female individuals producing eggs which are never fertilized and that hatch out into other females. However, at certain times, usually when environmental conditions are less favourable, another type of egg is produced which is haploid and which will hatch into a male. Fertilization of the females by males then becomes possible, and the fertilized eggs develop thick resistant shells which may lie dormant for some months before hatching. The generation time of these small animals can normally be very rapid, perhaps between 1 and 7 days.

The Cladocera can also be parthenogenetic and can therefore produce, without fertilization, several female generations, each of which may be from 5 to 24 days' duration. The young are produced from eggs which are held in a brood chamber within the dorsal region of the carapace. When adverse conditions occur males are produced followed by the production of fertilized eggs which are larger than the parthenogenetic ones. These become enclosed within the brood chamber which is transformed into a resistant capsule, the ephippium, which is shed at the next moult and protects the eggs until they hatch.

Reproduction of the copepods is entirely sexual and, unlike the Cladocera where the young are small editions of the adults when they hatch, the young copepods go through several larval stages, a procedure which is commonly found in crustaceans. The eggs are held externally in two sacs (Fig. 4.6) and hatch to produce the first-stage nauplius larvae which is a common feature of the life histories of many crustacea (Fig. 4.6). This larva is rather pear-shaped, with three pairs of limbs and a single eye and subsequent moults take it through four or five more naupliar phases or instars until it enters a second larval stage, the copepodite, which resembles the adult except that the abdomen is not segmented and there may be only three pairs of thoracic appendages. Typically the adult form is achieved after six nauplius and six copepodite instars, although the whole procedure may be completed within a week to 30 days.

4.6.4 Weight and Numbers

(a) Sampling

The numbers of zooplankton can be counted in samples concentrated from known volumes of water in the same way as phytoplankton. In fact, the same samples may often be used for estimates of both. Numbers may be converted to weight by measuring the approximate dimensions of individuals under the microscope and converting these to volumes using the geometric formulae most appropriate to the shape of the animal. By converting the volumes to weight per individual, then numbers per ml of each species can be used to give weight per ml. Although very small, zooplankton can actually be weighed on a very sensitive balance. A number of dried individuals of the same species at the same stage of development can be gathered on a sliver of foil to obtain their combined weights from which a mean can be derived. A relationship between weight and some readily measurable dimension such as total length can be obtained so that once such a relationship has been established only the linear dimension need be taken to obtain an estimate of weight. However, most laboratory balances are insufficiently sensitive to make this method practicable.

(b) Zooplankton biomass in the tropics

The difficulties involved in obtaining zooplankton biomass explain why few estimates have been made in the tropics. The most thorough study has been in Lake George where a mean annual standing crop of 248 mg dry wt m^{-3} or 559 mg dry wt m^{-2} has been found for the predominant zooplankton the cyclopoid copepod *Thermocyclops hyalinus* (Burgis, 1974). The number of species of crustacean zooplankton in Lake George was not very high and inclusion of the other elements, largely three species of cladocera *Daphnia barbata*, *Ceriodaphnia cornuta*, and *Moina micrura*, produced a total standing crop of 828 mg dry wt m^{-2}. The biomass tended to be rather higher towards the centre of the lake, which also reflected the phytoplankton pattern, and it was thought this could be due to the internal current system which tended to gravitate suspended particles into the central region reinforced by the more intense predation by fish on the zooplankton in marginal areas.

The notable feature of the Lake George zooplankton biomass was its small size compared to the phytoplankton standing crop which can be 100 times greater on occasions. Moreover, they are also low compared to shallow temperate lakes. Loch Leven in Scotland, for example, has a summer biomass four times the peak biomass shown by Lake George. Whether this is a general feature of tropical plankton or confined to Lake George remains to be seen.

Zooplankton biomass estimates from the altiplano Lake Titicaca in the Andes gave a range of 44–200 mg m^{-3} (from Widmer *et al.*, 1975) which is a similar order of magnitude to Lake George although slightly lower and may be linked to the high altitude and lower temperatures. The standing crop of zooplankton, whilst up to five times less than that of the phytoplankton, was not so much less than that in Lake George.

Standing crops from three different localities in Lake Chad were also of a similar magnitude with means of 105, 240 and 350 mg dry wt m^{-3} (Carmouze et al., 1972). In the oligotrophic, subtropical Lake Sibaya in South Africa, however, the mean biomass of one of the commonest zooplankters, *Pseudodiaptomus hessei*, was only 5.7 mg dry wt m^{-3}, which was probably a reflection of the low primary productivity of the lake (Hart and Allanson, 1975). By contrast, the biomass of the only planktonic crustacean in Lake Nakuru, the copepod *Lovenula* (= *Paradiaptomus*) *africana*, could be extremely high with a mean, one year, of 1500 mg dry wt m^{-3} (Vareschi and Vareschi, 1984). To this could be added a further 1400 mg dry wt m^{-3} of rotifer biomass which, on occasions, reached peak densities as high as 7000 mg dry wt m^{-3} in the lake. These enormous quantities of zooplankton are matched by extremely high phytoplankton standing crops in this shallow soda lake (Section 4.6.3). It does appear, therefore, that zooplankton biomass is broadly linked to phytoplankton density or primary production although the Lake George situation shows that this is by no means invariably so. Zooplankton numbers or biomass, however, often fluctuate markedly within a lake, as will be shown in Section 6.4.

4.6.5 Vertical Distribution and Diel Migration

In deeper lakes the zooplankton are not evenly distributed and may vary with depth and time of day. In Lago Tupé, a black-water lake of the central Amazon, the zooplankton numbers reached a sharp peak at 1 m depth which coincided with the peak of chlorophyll *a* concentration from the phytoplankton (Rai, 1978). This peak occurred in the epilimnion and it is not clear to what extent the anoxic conditions of the hypolimnion below the well-established thermocline restricted the distribution of the zooplankton.

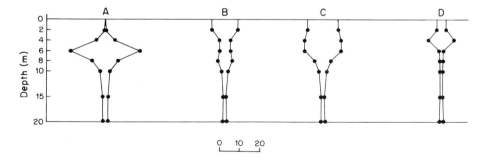

Fig. 4.7 Diel changes in the vertical distribution of cladoceran zooplankton at a sampling point on Lake Kainji, Nigeria when the thermocline was present at between 5 and 10 m depth. (*Reproduced by permission of Schweizerbart'sche Verlagsbuchhandlung from Adeniji, 1981.*) (A) 12.00 hours, (B) 18.00 hours, (C) 00.00 hours, (D) 06.00 hours

It is probable that in most tropical lakes there exists a diel migration of zooplankton and that their abundance becomes redistributed over a 24-hour cycle (Fig. 4.7). Such movements are almost ubiquitous in both marine and freshwater habitats and have been found in the few instances where they have been looked for in the tropics. In the shallow, unstratified Gatun Lake in Panama, where zooplankton was dominated by the calanoid copepod, *Diaptomus gatunensis*, the majority of the animals were found just above the sediment during the day, and rose up towards the surface during the evening until they accumulated within the top 4 m just after sunset. During the night the copepods became more randomly distributed down through the water column, but rose again towards the surface at dawn only to sink as the sun rose higher in the day (Zaret and Suffern, 1976). This is a cycle typical of zooplankton in most aquatic environments. The animals appear to swim upwards under the stimulus of the rate of change of light intensity and so are closest to the surface at dusk and dawn when the light intensity changes most rapidly. Such a widespread phenomenon must presumably have some adaptive significance. Retreat from surface waters to deeper, dimmer waters during the day may give protection from predators or the synchronized raising and sinking may enable the animals to exploit different current systems in surface and deeper waters, which often move in different directions, and thereby increase the area which can be covered by zooplankton in the search for food. One further suggestion is that by sinking into deeper, cooler water for part of the day the energy required for metabolism will be reduced at the lower temperatures and can therefore be used to maximize reproductive potential (McClaren, 1963). However, in the example of Gatun Lake there was no stratification and no cooler water so this explanation would not be possible. In many tropical lakes, marked diurnal stratification occurs and so the migration to deeper waters will allow the zooplankton to avoid the high temperatures of the surface waters as they heat up during the day.

The presence of a thermocline seems to telescope the diurnal movements of the zooplankton. In Kainiji Lake, Nigeria, where the thermocline was between 10 and 15 m depth, the largest part of the zooplankton populations remained above the thermocline (Fig. 4.7) and showed a rhythm of movement similar to that described above but largely within the epilimnion (Adeniji, 1978, 1981). Some migration of copepods and ostracods into the deoxygenated hypolimnion did occur, but the rotifers tended to remain close to the surface even at noon. Some separation of zooplankton types by a well-developed thermocline was also apparent on Subang Lake in Malaysia where rotifers, cyclopoid copepods, cladocerans, and the testate amoeba *Difflugia* were confined to the epilimnion whilst only a few types which could tolerate the anaerobic conditions, such as ciliate protozoa and the phantom-midge larva *Chaoborus*, were found in the hypolimnion (Arumugam and Furtado, 1980). The presence of a thermocline, especially when this is sufficiently well established to create a marked stratification in oxygen and other factors, can have considerable effects upon the distribution of species and biomass within the water column as well as restricting the scope for migration.

In Lake Kariba the cyclopoid copepod, *Mesocyclops leukartii*, shows marked diel migrations and will descend into the hypolimnion providing that there is at least $1.5\,mg\,l^{-1}$ oxygen there. The cladoceran, *Bosimina longirostris*, is much more readily affected and will no longer move into the hypolimnion once the oxygen falls below $2.5\,mg\,l^{-1}$ (Begg, 1976). However, whilst both of these crustaceans carry out significant vertical migrations over a 24-hour cycle, the rotifers, which are the main prey of *M. leukartii*, remain at the surface, as was noted in Subang Lake.

4.7 THE BENTHOS

4.7.1 Composition

The animals which inhabit the bottom regions of lakes and rivers belong to rather more diverse groups than those of the plankton. For example, a typical grab-sample from a lake bed might produce oligochaete and nematode worms, turbellarians, molluscs including both gastropod snails and bivalves, crustaceans, and almost certainly some insect larvae. The presence of insects both in adult and larval stages is perhaps the most characteristic feature of fresh waters and they are rarely of importance in marine or estuarine environments. Some insects are entirely aquatic, such as the water-beetles, and spend both larval and adult stages in water, but a great number such as the dragonflies (Odonata) live in water only during their pre-adult phase. In many cases the larval or nymphal stage, during which several intermediate moults take place as the young grow and change, is the longest phase of the animal's life history and, during this time, the energy and resources required for development are obtained exclusively within the aquatic system. Following the final moult the adults emerge and take to the air to become part of the terrestrial ecosystem and in some cases such as the mayflies, which do not even feed as adults, this may last for only very brief periods of a day or two. Mating takes place during the aerial or terrestrial phase but the eggs are then laid in water to complete the cycle. During life, the adults frequently live close to water and in death often fall on to the surface to add their remains to the resources of the aquatic system.

Within the aquatic insects, the diversity of shape and habit is very great. Not all insects are truly part of the benthos; many of the free-swimming beetles and hemipteran bugs can distribute themselves throughout the whole water column but, because they tend to take refuge in submerged vegetation, they are often taken along with truly benthic types when searching shallow, weedy, or marginal areas. These will be discussed further in Section 4.9. Most of the major families of aquatic insects are not confined to the tropics and, in fact, the main differences between temperate and tropical rgions lie mainly at the species level. The commonest orders include Ephemeroptera (mayflies), Plecoptera (stone-flies), Trichoptera (caddis-flies), Odonata (dragonflies), Diptera (midges, mosquitoes, etc.) and Coleoptera (beetles). The mayfly and stone-fly nymphs are relatively uniform in appearance and are characterized by having three and

76

two 'tail' filaments (cerci) respectively at the end of the abdomen. Mayflies (Fig. 4.8) also have gills along the abdomen ranging in form from simple plates to extensive feathery plumes. Trichoptera nymphs are most commonly distinguished by having a case of vegetation or grains of sand or mud intricately constructed around the body. A few, however, are caseless and have the two appendages at the rear of the abdomen adapted to clasping objects to anchor the animal rather than securing it within its case. These are most often found in rapidly flowing streams.

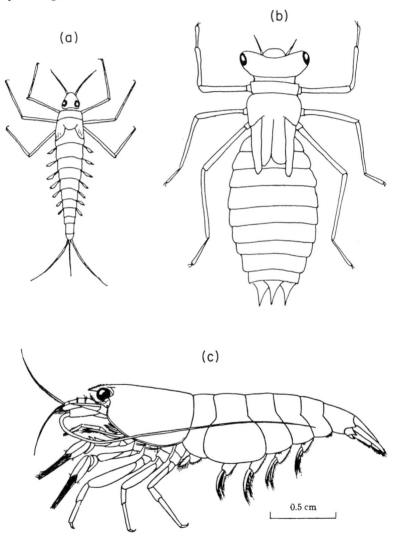

Fig. 4.8 Examples of some benthic macroinvertebrates: insect nymphs—(a) mayfly (Ephemeroptera), Baetidae, (b) dragonfly (Odonata), Libelluidae, (c) freshwater prawn (Atyidae), *Micratya*. (Fryer, 1977b)

The nymphs of the Odonata occur in basically two forms which reflect their division into two suborders—the Anisoptera and the Zygoptera—with the latter sometimes being referred to as damsel-flies in the adult form owing to their generally more delicate build and their frequently striking coloration. The nymphs of the Zygoptera also tend to be slender and their most readily distinguishable feature is the three rather feather-shaped gill plates at the end of the abdomen. This is in contrast to the much more rugged larvae of the Anisoptera, which are often dark brown and in some species up to several centimetres long, where the abdomen terminates in a triangle of three short, pointed structures. Both types of nymphs, however, have the powerful jaws which are borne at the end of the extensible labium normally held folded under the head to give the characteristic 'mask' of all the Odonata. The larvae of the Coleoptera also tend to have large jaws, although they are terminal unlike those of the dragonflies. They can be quite variable in body shape, the abdomen sometimes ending in one or two fine filaments.

Whereas each of the previous orders contained pre-adults of a reasonably uniform type, amongst the dipteran flies larval body shape can vary quite considerably. The only feature they share is that they have no true legs but some do have the muscular, unjointed 'false legs' similar to those found on the abdominal segments of lepidopteran larvae. Beyond this they can vary from larvae which are similar in shape to the familiar blowfly maggot, with a hook at the anterior end and two spiracles opening at the rear, as seen in the Tipulidae or crane-flies, to the more specialized and intricately differentiated mosquito which can be seen swimming to the surface of the water where it hangs upside-down from the surface film whilst it recharges its tracheal air via the siphon. Dipteran larvae are important because not only do they often occur in large numbers but several of them are vectors of major tropical diseases, such as the mosquitoes which carry malaria and yellow fever as well as other diseases, and the black-fly, *Simulium*, which carries the filarial worm causing 'river blindness', *Onchocerca* (Section 8.5).

Amongst the most widespread of the dipteran larvae are those of the midges belonging to the genus *Chironomus*, of which there are a large number of species. They grow up to a size of 1–2 cm and may be green in colour, but they can also be bright red owing to their capacity to synthesize the respiratory pigment haemoglobin facultatively under conditions of low oxygen. A head capsule can be distinguished at the anterior end, and just behind this is a small organ for spinning a proteinaceous thread whilst a gill system can be seen at the posterior end (Fig. 4.9). The larva tends to live in or on fine sediments and constructs a tube around itself composed of mud particles, held together with a salivary secretion. The ability to produce haemoglobin means that chironomid larvae can live in regions of low oxygen concentrations such as those which occur in the hypolimnion since they are able to regulate their oxygen uptake to maintain a constant respiratory rate down to less than 1 mg $O_2 l^{-1}$.

Another type of dipteran larva, that of the black-fly, *Simulium*, is typically found in faster-flowing waters of headwater streams or in broken-water turbulent regions of larger rivers. It tends to be anchored by silken threads, secreted by the mandibular glands, to the underside of stones. It is small, perhaps up to 5 mm long and rather swollen around the base. A close inspection will reveal a head capsule incorporating a rather feathery structure which is the main food-collecting organ (Fig. 4.9).

Finally, amongst the more notable of the dipteran larvae is the curious phantom-midge, *Chaoborus*. Its peculiarity lies in the fact that although it is effectively an open-water predator of zooplankton it also can be very abundant in the benthic sediments of lakes, with the third and fourth instars burrowing into the mud during the day and emerging into the water to feed at night. They can actually be found as deep as 25 mm down in the mud (Burgis *et al.*, 1973). Like its zooplankton prey, *Chaoborus* is transparent with the exception of two black eye spots and dark air sacs at the anterior end (Fig. 4.9).

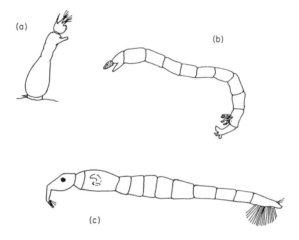

Fig. 4.9 Examples of some dipteran larval types: (a) the suspension feeding larva of the black-fly, *Simulium*; (b) larva of a chironomid midge; (c) larva of the phantom-midge, *Chaoborus*, which spends part of its time swimming in the open water and part in the bottom sediments

The non-insect component of the benthic fauna can also be very varied. The annelids are represented by both oligochaetes and leeches (Hirudinea). Amongst the gastropod snails there are the pulmonates which are derived from the primarily terrestrial group some of which have reinvaded fresh water, and then there are the operculate prosobranch snails which are an older branch with their main radiation in marine and fresh water habitats. The presence of the operculum, the horny plate which can be lodged into the opening of the shell to protect the retracted animal, is the main distinguishing characteristic of the prosobranchs. Since the pulmonates were essentially a terrestrial group the gills were lost to facilitate the use of air for respiration. In the freshwater pulmonates some families such as the Lymnaeidae and Physidae still use air and therefore must return to the surface to replenish their supply, although some species are able to extract all their oxygen requirements from the water. Members of the Planorbidae, the flat, ram's-horn snails, and the Ancylidae, the freshwater limpets, have evolved secondary gills. All of the operculates have retained their primary gills.

In addition to the snails which largely live on firmer surfaces, including rocks and vegetation, there are also representatives of another branch of the molluscs —the bivalves—which live in the sediment, particularly the pea clams (Sphaeridae) and the larger freshwater mussels (Uniondae). There are, in fact, examples of a wide range of invertebrate types to be found in, on or under sand, mud, or rocks of freshwater environments, including coelenterates, sponges, water-mites, and spiders. There is, however, one group the distribution of which characterizes tropical fresh waters—the crustacean Decapoda which includes shrimps, prawns, crayfish, and crabs. In temperate rivers and lakes a few species of crayfish and crabs do occur, but in tropical environments they are both diverse and widespread. The crabs of the family Potamonidae are truly freshwater animals and occur throughout the tropics and subtropics. Although crabs as a whole are generally regarded as being marine or estuarine, thriving populations of *Potamonautes* can be found in demineralized waters such as streams of the West African forest areas where the conductivity of the water is as low as $16 \, \mu\text{s cm}^{-1}$. Such streams can also hold freshwater prawns ranging from the palaemonid *Macrobrachium* which in some species such as *M. rosenbergii* of South-east Asia can grow up to 30 cm long, to some of the members of the Atyidae which may be only 1–2 cm in length. The atyids, in particular, are an important group in tropical fresh waters because they fill the roles played by isopod and amphipod crustaceans, such as *Asellus* and *Gammarus*, in the equivalent temperate environments. Both *Macrobrachium*, which is normally distinguished in the male by its pair of long claw-baring limbs (chelipeds), and the atyids are found in fresh waters throughout the tropics and contain a large number of species (Fig. 4.8; Fig. 8.8).

4.7.2 Feeding

The sources of food available to benthic animals include organic material accumulating on the bottom or suspended in the overlying water, the algae and other micro-organisms of the aufwuchs and, of course, the animals themselves which are the food source for the predators of the community. The material which accumulates can itself be quite diverse and include fragments of leaves, or terrestrial insects, smaller more finely particulate detritus, live benthic algae such as pennate diatoms, along with phytoplankton both live and dead which has been deposited from above. In addition there will be a bacterial and fungal flora which is engaged in breaking down the dead material. Not all of these sources will be equally as abundant in all fresh waters and there may be rather fundamental differences between lakes and rivers as can be seen from Figs. 4.3 and 4.4. In lakes, with their prominent phytoplankton populations and slow circulation, there can be a continual rain of plankton on to the lake bed providing organically rich sediments. In streams and rivers the rapid flow can ensure that the light organic material, often derived from the land, remains suspended. In this respect, however, rivers vary down their course, since the flow tends to be more rapid and turbulent in the headwaters and upper reaches but slows down as the river becomes wider and deeper as it approaches the sea. The regions of high flow rates where all but the heaviest fragments and particles are swept along and kept in suspension are considered to be eroding environments, and those of slow flow where the finer particles can begin to sink are depositing systems. Under eroding conditions most of the algae and micro-organisms are present as periphyton or aufwuchs although a rich periphyton can also be present in the shallowest parts of lakes where there is sufficient light.

The distinction between suspended material and material which sinks on to the bed of the water column as sources of food for benthic animals opens the way for two strategies for collecting fine organic particles. Suspension feeders are those which collect the particulate material from the water itself, and so it is not surprising that this approach is frequently found in animals from eroding environments where much of the fine material is kept in suspension. The larvae of *Simulium*, therefore, which are very characteristic of fast-flowing streams, have fan-like appendages on the head which are held into the current to filter out the suspended material (Fig. 4.9). The filtering of organic materials is not confined to animals from fast-flowing rivers; bivalve clams, which burrow beneath the surface of soft sediments for protection and tend to be found at the bottom of depositing environments including lakes, also have a highly effective filtering mechanism. Most of the cavity within the shell of these animals is taken up with two pairs of very large complex gills well supplied with ciliary tracts. The beating of the cilia draws an inhalent current through an opening in the fleshy shell-lining or mantle to the posterior of the animal which is held just at the surface of the mud. This inhalent current does a U-turn through the gills, where suspended material is filtered out, and emerges as an exhalent current close to the inhalent. In some bivalves, including the freshwater Sphaeridae, the inhalent and exhalent currents are kept separate by a fusion of the mantle to form tubes or siphons. The U-shaped current system enables bivalves to retain a free circulation of water across the gills for food collection, gas exchange, and elimination of nitrogenous waste whilst the animal itself remains entirely below the surface of the mud. Although the mechanism is primarily for filtering out suspended material from the water, some of the rich surface sediments will also be sucked down. Once stuck to the gill the particles are sorted into the potentially edible and the unacceptable by complex ciliary tracts on the gills, on the palps around the mouth, and in the stomach. For an animal living in such a close relationship with the sediments, which contain a high inorganic component, sorting of this nature is a major task.

Some of the animals living in the soft deposits use the organic material which accumulates within the sediments as their sole source of food. These deposit feeders include the aquatic oligochaetes which, like their larger terrestrial relatives the earthworms, burrow through the mud consuming it as they go and assimilating the most readily digestible portion as it passes through the gut. The frequently abundant larvae of *Chironomus* (Fig. 4.9) browse upon the entire algal cells which have sunk to the bottom from the phytoplankton or from the populations of benthic algae. They can also produce a sticky salivary secretion which can be spread over the surface of the mud to which particles sinking to the bottom will adhere. The secretion with its adhering organic particles can then be rolled up and consumed periodically.

Under eroding conditions in faster-flowing streams and rivers deposited material tends only to accumulate in sheltered places behind rocks, sunken trees, or in areas protected by the river bank. In the upper reaches of rivers much of the potential food material is often in rather large pieces such as leaves or seeds dropping into the river from the overhanging trees. Being heavy, these sink to the bottom fairly readily, particularly during the dry season when the current flow slackens. For this type of food, fairly robust biting mouthparts are needed to shred the large pieces into a manageable size. Several types of insect nymphs fall into this category including those of some stone-flies (Plecoptera) and case-building caddis-flies (Trichoptera). The finer particles are consumed by other types such as the larvae of mayflies (Ephemeroptera).

Most rocks, plant stems, or other submerged hard surfaces present a suitable substrate for the micro-organisms of the aufwuchs. To deal with these, which form a thin film over the surface, requires some form of scraping mechanism. The gastropod snails are the most abundant members of the benthos with this style of feeding. They possess a horny, file-like organ, the radula, which can be protruded through the mouth and rocked backwards and forwards so that its many small backward-pointing teeth rasp away the surface to which it has been applied. It can be effective on both the aufwuchs or on soft plant material, and since these can be common food sources in both eroding and depositing environments, the gastropods are widely distributed in both rivers and lakes.

Some animals are extremely versatile with regard to the feeding mechanisms and this is certainly true in the case of the atyid prawns. Detailed observations on species from streams in the Dominican rain forest have shown that several genera, including *Atya* which is widely distributed in the tropics, can either scrape up fine particles from the bottom or filter them from the water (Fryer, 1977a,b). The basis of their flexible feeding mechanism is the dense brush of bristles lining both segments of the pincers carried on some of the anterior thoracic limbs (chelipeds, Fig. 4.10). The pincers are opened, applied to a surface, and then closed, causing the opposing fans of bristles to sweep up small particles between them which can then be transferred to the mouth. Considerable variations can occur in the structure of the brushes so that, for example, in *Jonga serrei* some bristles are modified as scrapers and some are brushes so that as the claw closes the scrapers dislodge material from the surface and the brushes sweep it up. The bristles in *Atya* and its relatives can also be used in a completely different fashion; they can be expanded as a series of fans which, when held into the current, can act as a filtering mechanism to catch suspended

Fig. 4.10 A front view of *Atya innocuous* showing all four chelipeds fans expanded to form an almost continuous filtering surface to collect drifting particles. (Photo: Dr Geoffrey Fryer, FBA)

particles in a way which presents a striking parallel to *Similium* larvae (Fryer, 1977a,b). The prawns may filter-feed in this way for several hours, although the proportion of time allocated to using the various methods available to the atyids may vary from species to species. It is probable that the diversity of feeding mechanisms and, therefore, food sources available to these prawns has contributed to their widespread distribution in the tropics.

Finally, within the benthic communities there are the predators. Considerable predation of benthic animals is due to bottom-dwelling, demersal fish such as the catfishes or the mormyrids (see Section 4.8), but some invertebrate groups are also carnivorous. Amongst these are some beetle and stone-fly young and most dragonfly nymphs. Predatory beetles and stone-flies are generally active animals with a well-developed but simple jaw system. Dragonfly nymphs, by contrast, have an elaborate extensible jaw system which is normally held folded under the head giving the characteristic 'mask' of the Odonata. However, if an unwary animal moves into the vicinity the mask, with the jaws at the tip, are flicked out to snap up the prey. In calmer waters swimming insect predators such as the large notonectid water-boatmen, although not part of the benthic system, will dive down and take bottom-dwelling animals. Conversely, often the commonest predator to be found in mud is the phantom-midge larva *Chaoborus*; this, however, emerges into the water actively to hunt the small crustaceans of the zooplankton rather than prey on benthic animals.

4.7.3 Sampling

The simplest way of estimating numbers from the bottom of a stream or pond, at least in shallow waters, is by placing a quadrat over an area and collecting all the animals within the quadrat by sieving or close examination of stones or other inclusions within the quadrat area. To minimize loss of animals during the collecting procedure a fine-mesh net can be positioned immediately downstream of the quadrat or the quadrat itself can be in the form of a metal box without a top or floor which can then be pressed down into the sediment to more or less isolate the animals enclosed in this way. In deeper water, where the deposits are soft, some form of grab can be used which, upon being dropped down on to the bed from a boat, will enclose a known amount of sediment in its jaws when an automatic catch is released. There are a variety of patterns with varying sampling characteristics (Brinkhurst, 1974; Elliott, 1977; Elliott and Drake, 1981), although the Eckman grab is probably the most well known.

A different approach used particularly in gravel or stony areas which can be difficult to sample by the above methods, is to introduce a basket with a wide metal mesh enclosing a cleaned sample of the material characteristic of the bottom type. After leaving the basket for a sufficient time for colonization to take place, usually several months, it can be retrieved and all the animals which have become established, counted and weighed. If it is to be assumed that final numbers and species composition reflect the situation in undisturbed bottom material, then the artificial substrate must be the same as the surrounding environment.

Benthic organisms do not always remain on the bottom and methods for estimating the more mobile phases of the life history have been devised. For example, many bottom-dwelling invertebrates in streams leave the bottom periodically and enter the 'drift' of animals downstream. Normally most types which live in rapidly flowing water have some adaptation for either holding on to rocks or other hard surfaces or to exploit the 'boundary layer', perhaps a millimetre or two thick, close to the surface of rocks where the drag from the hard object slows down the current quite considerably, providing a very narrow zone where the prospect of an animal being washed away is severely reduced. However, animals do become dislodged from time to time and drift downstream, sometimes for only a metre or two but sometimes for much longer. The numbers in the drift can be estimated by suspending a net with a frame of known dimensions with its mouth facing the current. The numbers caught per unit time can then readily be assessed and if the current speed is known as well as the area of net frame the numbers per unit water volume can also be estimated. The phenomenon of drift is probably not entirely accidental, since some animal types appear more regularly than others and numbers can vary diurnally, seasonally, or even in relation to food availability. In basically sedentary animals it does convey the advantage of a greater potential rate of colonization of downstream habitats whenever new opportunities arise, or it can provide a route of emigration when local conditions become adverse.

A second route of emigration for benthic animals, which represent the larval phase of terrestrial adult is through flight. Dragonflies, mayflies, and diptera, amongst others, all possess nymphs or larvae which eventually pupate and emerge as flighted adults. The rate of this occurrence can be estimated using emergence traps, usually in the form of funnel-shaped nets inverted to cover a known area of water. As the adults emerge they fly up to the apex of the funnel to be trapped in a container attached for the purpose. These adults represent a loss of energy from the aquatic to the terrestrial system, but their mobility between emergence and laying eggs in the water once more allows them to influence the distribution of the larvae in lakes or rivers.

4.7.4 Biomass and Assemblages in Natural and Man-made Lakes

In a few instances the benthic communities of tropical waters have been examined quantitatively as well as qualitatively and biomass estimates have been made. In the well-established reservoir of Bung Borapet in central Thailand, particularly high densities of bottom-dwelling animals were found with up to 223 g dry wt m^{-2} occurring (Junk, 1975). There was, however, considerable variation in the biomass and the composition of the fauna which was related to the nature of the sediment and other environmental conditions. In this reservoir the majority of most of the samples constitued bivalve molluscs, notably the unionid *Corbicula* and to a lesser extent *Scaphula*, which sometimes contributed up to 99% of the total biomass, shell weights being included in the biomass. Second in prominence to the bivalves were the insect larvae *Eatogenia*, an ephemeropteran, and *Dipseudopsis*, a trichopteran, although in terms of biomass they never contributed more than 0.95 and 0.7 g dry wt m^{-2} respectively. Although the total biomass was generally high at Bung Borapet there was a considerable range of estimates from 0.116 to 223 g m^{-2} with some variation between the major areas; for example, the mean for the outflow region was 31.1 g m^{-2} whilst for the inflow it was only 0.32 g m^{-2}. The magnitude of the biomass was largely dependent upon the suitability of the site for bivalves. *Corbicula*, by far the most common of these, tended to occur most frequently in shallow-water areas kept free of superficial sediment by wave action (Fig. 4.11). This is presumably because the bivalve needs relatively stable bottom deposits into which it can burrow; consequently the coarse organic material associated with the marginal vegetation is particularly unsuitable (Fig. 4.11) and the thick decaying layers of organic detritus near the inflow have also been avoided. Nevertheless, when the sediments are suitable bivalves can be expected to do well in moderately productive lakes since their feeding mechanism is well suited to filtering out the rain of organic material sinking from the overlying water (Section 4.7.2).

The other components of the Bung Borapet bottom fauna also show considerable variation in distribution in relation to environmental conditions. The mayfly larva, *Eatogenia*, occurs most prominently where there are fine sediments suitable for its burrowing habit whilst the caddis, *Dipseudopsis*, is more common amongst the coarse organic material associated with the marginal

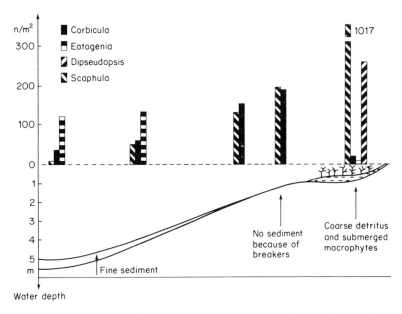

Fig. 4.11 Distribution of benthic organisms in Bung Borapet Reservoir, Thailand. (*Reproduced by permission of Schweizerbart'sche Verlagsbuchhandlung from Junk, 1975*)

vegetation (Fig. 4.11). Also characteristic of this zone were the bivalves *Scaphula pinna* and *Limnoperna siamensis* which were frequently attached to the roots of the plants. The marginal vegetation itself harbours a great diversity of types including gastropod snails, dragonfly larvae, and aquatic coleoptera but these are types associated with the vegetation itself rather than the lake bed.

In relation to other lakes Bung Borapet appears to be similar to Lake Chad where the range of mean mollusc shell-free biomass is 0.02–7.2 g dry wt m^{-2} (Lévêque and Saint-Jean, 1983) compared to Bung Borapet where an equivalent range of 0–6.3 g m^{-2} was found (Junk, 1975). The mean biomass for all benthic animals in Lake Chad was 3.7 g m^{-2} in 1970 of which 3.3 g were due to molluscs, 0.29 g to oligochaetes, and 0.12 g to chironomids. Both of these lakes, therefore, support a high biomass of benthic animals the great majority of which are molluscs. The standing crops of non-molluscan animals in the two lakes are of a similar order with oligochaetes providing the greatest proportion in Lake Chad and insects in Bung Borapet.

In other lakes the benthic biomass can be considerably lower; for example in Lake George it averaged 0.74 g dry wt m^{-2} (Greenwood, 1976) and a selection of the varzea lakes from Amazonia proved to have values ranging from 0.14 g m^{-2} in the black-water Lago Tupé (Reiss, 1977) to 2.65 g m^{-2} in Lago Muratu with only the Lago Jacertinga possessing rather elevated levels of up to 6.2 g m^{-2} in the central body of the lake (Fittkau *et al.*, 1975). The low biomass levels from the varzea lakes are partly due to the deoxygenated conditions and production of hydrogen sulphide in the hypolimnion which occurs during the stratification cycle (Section 3.5). In addition the low mineral content of the waters (25–70 µS cm^{-1}), including low calcium and low pH, is not suitable for molluscs which require calcium for shell formation. The bivalvia, so predominant in Bung Borapet and Lake Chad, are therefore either absent or present only at low densities in the varzea lakes. In Lago Tupé, which has the lowest biomass, the major inhabitants of the benthic region are the chaorbids which, although they may live in burrows in the mud, actually actively hunt in the open water (Section 4.7.2); the major detritivores are the ostracod crustacea whilst the hydrachnella (water-mites) are the principal predators (Reiss, 1977). In other Amazonian lakes the major components are either oligochaetes or chironomid larvae.

In many ways the benthic fauna of Lake George is similar to that from the varzea lakes; the biomass is of a similar order and bivalves make a minimal contribution. In the central lake area, as in the Lago Tupé, over 30% of the biomass consists of the larva *Chaoborus*, whilst a further 40% are due to other dipteran larvae of the chironomid midges, *Chironomus* and *Procladius* (Burgis *et al.*, 1973). Both of these are essentially detritus feeders although *Procladius* can also be predatory. Other detritivores such as oligochaetes and ostracods also make a significant contribution. The role of *Chaoborus* can be variable, however, since the first and second instars are wholly planktonic whilst the third and fourth instars are mainly benthic during the day but migrate into the plankton at night (McGowan, 1974). It seems that the diurnal retreat into burrows is a negative response to light and that in deeper lakes such as Lake Edward all instars of *Chaoborus* remain planktonic but sink to depths of very low light intensity during the day (Verbeke, 1957). The restricted range of types and low biomass found throughout the main central portion of Lake George is mainly due to the soft, unstable, deoxygenated nature of the mud there where the top 20 cm is prone to disturbance which, in itself, would tend to discourage the colonization by bivalves. The range of types does increase in inshore areas of Lake George where mayflies and caddis-flies can be found along with molluscs and increased proportions of oligochaetes. It is evident, however, that the greatest source of variation in the biomass of the benthic fauna of these lakes is the contribution of molluscs, particularly bivalves. With regard to the other constituents, notably the insects, oligochaetes, and ostracods, the values available so far suggest that their densities are more immediately comparable whilst their role in the functioning of the community may be more significant than that of the molluscs.

In the same benthic communities, shrimps can be the dominant organisms. In Lake Sibaya, South Africa, for example, the atayid *Caridina nilotica* contributed a standing crop of 2–3 g dry wt m^{-2} (Hart and Allanson, 1975). In Tarek Bera swamp, Malaysia, the extensive mats of bladderwort, *Utricularia*, contained a mean of 1.49 g dry wt m^{-2} shrimps belonging to *C. thambipalli* and *Macroproachium trompi* (Mizuno *et al.*, l982).

Under the extreme conditions of Lake Nakuru, a Kenyan soda lake, virtually the only benthic organisms were two chironomids, *Leptochironomus deribae* and *Tanytarsus horni*, giving a mean biomass of 0.4 g dry wt m^{-2} which is quite low compared to most of the values given above (Vareschi and Vareschi, 1984).

The richest North American lakes in the temperate zone can have benthic standing crops of around 8.5 g dry wt m^{-2} (Cole, 1968) which is equivalent to the higher values recorded in tropical regions although falling well short of the highest such as those occurring in Bung Borapet. Moreover, values from temperate lakes are generally summer ones, when biomass is at a peak whilst those for the tropics are often annual mean values (Vareschi and Vareschi, 1984).

Man-made lakes provide a rather special case for the benthic fauna. The inundation of large tracts of land should provide a wealth of new opportunities for those animals adapted to live on the lake bottom but there are, however, limitations to this. We have seen above how important the nature of the sediment is to the abundance and distribution of benthic animals; consequently the nature of the soil soon after flooding is not suitable for most of these. Moreover, during the early stages of the filling process a good deal of decomposition of terrestrial vegetation can take place which causes deoxygenation of the bottom waters, particularly if stratification of the water body occurs. In this situation a habitat of particular importance to the benthic animals, which may have been denied their more typical habitats, is the trunks and branches of the trees which have become submerged. Often these extend up into the oxygenated waters close to the surface. In the Volta Lake, for example, the trees rapidly became colonized by an ephemeropteran nymph, *Povilla adusta*, capable of burrowing under the bark, and they were able to build up enormous numbers to the extent that biomass ranged up to 10 g m^{-2} (Petr, 1970). In the Volta, however, most of the submerged trees were softwoods but in other lakes, such as Kariba, the majority were hardwoods. In this situation burrowing types such as *Povilla* were unable to penetrate the trees directly and, to begin with, only the bark could be colonized by a few oligochaetes and chironomids which attained a biomass of 0.022 g m^{-2}. However, when the annual drawdown occurred the dead trees were exposed and attacked by the terrestrial wood-boring beetle *Xyloborus*; the tunnels formed by these could then be colonized by *Povilla* and the trichopteran *Amphipsyche senegalensis* when the trees were again submerged and the biomass subsequently increased to 0.097 g m^{-2} (McLachlan, 1974). The presence of hardwoods, therefore, tends to delay the colonization of the trees by benthic animals.

Eventually, however, the bark decays from the trees and the trees themselves become broken down and of less value for colonization. By this time, however, more typical aquatic sediments have started to form through sorting of particles by turbulence and wave action as well as through the inclusion of organic material derived from the river inflows and from detritus originating from autochthonous plant production within the lake. These sediments should provide a suitable medium for invertebrate colonization but there is, in addition, the annual drawdown which can expose great areas of the benthic region to desiccation. Many animals cannot deal with this and, therefore, the fauna at the bottom of man-made lakes is rarely diverse if the drawdown is great. Most often chironomids are the major if not the sole member of the fauna although high densities can still be attained. Biomass of chironomids in Lake Volta generally ranged from 0.7 to 2.5 g m^{-2} although unusually high values up to 24 g m^{-2} were found in sheltered bays where blooms of *Microcystis* were occurring (Petr, 1969). The chironomids can apparently remain plentiful even during the drawdown period owing to the adults continually laying their eggs in water as it recedes (McLachlan, 1974).

4.7.5 Biomass and Assemblages in Flowing Waters

The linear course of a river from source to estuary generally involves loss of altitude and changes in slope, cross-section, and current (Section 2.2). This leads to gradual changes in the fauna although it has been possible to identify a point of transition in rivers from a wide range of latitudes whereby the upstream 'rhithron' zone can be demarcated from the downstream 'potamon' zone (Illies, 1961; Illies and Botosaneanu, 1963). These categories can be defined as (from Hawkes, 1975):

(i) *Rhithron*—that part of the stream from its source down to the lowermost point where the annual range of monthly mean temperatures does not exceed 20 °C. The current velocity is high and the flow volume small. The substratum may be composed of fixed rock, stones, or gravel and fine sand. Only in pools and sheltered areas is mud deposited.

(ii) *Potamon*—the remaining downstream stretch of river where the annual range of monthly mean temperatures exceeds 20 °C, or, in tropical latitudes, with a dry season maximum exceeding 25 °C. The current velocity over the river bed is low and tends to be laminar rather than turbulent. The river bed is mainly of sand or mud, although gravel may also be present. In the deeper pools oxygen may be depleted, light penetration limited, and mud deposited.

In effect the transition represents the change from the upland phase of the river when the source is at altitude, to the lowland plane where the river becomes wider and deeper. The reference to temperature ranges made in the original classification may not be entirely justified throughout the whole of the tropics and warmer regions of the world. In South Africa, for example, the physical differences between rhithron and potamon stretches were found to be suitable, but the minimum seasonal water temperature was found to be a better criterion than temperature range (Harrison, 1965). On the mountainous Caribbean island of Trinidad the rhithron zone was found to extend from 500 m altitude down to between 30 and 17 m, restricting the potamon section to the narrow coastal strip (Hynes, 1971), and gradient appears to be an important factor. Equivalent changes in species zonation occurred over only a 500 m change in altitude here compared to a change of 1000 m or more in the Andean headwaters of the Amazon (Illies, 1964).

The concept of the rhithron and potamon zones appears to be useful in many parts of the world and many of the invertebrate families inhabiting these are common to both temperate and tropical regions. For example, amongst the dipteran larvae the Simulidae are characteristic of the rhithron and the Chironomidae of the potamon, a difference which is related to their feeding mechanisms and other habits (Section 4.7.2). The slower flow of the potamon area also allows swimming types to survive, such as corixid and notonectid water-boatmen, whilst the silt which can be deposited provides a suitable medium for bivalves to burrow into. Nevertheless, species diversity is often highest within the rhithron (Bishop, 1973; Hynes, 1971).

The standing crop of benthic animals in tropical rivers has rarely been estimated. Bishop (1973) determined abundance and biomass along the course of a Malaysian stream, Sungai Gombak, and found that both were higher in stony riffles where eroding conditions occurred compared to areas of deposition. Numbers and weight of organisms in riffle areas also tended to increase downstream, but this was emphasized in this particular instance by a source of organic pollution at the lowermost station which enriched the waters and led to the development of large populations of turbificid oligochaetes and chironomids which thrive upon the organic detritus. In upstream samples oligochaetes could be numerically important, but the main bulk of the biomass was made up of Insecta such as Ephemeroptera and Plecoptera. The range of biomass determination for the unpolluted stretches of the river was 0.2–1.4 g dry wt m^{-2}, although this excluded the few molluscs which occurred in the two lower stations which would have considerably inflated the estimate of biomass owing to their large size. The values are of the same order as those obtained for the non-molluscan components of tropical lake-bottom faunas (Section 4.7.4), although as Bishop observes care must be taken in comparing values since sampling under different conditions by different methods, even to the extent of the mesh size of the dip net used, can influence the efficiency of sampling. In the polluted or enriched area of the river the biomass was considerably higher, up to 4.1 g dry wt m^{-2}, indicating the extent to which the deposit feeders can utilize the organic effluent. In addition particular microhabitats yielded particularly high levels of biomass such as the *Saraca*-root habitat, largely dominated by decapods, where weights in excess of 10 g m^{-2} were found, although, as with a number of specialized stream habitats, exact quantitative sampling was too difficult. In the Black Volta, Ghana, a biomass range of 4.5–43 g wet wt m^{-2} was found (Petr, 1970). Allowing for a 20% conversion to dry weight this would put the biomass of the Volta river bed rather higher than that typically found in Sungai Gombak which might be attributed to the fact that the Volta is a savanna river and is also chemically richer (Bishop, 1973).

In addition to the actual biomass of the fauna at any one time the magnitude of the drift must also be considered since this is probably the most significant component when colonization and availability of bottom-dwelling organisms to predatory animals such as fish are concerned. In an Andean tributary of the Rio Matadero in Ecuador (Turcotte and Harper, 1982b), the range of types found in the drift coincide with the type of animals occurring in the bottom fauna, but in terms of percentage contribution platyhelminth flatworms, nematodes, and oligochaetes were under-represented owing to the fact that they live either tightly applied to or within the substratum and are less likely to be washed away. Most numerous were insects, particularly Ephemeroptera and caseless Trichoptera, and aquatic insects altogether accounted for 30–46% of numbers in the drift. The caseless net-spinning caddis types such as the Hydropsychidae are typically found in the rhithron portion of streams and consequently appear in the drift. They were also found in the drift of the Black Volta (Petr, 1970), a forest stream on the West Indies island of St. Vincent (Harrison and Rankin, 1975), but not in Sungai Gombak in Malaysia (Bishop, 1973). In a hillside forest stream in Sierra Leone, Ephemeroptera were prominent in the drift, although in equal proportions to small decapod atyid shrimps which were also common in the benthos (Payne, 1975). Terrestrial insects frequently appear in the drift and in the Ecuador Andean stream 13–23% of the animals were from this source and most commonly included spiders, Collembola, or Homoptera blown from the surrounding grasses and bushes of the paramos vegetation (Turcotte and Harper, 1982a,b). Terrestrial insects falling into the stream from the dense overhanging trees also made up a significant proportion of the animals being carried along in the Sierra Leone forest stream, amounting to some 26% of the total (Payne, 1975), but in a Ghanaian stream only 1% of the numbers taken were terrestrial (Hynes, 1975). In such circumstances, however, any disturbance of the surface by a small object brings immediate attention by small fish which often inhabit the tiniest stream and most insects falling into the stream are subject to rapid predation and are, therefore, likely to be under-estimated in the drift. The drift organisms may also be subject to this same intense predation which is probably more considerable than in temperate streams, judging from observations on the activities of fish in streams from both regions, and this should be borne in mind when comparing drift estimates from the two.

In the Andean stream, 1898–7923 animals day^{-1} were caught compared to 7000–40 000 in Sungai Gombak. In relative terms the values were 0.85–3.38 animals m^{-3} day^{-1} and 1.56–1.79 m^{-3} day^{-1}, which is perhaps slightly higher than the equivalent value for the Ghanaian stream of 0.1–1.9. Temperate drift rates can be higher than this although they rarely exceed 10 m^{-3} day^{-1}.

The numbers in the drift of the Ecuador stream were also fairly stable, showing little diel variation except on moonlit nights when the numbers fell. This was not true of a forest stream in Hong Kong where a marked burst of animals entering the drift just after nightfall was observed (Dudgeon, 1983).

4.8 THE NEKTON

4.8.1 Composition

The nekton is that part of the community which possesses sufficient locomotive powers to determine its own distribution. In fresh waters this largely means the fishes. There are a number of insect types such as the diving beetles and hemipterid water-boatmen which can swim very actively, but they nevertheless tend to remain in the shelter of marginal vegetation rather than become widely distributed in open waters. The only truly pelagic insect is the phantom larva *Chaoborus* but, by virtue of its size, it still tends to be at the mercy of water currents and therefore belongs to the plankton.

The fish of tropical waters present a completely different situation from that seen in the plankton and benthos with respect to composition. The type of plants or animals found in the phytoplankton or zooplankton, or living on the bottom, tend to be basically similar to those found in the temperate regions, at least with respect to families and even down to generic level in some cases. Amongst the fishes, however, not only are there considerably more species—for example Brazil alone has more than 1400 compared to 192 in Europe—but also many belong to exclusively tropical families or even in some cases such as the Characoidea and Osteoglossiformes, to suborders and orders. It is with the fishes, therefore, that the structure of tropical aquatic communities becomes most immediately distinctive. Moreover, the three major areas of the tropics, South and Central America, Africa, and southern Asia show considerable differences from each other.

Tropical fish faunas are dominated by members of the Ostariophysi, a tremendously diverse taxonomic grouping which includes fish as different as the catfishes and the carps, but which share a common feature in the possession of the Weberian ossicles. These are modifications of the first four vertebrae which connect the swim bladder to the inner ear. The swim bladder itself has an opening to the anterior alimentary canal.

In southern Asia the predominant groups are the carp family (Cyprinidae) and the catfishes (Siluroidea). The carps alone constitute more than a third of the known species whilst, together with the catfishes they contribute almost 60%. In Africa both cyprinids and catfishes are important whilst in addition a further group of ostariophysians, the Characoidea, also plays a significant part. Non-ostariophysian fishes do, however, contribute to the African ichthyofauna, in particular members of the percomorph family Cichlidae which have come to dominate the faunas of many of the large lakes in the continent. These typically have the anterior portions of the dorsal and anal fin supported by spiny rays, and their swim bladder is closed so that all regulation of the gas content is carried out physiologically. Africa also contains several unique families endemic to the continent including the ancient Polypteridae, the Mormyridae, which includes the 'elephant trunk' fishes with their long protruding snouts, the Gymnarchidae which, like the mormyrids, possess electric organs, and the Pantodontidae or butterfly fishes with their greatly expanded pectoral and pelvic fins which are probably useful when the fish jumps from the water during its hunting.

Tropical South America shares the Characoidea and Cichlidae with Africa as dominant members of the fish fauna whilst, again, the catfishes are very diverse and have produced a number of unique families amongst which are to be found the 'armoured catfishes' (Loricariidae) in which the body is covered in hard dermal plates, and the most widely distributed family, the Pimelodidae. The cyprinids are entirely absent from Central and South America and in many cases their place is taken by the characoids which have diversified to an incredible extent in this area to produce both unique forms as well as those which parallel types found in Africa and Asia; over a 1000 species have so far been described from South America. Parallel evolution is also shown in a number of ways by other groups—for example the development of electric organs for communication, navigation, and defence in totally unrelated groups of South American and African fishes, the Gymnotids and *Electrophorus*, the electric eel, in the first case and the mormyrids, gymnarchids, and an electric catfish, *Malapterurus*, in the second.

Within the faunas of tropical fresh waters are relic species of the early evolution of the fishes and these are generally widespread. The lungfishes or Dipnoi have representatives in South America, *Lepidosiren paradoxa*, in Africa with four species of *Protopterus* and also in tropical Australia, with *Neoceratodus fosteri*. Amongst the oldest teleostean or bony fishes are the Osteoglossidae which derive their scientific name from the hard tongue-like organ in the mouth. Several species still survive in the tropics such as the South American osteoglossid *Arapaima*, one of the largest freshwater fishes which can attain a length of 3 m and a weight exceeding 100 kg. There is one species in Africa, *Heterotis niloticus* and representative species of *Scleropages* in South-east Asia and northern Australia. In fact, *Scleropages leichardti* and the lungfish *Neoceratodus* are the only primarily freshwater species in the fresh waters of tropical Australia, the rest are either close relatives of marine forms or are estuarine species. In this respect the freshwater fauna of Australia is very impoverished and represents the long separation of Australasia from south-eastern Asia. In common with many organisms a division representing Wallace's line can be discerned for freshwater fishes between these two biogeographical zones which runs to the west of the Philippines between Borneo and Sulawesi and continues between Java and Lombok (Myers, 1951). On the western side of this line are the rich ichthyofaunas of the oriental region and to the east the considerably sparser Australasian faunas.

Although the major groups of fishes do show a considerable range of species owing to the evolution of form to suit function, each group does often possess certain common characteristics (Fig. 4.12). The Cyprinidae, for example, are frequently of a relatively elongated shape and are morphologically quite conservative. They are primarily fish of streams and rivers, often being found in the smallest headwater streams as well as the largest rivers of Africa and Asia although some have become adapted, at least partially, to life in lakes. No cyprinid has true jaw teeth but they do have well-developed pharyngeal teeth on the gill arches, and it is modifications of these that contribute to any feeding specializations that individual species evolve. Most are, in fact, foragers on plant or insect material with only a few, like the small African *Barilius*, being predatory, principally no doubt due to the lack of jaw teeth. A common modification amongst some cyprinids, such as the very widespread genus *Barbus* (sometimes referred to as *Puntius*), is the presence of filamentous barbels from the maxillary region (upper jaw), often close to the point of articulation. Another distinctive development, at least in the genus *Labeo*, is the development of the lips into a series of complex folds to produce a rather protrusible, suctorial mouth which can be used to suck up fine sediments. This design is also effective when applied to the aufwuchs and chains of semicircular feeding marks down the surface of rocks are often a sign that *Labeo* are present.

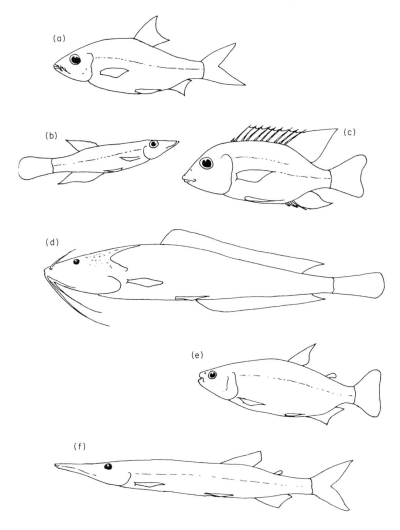

Fig. 4.12 Examples of some of the principal types of fish from tropical waters: (a) Cyprinidae, *Barbus*; (b) Cyprinodontidae, *Epiplatys*; (c) Cichlidae, *Tilapia*; (d) catfishes, Siluroidei, *Clarias*; (e) Characoidei, *Brycon*; (f) Characoidei (predatory), *Ichphyoborus*

The Siluroidea are considerably more diverse in body form than the cyprinids and they are also the most widespread of all the major tropical groups. Commonly referred to as 'catfishes', this is due to the development around the mouth of a variable number of 'whiskers' reminiscent of their feline counterpart. These are present in the majority of types, although by no means all, and they can be associated with upper or lower jaw or the nostrils. Almost universally there are no true scales and most have only an unprotected, mucus-covered skin although some, particularly in South America, have developed a system of hard external plates and are therefore termed 'armoured catfish'. In many, however, such as *Clarias*, a common genus of Africa and Asia, the bones of the skull have become fused and reinforced so that the head at least is well protected. The head is also often dorsoventrally flattened as a general adaptation to living on the bottom. This feature is also retained in some catfishes which have become secondarily adapted in other features to life in mid-water, such as the Schilbeidae. Some catfishes have developed spines in association with their fins as a means of protection. The most outstanding example of this is probably the African genus *Synodontis* in which three stout and often serrated spines are found on the leading edge of the dorsal and each of the two pectoral fins. When alarmed these can lock, pointing outwards to produce a rigid triangle of spines which presents a formidable mouthful for any predator. This same response is the bane of fishermen since it ensures that the fish become badly tangled in the net. The spines are widely regarded as being poisonous and whilst this is not true, the mucus on the surface of the spines appears to ensure that wounds readily go septic.

Probably the most morphologically plastic group of fishes is the third major branch of ostariophysian fishes—the Characoidei. These have produced examples of almost every conceivable life-style and many ways of accomplishing each specialized way of life. The readiness to diversify and evolve new forms has led to considerable parallel evolution of comparable types in South America and Africa, from the fairly generalized *Brycon* and *Alestes* to the specialized highly streamlined predators *Boulengerella* and *Ichthyoborus*. Types can range from the voracious tiger-fish of Africa, *Hydrocynus*, to the tiny, brilliantly coloured tetra, *Hemigrammus*, much in demand as ornamental fish. There are also those which mimic others or objects such as leaves or twigs. Although they are so diverse they do share one common obvious feature and that is the presence of a tiny adipose fin between the dorsal fin and the tail.

The Cichlidae are a family of percomorph fishes which have the anterior portion of the dorsal and anal fins supported by spiny rays whilst the posterior section has softer, more flexible rays. The range of body forms evolved by the cichlids is not nearly so diverse as those of the characoids; nevertheless, certain types have speciated extensively. For example, there are more than 250 closely related haplochromine species recorded for Lake Victoria alone, most of which are found nowhere else (Greenwood, 1974; Greenwood and Gee, 1969; Barel *et al.*, 1977), and the genus *Cichlosoma* has produced more than seventy species in Central America. Many of the specializations of the cichlids do not involve major morphological differences but depend upon behavioural adaptation supported by subtle modifications for different types of feeding, which is possible because of the great evolutionary flexibility of their dental arrangements. In addition to teeth in the upper and lower jaw, which themselves can become modified to differ with various roof types, the lower pharyngeal bones at the back of the gill cavity have fused to form a single, strong, triangular-shaped plate which bears large numbers of small teeth. This tooth-encrusted plate can be raised to work against another tooth-bearing plate on the roof of the pharynx. The pharyngeal bones can easily be removed for examination and indeed are an important feature in classifying cichlids, many of which can look superficially very similar. The nature and number of pharyngeal teeth also give an indication as to the diet of the species. For example, amongst the African tilapias, members of the genus *Tilapia* itself feed by chewing leafy macrophytes and so have a few large pharyngeal teeth for grinding the often fibrous plant material. Two further genera *Oreochromis*, from East and Central Africa and *Sarotherodon* from West Africa, are phytoplankton or generally microphagous feeders and these have large numbers of very fine teeth on the pharyngeal bones for dealing with much smaller particles (Trewavas, 1983).

In some cichlids which eat hard particles such as insects or seeds the central teeth on the pharyngeal bones become enlarged and often shaped like human molars, for crushing. Even so the most unusual and even bizarre feeding techniques have been evolved with no major physical adaptations; for example amongst the species flocks of cichlids occurring in the African Great Lakes are types which feed by carefully approaching other fish and gnawing scales from close to their caudal fins. This would appear to be a rather unpromising food source, yet in Lake Malawi several species have become adapted to this way of life (Fryer and Iles, 1972). Others feed by sucking the eggs from mouth-brooding cichlids, whilst other species defend their own algal 'gardens' (Ribbink *et al.*, 1983).

The tilapias were originally restricted to Africa and the Levant, but their use for culture has led to them being widely introduced to many warm-water regions of the world and there are few parts of the tropics where they are not found, at least on fish farms. Initially the tilapias were grouped together with the single genus *Tilapia* but now their three significantly different reproductive habits, as well as feeding methods, have indicated that they should be split into three genera (Trewavas, 1983) and are a further indication of the variety of behavioural differences within the cichlids. Most cichlids show some form of parental care for their young; in some cases this can be by guarding the nest and young, but others brood the young in the mouth by one or both parents.

The variety of niches which can be exploited by fish within communities is immense and the specializations which have been developed to suit each way of life are equally varied, as much within families as between them. It is, however, often possible to deduce quite a lot about where and how a particular species lives from a careful inspection. The position of the mouth can give some indication as to where the fish feeds; those in which the mouth is ventral with the snout overhanging the lower jaw are usually bottom feeders, those with terminal mouths may feed in mid-water, whilst those with the mouth angled upwards usually feed near the surface. The last situation can be accentuated in types, such as many of the cyprinodonts, which live right at the surface when the head itself can be flattened. Fish which habitually live close to the bottom can be flattened dorsoventrally whilst active mid-water swimmers tend to be laterally flattened. Most predators are highly streamlined for rapid acceleration; those which tend to lurk in wait for their prey, however, are often more deep-bodied than the more continually active silvery hunters from open waters. Rather, rounded deep-bodied types are characteristic of fish which browse in still or slow-flowing waters. Fish inhabiting cracks or crevices are often sinuous, flexible, and eel-like or very compressed from side to side. By contrast, fish with electric organs tend to hold their body very still and obtain propulsion from fin movements only in order to maintain an undisturbed electric field around them. These, of course, are just some of the general indicators to which can be added more specific pointers, particularly from jaws and teeth.

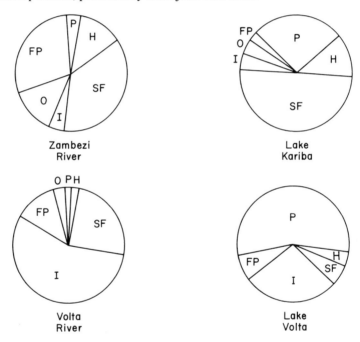

Fig. 4.13 Relative abundance of feeding types in terms of percentage biomass amongst fish assemblages of two rivers and two man-made lakes resulting from their impoundment. (P) plankton feeders, (H) herbivores, (SF) substrate feeders, (I) invertebrate predators, (FP) fish predators, (O) omnivores

The methods by which fish exploit their resources are rather too diverse and complex to treat systematically. However, the resources themselves, particularly food or energy supply, can be grouped into fairly well-defined categories, as demonstrated initially in Figs. 4.3 and 4.4. Fishes are important because they can be found at any trophic level from feeding on phytoplankton to eating animals that are themselves predators; they also contribute to the saprophytic chain (Fig. 4.1) when they utilize detritus or benthic invertebrates. Those fishes which do normally live and feed in the benthic areas are termed 'demersal fishes'. By dividing up the fish in the community in terms of their food sources a picture of the relative role of the various fish types can be constructed which, owing to the fact that fish exploit most trophic levels, can also show something of the structure of the community itself. Petr (1970) adopted a very simple scheme and separated the fish from the River Volta and Lake Volta into the six feeding groups shown in Fig. 4.13. The percentage biomass of fishes give an indication of the availability of the different food resources within each category, and Fig. 4.12 includes a second pair of examples taken from the Zambezi River and its resulting lake, Kariba. These indicate, for example, that the plankton only supports a small percentage of the total biomass in a river whilst in the lakes plankton feeders dominate, which coincides with the relative status of plankton in lakes and running water (Section 4.5.3; Figs. 4.3 and 4.4). In the Zambezi, the most prevalent groups are those feeding on the mud, whilst in the River Volta insect predators predominate. The difference between the rivers will reflect the nature of the places where the samples were obtained. For example, muddy deposits are typical, as we have seen, where a river is wide and slow flowing, whilst the insect fauna will become more important when the river is swiftly flowing or more rocky, but both could be found in the same river at different points. Substrate feeders could also be important in lakes, as they are in Kariba, since the lake bed presents a large area over which the sinking organic material can accumulate. There are, however, limiting constraints within lakes. Stratification may make large areas of the lake anaerobic and therefore inaccessible to most fish; moreover, many fish in tropical lakes rarely go below 20 m depth which again restricts the area available. The substrate feeders are, therefore, by no means always prevalent in lakes. In both Volta and Kariba, the resources available to macrophyte feeders show only a modest increase over their inflowing rivers. This is because all macrophytes are confined to the margins either because they are emergent or, even though submerged, are limited to the shallow areas within the euphotic zone, where there is sufficient light for phytosynthesis.

The divisions used by Petr (1970) are rather simplified (Fig. 4.13) and could be refined, but in the broadest sense they do present a reflection on the structure of the communities. The separation into groups by food source does ignore the point that many species show variations in their diet throughout the year. Indeed Vaas et al. (1953) when categorizing the fishes from the Kapaus River in Borneo divided them into stenophages, those with a precisely defined diet, and euryphages, those in which the diet is more variable. The term 'substrate feeder' used above is also rather vague since as well as detritus, live benthic algae or invertebrates may be ingested. Some can also suck the aufwuchs off hard surfaces. Distinctive sources such as the aufwuchs and detritus really require their own categories (Welcomme, 1979). Nevertheless, there are evident differences in the opportunities provided for fishes in rivers and lakes and these differences are emphasized by the different types of fishes which are found there.

4.8.2 River Fishes

It has been apparent from previous sections that rivers present a series of habitats along their length although they do fall into two major categories, the generally eroding rhithron upper reaches and the more depositional potamon stretches downstream. The rhithron zone is only present when some significant change in altitude is involved and the difference between the two zones is reflected in the benthic animals (Section 4.7.5). Fish are more mobile than the benthic invertebrates and are therefore able to move up or downstream; however, many are restricted to specific parts of a river system. Close to the source, streams are generally narrow and shallow but, in the tropics, often contain one or two species of small fish. When streams arise in upland areas where there is a recognizable rhithron section, the fish most commonly reported in the uppermost, often forested regions, in Africa and Asia, are *Barbus* species. This is the case, for example, in the Fouta Djalon massif (Daget, 1962), the mountains of Sierra Leone (Payne, 1975), central Nigeria (Sydenham, 1975), Sri Lanka (Kortmulder, 1982a,b), and Malaysia (Bishop, 1973). In South America, where there are no cyprinids, the small streams of the Andes tend to be populated with small, torrent-dwelling catfishes, characoids such as *Hemibrycon*, cyprinodonts, particularly *Orestias*, and some cichlids including *Aequidens* and *Cichlosoma*.

Although there may be only one or two species close to the headwaters of a stream, lower down the number of species tends to increase. The gradient itself may also influence the types of fish found closest to the source. In the streams descending from the mountains of the Freetown peninsula in Sierra Leone, for example, there is a distinct difference between the steeper seaward slopes where only *Barbus liberiensis* is found in the uppermost reaches and the less steep, although still rapidly flowing, streams of the landward side where the cichlid *Hemichromis bimaculatus* and the cyprinodont *Epiplatys fasciolatus* are always found. In both cases as the streams descend, becoming wider and deeper although still essentially rocky and boulder strewn, more species appear in the community (Fig. 4.13). Most of these particular streams do not level out into a distinct lowland potamon section before reaching the sea but in one that does, to produce a broader stream of 3–4 m wide and 1 m or so deep, with a more linear flow and silty bed, sixteen species occurred including specialized fish predators and scavenging catfishes. There are still major gaps in the composition compared to rivers—for example a lack of significant numbers of mud feeders and fast-swimming open-water carnivores—but the basic structure of this community is recognizable in those of other streams of similar dimensions in the lowland tropics (Table 4.1). The characins or their Asian counterparts, the cyprinids, tend to be active mid-water swimmers whilst the small cypri-nodonts and their South American equivalents the Poecelidae tend to stay in the more marginal areas as do the climbing perches. Specialized fish predators, often with elongated jaws, can also be found in such streams, such as *Hepsetus odoe* in West Africa and the garfish *Xenetodon cancila* in Sri Lanka. Most of the remaining fishes live close to the bottom and some are nocturnal such as many of the catfishes and the eel-like types, *Mastecembalus* and *Synbranchus*. Judging from the number of species found in streams of this order the number of niches in these lowland streams is limited but still greater than those of the upland sectors (see Fig. 4.14), although different families may fulfil equivalent roles in the different continents depending upon their zoogeographical history. In the African streams the relatively large number of endemic higher taxa are represented by *Polypterus*, *Papyrocranus*, and the mormyrids. In Sri Lanka the cyprinid species are plentiful in the absence of the charocoids and cichlids. Moreover, the *Barbus* species themselves show zonation as conditions change downstream (Kortmulder, 1982b). Even in the small headwater streams one species *B. titteya* occurs if the brook is marshy, or another *B. bimaculatus* if it is rocky. Combinations of others occur further down depending upon turbulence and current speed whilst yet others take their place in the potamon, lowland stretches.

The increase in species down a stream represents an increase in the nature and availability of resources including food, places to swim and places to hide. Right at the top of the system there may be just one species. Where *Barbus liberiensis* occurred alone in the upland forest streams in Sierra Leone (Fig. 4.14), a detailed examination of its diet showed that by far the most important item of the stomach contents was terrestrial insect fragments whilst the second most common was fragments of vascular plants (Payne, 1975). A few contained benthic invertebrates or algae but these were minor categories. These streams are close to the source and so have little accumulated organic material; they are also overhung by several tree layers, which causes considerable shading thus discouraging primary production of autochthonous material. Instead the stream receives a rain of whole insects, insect exuviae and insect faecal debris containing chewed-up leaf fragments from the overhanging trees. The fish rely upon taking the particles almost as soon as they hit the surface and swim in shoals with great speed to the source of any surface disturbance. Higher plant fragments and insects were also found to be the major components of the diet of the Sri Lanka upland *Barbus* species, although the smaller fish of the species found lower down in calmer waters also took crustaceans, such as copepods, which would occur in such areas of slower current (De Silva *et al.*, 1980).

As the streams descend from the headwaters they become larger, and active mid-water swimmers such as *Alestes* or other characins are able to feed on the increasing amount of debris from the forests which comes down the stream, including insects, seeds, and fragments of vegetation, much of which is left in suspension by the rapid flow and turbulence.

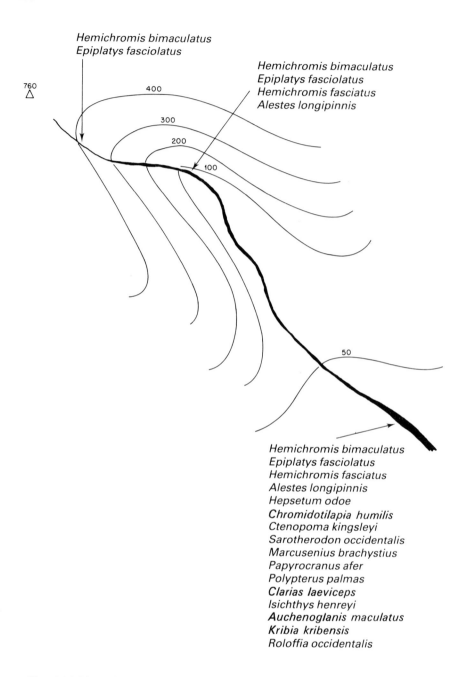

Fig. 4.14 The effect of slope on species number and composition of the fish assemblages in streams on the landward side of the Peninsula Mountains, Sierra Leone. The contours are in metres

Table 4.1 Species totals of fish assemblages from some tropical streams

Suborder	Family	Sierra Leone	Nigeria[a]	Panama[b]	Sri[c] Lanka	Malaysia[d]
	Polypteridae	1				
	Notopteridae	1				
	Mormyridae	2				
Characoidei	Characidae	1	2	4		
	Hepsetidae	1	1			
	Cyprinidae		1		7	5
Siluroidei	Five forms	2	2	2	1	4
Exocoetoidei	Belonidae				1	
Cyprinodontoidei	Poecelidae			2		
	Cyprinodontidae	2	1			
	Synbranchidae			1		
	Channidae		1			
	Cichlidae	4	3	1		
	Eleotridae	1		2		
	Anabantidae	1	1		1	
	Mastacemblidae		1			2
	Syngnathidae		1			1
		16	12	12	10[+]	12

[a] Sydenham, 1975.
[b] Zaret and Rand, 1971.
[c] Kortmulder, 1982a,b
[d] Bishop, 1973.

Once the stream reaches the lowlands and has become broader and deeper with a slower less turbulent current flow, then the whole water body can be used more efficiently and shared out between an increased number of species. This partitioning of the living space amongst the species was clearly shown to be the case in the forest stream (Table 4.1), of Rio Pedro Miguel, Panama (Zaret and Rand, 1971) in similar fashion to that generally outlined above. Insects were probably still the major food resource in that they constituted more than 60% of the diet by dry weight, often considerably more, in seven of twelve species found in the stream. In most cases terrestrial insects falling from the trees were a very significant part of the total insects consumed although they were only dominant in two species; the remainder consisted of benthic aquatic larvae of the types discussed previously (Section 4.7.1). Higher plant fragments did appear in the diet of some of the fishes, but did not play such an important role as in the headwater streams just mentioned. Two of the remaining species, the characins *Astyanax* and *Piabucina*, were fast-moving foragers taking a mixture of everything available, including algae, seeds, and a few insects, rather like *Alestes longipinnis* in the Sierra Leone stream, whilst two other types were almost solely algal feeders, the small poecelid, *P. spherops*, and the catfish, *Plecostomus plecostomus*, which grazes the algae off rocks with its sucking-type mouth. The presence of the last two species indicates some degree of primary production taking place in the river at this point. The remaining species in this Panamanian stream are fish predators, to some degree. Within lowland streams of this type it should be borne in mind that much of the finer detritus is able to sink to the bottom where it provides food for the invertebrates living there, as described in Section 4.7. This is particularly true of the prawns such as *Macrobrachium* and atyids like *Caridina* which, in many ways, act as part of the fish fauna.

In any sizeable river basin these streams of the potamon zone join together in a dendritic network to form progressively larger rivers. Here the niches multiply and the nature of the resources can change. As we have seen (Fig. 4.13), the sedimentation of finer organic particles may support the largest portion of fish biomass whilst, even in those rivers where they remain the most important food source, insects are overwhelmingly aquatic because at this point the river is probably broad and the fringing trees will have a negligible direct effect. Any insects or vegetation falling in or washed in upstream will have been already reduced to detrital particles. How much of this organic material supporting the fish populations is allochthonous and how much is autochthonous is extremely difficult to say, but since primary production in rivers is often low (see Section 5.2.3), the bulk may originate in the terrestrial system. It should be remembered also that even where aquatic insects form the major food source, they also most frequently depend upon this organic detritus.

As the streams join to form rivers so the number of niches available increases, and the extent of this increase can be related to the size of the river basin. The larger the basin the more species there are, and this can even be described by empirical equations obtained from plotting basin area against number of species occurring there (Welcomme, 1979) to give the following. For Africa $N = 0.449A^{0.434}$ and for South America $N = 0.169A^{0.552}$ where N = species number and A = basin area (km^2). Whilst physical increase in river size may play a part in this, the most potent factor will be geographical since in large river basins, groups of tributaries may be effectively isolated from each other giving the opportunity for independent speciation to occur and equivalent niches to become occupied by different species.

Compilations of the biomass records from rivers has been made by Lowe-McConnell (1975) and of standing pools of river floodplains by Welcomme (1979). Of the rivers themselves estimates from the channel of the Kafue River gave the biomass at 33.7 and 55.1 g f. wt m^{-2} obtained by using fish poisons (chemofishing) and multiple seining respectively (Lagler *et al.*, 1971). During the construction of the Kainji dam a section of the Niger was isolated by cofferdams and drained; here the collected fishes give a biomass estimate of 5.9 g m^{-2} (Motwani and Kanwai, 1970). Owing to the difficulties in obtaining determinations of biomass from flowing waters of large rivers, since poisons are readily diluted and netting in restricted areas is difficult, these are the only observations made for main channels. They are certainly no higher than those of 40.2 g m^{-2} made for the small Nigeria streams (Sydenham, 1975) mentioned previously (Table 4.1; see also Table 7.2), and may suggest that biomass does not necessarily increase down rivers. The pools resulting from the seasonal drying up of the River Sokoto in northern Nigeria produced standing crops of $7.8–207 \, \mathrm{g \, m^{-2}}$ (Holden, 1963), but these obviously represented concentrations of fish into restricted areas. In fact the average biomass was $41.5 \, \mathrm{g \, m^{-2}}$ which, if applied to the whole floodplain area from which the fish had retreated, would amount to only $1.2–1.7 \, \mathrm{g \, m^{-2}}$. However, fish are subjected to considerable mortality as they come off the floodplain and this may therefore be an underestimate. Similar ranges of biomass have been found for floodplain pools of other tropical rivers (Welcomme, 1979); for example in south-eastern Asia the Mekong lagoons had $6.3–39 \, \mathrm{g \, m^{-2}}$ and the Parana in South America gave $126.4–221.6 \, \mathrm{g \, m^{-2}}$.

Tropical rivers do present certain problems as an environment to the fishes which live there. As we have seen (Section 2.3.2), many show seasonal changes of flow and some, like the Sokoto, cease to flow entirely and become a series of isolated pools. The low solubility of oxygen at high temperatures and the decomposition of organic debris can often combine to produce low oxygen concentrations in the rivers, particularly during the dry season. To augment the availability of oxygen from the water, numerous fish types have developed accessory respiratory organs which enable the fish to use air. Gills are not inherently incapable of functioning in air; it is just that, normally, out of water they collapse and there is no space for circulation between the lamellae. The mud-skipper, *Periophthalmus* has its gills stiffened so that they do not collapse and therefore can be used out of water. Most often, however, some non-branchial modification occurs. The simplest is the vascularization of the gill cavity wall, which is found in a few types such as *Synbranchus*, but many elaborate organs have been developed for more effective use of air. Catfishes such as *Clarias* and *Heteropneustes* have tree-like or sac-like chambers opening from the top of the gill cavity, and equivalent structures are also to be found in other groups such as the labyrinth organs above the bronchial cavity of the anabantids. Many South American catfishes and others modify parts of the gut, either the pharynx such as in the case of the electric eel, *Electrophorus*, or the stomach or even the intestine in the case of *Haplosternum*. The need to use respiratory modifications of the gut in these South American fishes often coincides with the period of low water in the dry season when the fish cease feeding (Lowe-McConnell, 1964). The most well-known of fishes capable of aerial respiration are the lung-fishes which have a single dorsal lung although the opening to the pharynx is ventral. The primitive African *Polypterus* is also regarded as having true lungs, with a pair placed ventrally, whilst others have similar organs derived from a modified swim bladder.

Many of these fishes with the ability to use air are thus able to live in warm, stagnant waters of isolated river pools or swamps and, in addition, some such as *Clarias* will move overland on wet nights which explains why these fishes are able to colonize apparently isolated waters. In such cases the copious secretion of mucus over the skin of fishes like *Clarias* and *Anabas* may restrict dehydration. There is also the possibility that, in the case of fishes without scales, such as the catfishes and *Periophthalmus*, some oxygen uptake can occur across the moist skin when the fish emerges into the air. In many of the fishes with modifications for utilizing air their use is not optional; the fish must be able to return to the surface periodically to replenish their air supply and will die of asphyxiation if prevented. In other types, air breathing remains an auxiliary method to aquatic gas exchange; this is even the case in one of the lung-fishes, *Neoceratodus*, from Australia.

Air-breathing fishes can have rather subtle modifications including a lack of sensitivity of the blood pigment, haemoglobin, to carbon dioxide. Normally the haemoglobin of fish living in clean waters shows a decreased ability to take up oxygen if the environmental concentration of carbon dioxide increases slightly. Carbon dioxide always tends to accumulate in organs where gas exchange using air takes place, and consequently the haemoglobin of air-breathing fishes shows a much reduced sensitivity to carbon dioxide picking up oxygen efficiently even when carbon dioxide concentrations are high. Since many of these fishes can live in swamps where the carbon dioxide concentration of the water can be high, this lack of sensitivity of the pigment will also be advantageous for aquatic respiration.

Air-breathing is a commom feature of tropical fish assemblages; for example, of the species referred to in the small streams in Table 4.1, slightly more than half have some means of enhancing their respiration by using air. There is, however, another source of improved oxygen supply apart from the air itself and this is the thin film of oxygen-enriched water at its interface with the air. Some fish, such as the cyprinodonts, are particularly well adapted to use this zone since the head is flattened for close approach to the surface and the mouth is directed upwards (Lewis, 1970). In a Panamanian stream, 93% of the non-air-breathing fish showed evidence of being able to conduct aquatic surface respiration (Kramer, 1983).

In tropical rivers where sediment loads can be high and visibility low and where a significant proportion of the species may be nocturnal, the fish often require ways of gathering information and communication other than by visual means. The catfishes, for example, often have rather small eyes but well-developed 'whiskers' which will aid their orientation and hunting in the bed of the river. The catfishes can have up to eight pairs, and one pair at least can sometimes be very long, up to half the body length as in the case of the Asian genus *Mystus*, or become finely branched as in the case of the mandibular processes of the synodontids. Some cyprinids, including *Barbus*, have one or two pairs of barbels but they are not so well developed as in the catfishes.

Perhaps the most sophisticated method of information gathering and communication, however, is shown by the several lines of fish with electric organs which have developed independently in Africa and South America. The source of their electrical activity is generally a large series of plates devised from modified muscle cells working in series in order to co-ordinate their discharges. Most often, as in the South American gymnotids and the African mormyrids, they are derived from the myotomal muscle blocks of the body but in the electric catfish, *Malapterurus*, they are formed from the dermal musculature. In some, such as the electric eel, *Electrophorus*, the discharge can be of great power, up to 600 V in this case, and can be used to stun prey or for defence. The electric catfish can also deliver a shock, even small fish giving a powerful jolt to the unwary, and fishermen generally cause them to discharge before handling since it takes some time for the potential to be built up again. The most common use of electric organs, however, is for navigation and signalling. The electric organs give off a low-voltage discharge at a frequency which may depend upon the species. The fishes create an electric field around themselves and can detect disturbances in this field. The need to keep this field steady to detect disruptions means that the body itself, or at least the part bearing the electric organs, should be kept relatively rigid. Thus, in *Electrophorus* and *Gymnotus* where the electroplate organs are aligned along a large area of their elongated bodies, the body is held still and propulsion is by rippling of the dorsal and anal fins. In the mormyrids the electric organ is in the caudal peduncle which is consequently stiff and not flexible as in most fish. The discharge patterns can be detected in the field by dipping two electrodes into the water attached to an amplifier. Amongst a group of eleven gymnotids occurring together naturally in a creek in the Rupununi district of Guyana, each had its own distinctive signal and in at least one, *Sternopygus macrurus*, males and females had their own patterns (Hopkins, 1972). The male would respond with a characteristic pattern when a female approached but not a male, and would also respond to a female play-back signal. The system of communication which is possible is quite elaborate and, in at least one case, *Eigenmannia*, there is a jamming avoidance reaction in that if the fish detects a frequency close to its own it shifts its frequency (Hopkins, 1973). Most gymnotids have very regular discharge patterns which may be more concerned with orientation than communication although the mormyrids, in at least some cases, do produce more regular bursts when others are in the vicinity. The irregular, low-frequency discharges of the predatory electric eel may allow it to blend in with the 'noise' of the electrical background whilst its potential prey emit their more regular distinctive signals.

4.8.3 Fish in Lakes

In many ways conditions in lakes are considerably more stable than those in rivers. The hydrological stability means that autochthonous production in the form of phytoplankton is encouraged and the temporal stability appears to allow considerable specialization to occur. The production of the phytoplankton tends to be reflected in a high relative biomass of planktivorous fish, as exemplified in Fig. 4.13. Phytoplankton feeding, however, is a rather specialized technique amongst fishes and has only originated in a few groups. In Africa, within the Great Lakes, the Cichlidae dominate this feeding niche. Virtually the whole of the biomass of phytoplankton feeders referred to in Fig. 4.12 belong to this family and most notable amongst these are the tilapias. Phytoplankton feeding is just one of the methods to which the rather flexible cichlid feeding mechanisms can become adapted (Section 4.8.2). The specific feature which adapt tilipias for feeding on phytoplankton was shown in *Oreochromis esculentus* from Lake Victoria to be due to the copious secretion of mucus along the gill arches to which phytoplankton cells then stick as they pass the arches with the respiratory current (Greenwood, 1953). The mucus and algae are then scraped off the arches by the fine teeth on the pharyngeal bones to be passed back down the oesophagus for digestion. As well as the capture of the algae there is also the question of digestibility. Algae such as diatoms are readily digestible and their empty frustules are common in faecal material, but other types with chemically complex cell walls, such as blue-green algae, can be more difficult since the walls contain celluloses and fish do not produce cellulase. However, the cell walls of blue-green algae and bacteria can be split by low pH (Moriarty, 1973; Bowen, 1976) and the gastric pH of cichlids is sufficiently low at 2 or less to cause these cells to rupture and their contents to be exposed to the digestive enzymes (Moriarty, 1973; Payne, 1978). This is significant since blue-green algae are often common or dominant in lake phytoplankton.

The microphagous feeding habits of the cichlids probably originate in the rivers, but here these fish make up a very small percentage of the standing crop. Within the rivers, cichlids make their nests in quiet backwaters and are not favoured by rapid currents. Once fish with these attributes colonized lakes, their feeding and reproductive habits were well suited to the more stable conditions and, in this sense, the cichlids were pre-adapted to the lacustrine way of life. Examples of this happening have been provided by the damming of major rivers to make man-made lakes where, almost without exception, the cichlids, principally the tilapias, came to dominate the fauna. Most of the natural lakes of Africa also have tilapias as their most significant phytoplankton feeders; for example, Lake George has *Oreochromis niloticus*, Lake Victoria *O. variabilis* and *O. esculentus*, and Lake Malawi has six species. Such were the original cases at least, but fishing pressure and introductions have complicated the situation particularly in Lake Victoria (see Chapter 8). Other cichlids as well as tilapias do feed upon phytoplankton; in Lake George *Haplochromis nigripinnis* also utilizes this as well as *O. niloticus*.

Some non-cichlids are also planktivorous and therefore do well in lakes. Amongst these the African freshwater herrings are perhaps the most prominent. They do occur in rivers, but where they have colonized lakes they are very successful. Lake Tanganyika, for example, has two endemic sardine species, *Limnothrissa miodon* and *Stolothrissa tanganicae*, both of which feed on a mixture of phytoplankton and zooplankton when young whilst the former, although still essentially planktivorous when older, does take a more mixed diet than *S. tanganicae*, which eats mainly zooplankton. The principal fishery on the lake is for these sardines, or *dagaa* as they are known locally, which indicates a substantial standing crop and production. Similarly, since *L. miodon* was introduced into Lake Kariba it has come to make up 80% of the fish yield of that lake. In the Volta Lake a major component of the fauna and fishery, besides the cichlids, was a herring *Pellonula afzelius* which had previously been unrecorded in the river before its impoundment.

Amongst other non-cichlid fishes the cyprinids also include types which can be generally microphagous. Two of the Indian major carps, for example, *Catla catla* and *Labeo rohita* will take fine particles from the water column, the former closer to the surface and the latter closer to the bottom. In addition there are other carps, such as *Cirrhinus cirrhosa*, which take zooplankton but, on the whole, they lack the specialization of the tilapias.

To be truly successful in the lake environment fish need to break their dependence upon rivers. The cichlids have done this, but many non-cichlids still need to return to rivers to spawn. Some of the non-cichlids in Lake Victoria, for example, such as the *Barbus* and *Labeo* species are in this category and they migrate up the inflowing rivers during the spawning season which usually begins as the river level begins to rise (Corbet, 1961).

Table 4.2 Survey of trophic specializations and species numbers (from Kortmulder, 1982). (A) Lake Victoria cichlids (after Van Oijen *et al.*, 1977–79); (B) Lake Redondo fish (after Marlier, 1967).

(A) Diet	Species	(B) Diet	Species
Piscivores	80	Omnivores	12
Lepidophages	1	Non-specialist herb.	3
Pedophages	15+	Non-specialist carn.	8
Algal scrapers	17	Specialist herb.	6
Insectivores/detritus	53	Piscivores	5
Mollusc crushers	20	Insectivores	2
Crustacean eaters	5	Zooplankton eaters	2
Zooplankton eaters	4	Mud eaters	4
Parasite feeders	1		

The stability of lakes with time appears to encourage the occupation of a large number of niches and diversification within feeding types also occurs. This can be demonstrated by a breakdown of the major feeding niches in Lake Victoria (Table 4.2) as mentioned earlier (Section 4.8.1), but is not only confined to large lakes; even smaller lakes such as the 35 ha Lake Redondo in Brazil shows a considerable diversity (Kortmulder, 1982a,b). The temporal stability of lakes allows species time to adapt to specialized food sources whilst the hydrological stability means that, unlike fish in rivers which have to be prepared to feed generally on anything which is available during regular times of scarcity in the low-water season, fish in lakes have reliable specific sources of food all the year round to which they can become specialized to exploit. Not that river systems as a whole are necessarily poor in species, but this is due to their geographical diffuseness as much as to the number of niches which are actually available. Within the circumscribed area of a lake, however, there appears to be some pressure to diversify which has led, in many of the older lakes, to the evolution of species flocks, that is, groups of closely related species each minutely specialized to a different niche (see Chapter 7). We have already mentioned the haplochromines of Lake Victoria, but there are further examples in the other Great Lakes such as the cichlid flocks of Lake Malawi, amongst which the mbuna show many variations on living on and around rocks, and the utaka, which are plankton-feeding pelagic species. This is not a phenomenon restricted to large lakes; a range of endemic cichlids are found in Barombi Mbo, a small crater lake in Cameroon. Nor is it confined to African species; flocks of cyprinids occur in Lake Lanao in the Philippines and a flock of cyprinodonts in Lake Titicaca. Lakes are, therefore, rather more self-contained entities than rivers in that much of the production is autochthonous and their stability allows specializations to be developed.

Of the few estimates of fish biomass made for tropical lakes, the majority are of the same order of magnitude as those made for rivers although, as always, fish are rarely evenly distributed throughout a lake and there is considerable variability. For example, samples in Lake George with purse-seine nets which are used to encircle and trap fish in open water produced a mean estimate of 23.3 ± 5.6 g fresh wt m^{-2} although the range was between 6.3 and 90 g m^{-2} (Gwahaba, 1975). Chemofishing in two Cuban lakes showed ichthyomasses of some 32 g m^{-2} (Holcik, 1970), whilst for a lagoon in Venezuela the value was 100 g m^{-2}. In Lake Kariba, particularly high densities of 283 g m^{-2} were found (Balon, 1973) although, by contrast in another man-made lake, Lake Volta, the biomass was only 11.4 g m^{-2} (Regier and Wright, 1972). In Lake Nakuru, Kenya, which has a very high salinity, the biomass was quite low, with a mean of 2.1 g dry wt m^{-2} or 9 g f.wt m^{-2} (Vareschi, 1979), even though primary production here is very high. The lake does, however, contain only one species of fish, the introduced tilapia *Oreochromis alcalius grahami*, which is known for its tolerance of extreme conditions such as occur in this lake.

4.9 LITTORAL AND SURFACE VEGETATION

4.9.1 Composition

The shallow margins of lakes and rivers provide opportunities for rooted or attached vegetation composed of larger plants, collectively termed macrophytes, to exist. Many of these macrophytes are vascular plants although they may include a few large algae. Aquatic vascular plants occur in a number of diverse forms although, basically, there are four categories:

1. Emergents—these are rooted in the soil below the water surface whilst the reproductive organs of the shoots are held above the surface. This includes many of the reeds and rushes such as *Typha* (Fig. 4.15), *Phragmites* (Fig. 4.16) and *Cyperus*, as well as many grasses.
2. Floating leaved—these are rooted on the bottom, but their leaves float on the surface as shown by *Nymphaea* (Fig. 4.17) and other water-lilies.

Fig. 4.15 The reedmace *Typha*

3. Submerged—although rooted or attached to some hard surface the whole plant remains entirely under water except for the reproductive parts which emerge at the surface.

4. Free floating—the plant is not attached and roots may dangle in the water whilst the shoots appear above the surface, as in the water-hyacinth *Eichhornia crassipes* and the aquatic fern *Azolla*, or the plant may be submerged but entangled amongst other water-plants as is typical of *Ceratophyllum*.

Fig. 4.16 The reed, *Phragmites*, which can occur thoughout most of the tropics with the exception of the Amazon basin

Where conditions are suitably undisturbed there can be a typical gradation from emergents in the shallowest waters to floating leaved types and finally submerged plants in the deeper water. This can be seen in many of the shallower East African lakes where dense banks of papyrus (*Cyperus papyrus*) and grasses give way to water-lilies (*Nymphaea*) and water-chestnut (*Trappa*) and then to submerged *Ceratophyllum*, *Potamogeton*, and *Utricularia*. The depth to which submerged vegetation can occur is limited, like the phytoplankton, to the euphotic zone where sufficient light is available for phytosysthesis. It is inhibited, therefore, by high turbidity and is, to some extent, in competition for

Fig. 4.17 The water-lily, *Nymphaea*, in which the outermost petals are much longer than the sepals

resources with the phytoplankton. This is less true for other aquatic macrophytes since their leaves intercept light before it reaches the water surface, although they may share the same sources of essential nutrients such as phosphorus and nitrogen depending upon whether these are obtained from the soil or the water.

4.9.2 Littoral Vegetation

Whilst depth can limit the degree of colonization of lakes and rivers by macrophytes, nevertheless areas of permanent shallow water or where the soil remains saturated can provide suitable conditions for more or less continuous cover by emergent aquatic macrophytes leading to swamp formation. Swamps can occur around the margins of lakes; in Lake Chilwa, East Africa, the area of *Typha* swamp is equal to the area of the open water. They can also occur in river valleys and in floodplains. Some are of considerable size—the Plain of Reeds in the Mekong Delta in South-east Asia covers 1 million ha, whilst the immense Sudd region of papyrus swamp across the course of the White Nile in southern Sudan extends over 4 million ha which is considerably increased during flood periods. For the African continent, in fact, Beadle (1981) has estimated that the area under swamp was similar to that covered by all the lakes. Although many swamps associated with river systems occur in the lower regions, there are examples of upland swamps occurring in upland areas such as those found in the Kupas lake district of West Borneo and on the Sepik River in Papua New Guinea.

Extensive coverage of water by emergent plants tends to modify the environment to an extent rarely seen amongst other types of organisms. Their large size and three-dimensional structure result in a physical complexity which supplies an increased number of places for other plants and animals to live. The underwater parts of emergent plants as well as submerged macrophytes all provide surfaces for a flora of epiphytic algae and bacteria, as well as hard surfaces for such animals as gastropod snails to creep along as they browse on the epiphytes and decaying organic material on the macrophytes and associated debris. The tangle of underwater stems and roots becomes a refuge for swimming insects including water-beetles and water-boatmen, as well as providing cover for predators both large, such as fish, and small, for example, dragonfly nymphs. It is not surprising, therefore, that the shallow, marginal areas, or littoral zone, where rooted plants are common, often contain a greater diversity of animals than the benthic areas of open-water regions (Section 4.7.4). For example, in Lake George the vegetated littoral regions contained nineteen fish species compared to around ten in the mid-lake (Burgis *et al.*, 1973). The actual biomass of animals can also be higher. In the reservoir of Bung Borapet in Thailand, a case used earlier to consider the distribution of benthic animals generally (Section 4.7.4), the bottom fauna of the littoral vegetated region, which was dominated by the submerged *Hydrilla verticillata* and *Ceratophyllum demersum* growing between the water-lily *Nymphaea lotus* and *Nelumbo nucifera*, had an animal biomass which was the highest recorded in the lake at $25\,\mathrm{g}$ dry wt m^{-2} (cf. Section 4.7.4).

The complexity of their physical structure is by no means the only way in which macrophytes modify the environment for other organisms. The constant death and deposition of leafy parts tends to produce a covering of organic material for the sediments in the vicinity of the plants, and the organic content of the sediments can be greatly enhanced owing to the settlement of suspended organic material from the water in the calm protected bays and lagoons often formed by the water-plants. This can have the effect of stabilizing water margins and paving the way for colonization by more terrestrial vegetation in the frequently observed phenomenon of ecological succession as the terrestrial system encroaches, with time, on the aquatic. This stabilization can be particularly important in newly created environments such as man-made lakes. In Lake Kariba the colonization of the marginal area with larger water-plants led to the accumulation of autochthonously produced organic material in their vicinity. This encouraged the development of the mud faunas normally associated with lakes (McLachlan, 1974). Moreover, the invasion of shallow areas, particularly by the grass, *Panicum repens*, physically bound the bottom material together which reduced the erosive effects of wave action.

Large banks of water-plants can produce considerable chemical modifications within the environment. In relatively open areas where water-lilies or submerged types such as *Hydrilla* are plentiful, the high levels of incident sunlight during the day promote rapid photosynthesis leading to such high levels of oxygen output that by early evening the water can be more than 100% saturated with oxygen, i.e. supersaturated. During the night this is reduced as oxygen is used for respiration by both the plants and animals and the cycle begins again the next morning. In extensive areas of swamps dominated by emergent vegetations the water surface tends to be heavily shaded, which means that there is little light available for photosynthesis by phytoplankton or submerged macrophytes actually in the water. Dense shading is particularly true of papyrus canopies in African swamps, and only slightly less so for *Typha* and *Phragmites* canopies where the leaves are more linearly aligned.

The constant rain of dead vegetation into the water below the emergent plants produces an accumulation of organic debris which, under swamp conditions, can lead to a reduction or the complete absence of oxygen in the water under the dense growth of vegetation. The decomposition process also causes this water to be markedly more acidic than open water, perhaps as low as pH 6 in some cases. The deoxygenated conditions found in swamp waters are also a result of the difficulties of reoxygenation, since the plants themselves protect the water surface from the wind action and turbulence which would alleviate this effect. Their shade inhibits phytosynthesis within the water which would also promote reoxygenation.

These rather singular conditions prevail over large areas of water where swamps occur. Emergent vegetation, however, is not necessarily limited to shallow water. For example in the case of papyrus, the range of rhizomes and accumulated organic material can begin to spread from the shore over the water surface as a dense floating mat. Below this develop the same conditions generally characteristic of swamps, even though the underlying water may be several metres deep. Large fragments of such vegetation may break loose, particularly after storms, to form floating islands of vegetation.

In these situations the lack of oxygen presents very rigorous environmental conditions for animals. The quantity of animals associated with such dense vegetation can be rather low; the biomass of invertebrates under these conditions close to the inflow of Bung Borapet reservoir was $0.324\,g$ dry wt m^{-2} (Junk, 1975), close to a hundred times less than the highest value associated with the less dense marginal *Nymphaea* and weed vegetation mentioned above. Potentially anaerobic conditions in swamps might even be considered likely to eliminate animal life altogether but Beadle (1981) gives a list which includes representatives of most freshwater groups found deep in papyrus swamps in Uganda. Some, such as mosquito larvae, use oxygen from the air and therefore need to return regularly to the surface to refill their trachael systems, but there is also the possibility that some can undergo limited anaerobic metabolism. Amongst the fish it was noted in a previous section (4.8.3) that many have accessory respiratory organs for using air and also a blood pigment which has a high affinity for oxygen and a reduced response to high carbon dioxide levels. Fish with these modifications can survive quite well in swamps.

4.9.3 Plants of the Water Surface

The influence of macrophytes upon aquatic systems is not limited to marginal, littoral areas. The tendency of some emergent types to form floating mats over quite deep water has already been mentioned but, in addition, there are a number of water plants which float entirely free although these too can become aggregated into continuous mats. In fact some of these, including the aquatic fern *Salvinia*, the water-cabbage *Pistia stratiotes* (Fig. 4.18) and the water-hyacinth, *Eichhornia crassipes*, can become major nuisances by blocking waterways. During the filling stages of Lake Kariba, *Salvinia molesta* covered 22% (over 1000 km^2) of the lake's surface and became a major impediment to movement and fishing in the lake.

These floating plants can form the basis of important biotopes equivalent to swamps in many ways. The best documented examples of this type are the 'floating meadows' of the white-water sections of the Amazon and their dependent varzea lakes (Marlier, 1967; Junk, 1970). The basis of these meadows are rooted grasses, *Paspalum* and *Echinochloa*, with shoots which float on the surface. Often forming banks on the perimeter of the beds of grass are true floating plants including, *Pistia*, *Eichhornia*, and *Salvinia*. Sometimes these banks become very dense and compressed with a consequent depletion of oxygen and even the formation of hydrogen sulphide. These compact areas form floating islands which can be sufficiently well established for trees to grow on them (Sioli, 1975).

As with swamps, where the water beneath floating vegetation is devoid of oxygen, animal life can be completely eliminated or very sparse. In the floating meadows, for example, the denser forms supported an animal biomass of 0.3–4.2 g dry wt m^{-2}. By contrast, the more loosely structured mats contained a density of fauna of between 2.5 and 11.6 g dry wt m^{-2} which is equivalent to the more open type of littoral vegetation. When beds of floating vegetation drift along with the surface waters they often carry their associated animals with them, forming their own microhabitat. In Lake George, the drifting rafts of *Pistia* also carry with them shoals of small fish such as the cyprinodont, *Aplocheilichthys*, which swim amongst the roots feeding upon the various invertebrates associated with the plants.

4.9.4 Vegetation of Rivers

River floodplains often contain wet or marshy areas where large stands of emergent vegetation or swamps develop in rather the same way as they do in the shallow margins of lakes, except that the rise and fall of the river imposes a distinct seasonal cycle on them and they may be of a more temporary nature. Many rivers or streams, however, have well-defined channels with relatively small floodwater areas and, within these channels, the great changes in water level between dry and wet seasons, together with high current speeds, tend to discourage the establishment of marginal vegetation. Moreover, the intense

Fig. 4.18 The floating water-cabbage, *Pistia*. (Photo: Dr Jenifer Owen, Leicester)

exposure of sandbanks and marginal areas to sunlight during the dry season and the high turbidity of many rivers, which can severely reduce light penetration for submerged plants, are other features which help to create difficult and unstable conditions for littoral vegetation. As the rivers rise and fall, temporary vegetation may develop on sandbanks and a permanent vegetation on rocky islands, but these are often composed of basically terrestrial plants.

There is, however, one family of plants which has become adapted to the rather unstable and demanding life imposed by rivers and this is the Podostemaceae. These are small plants which live encrusted upon rocks and are amongst the most highly specialized of all water-macrophytes (Fig. 4.19). There are some 200 species, most of which are tropical with the greatest number in South America but occurring in most tropical areas (Grubert, 1975). Most of the species are annual, germination and vegetative growth beginning at some point when the rocks are covered with water and the river level is high. At this time the current speed is often rapid and the plants must be held firmly on the rocks. This is accomplished by chemical adhesion, the root system being effectively absent. The form is quite variable. In some species they can be rather thallus-like in others finely divided; often they look rather like mosses and liverworts which can fulfil a similar role in temperate streams. Flower buds begin to form as the water surface of the falling river reaches a critical point above them. As the river level declines further, more buds appear and flowering takes place just as the water level drops below the plant (Grubert, 1974).

Fig. 4.19 *Ledermaniella* flowering showing the prominent anthers (two here) and reduced carolla typical of the Podostemacea. (Photo: Dr P. J. Harris, Coventry Polytechnic)

Usually plants of a given species occur as a layer over the rock surface and consequently there is a narrow band of flowering just above the water-line. The mats of vegetation are rather absorbent and therefore, during this crucial flowering period, the plants can be kept moist. Since the river level is often dropping day by day, the timing and synchronization of bud formation and the brief flowering period must be very precisely controlled as reproduction must be completed before the exposed plant becomes desiccated. The varying degrees of responsiveness to current speed, light, or other factors leads to zonation of podostemacean species down the rocks (Fig. 4.20). The flowers, themselves, are often rather reduced with only the anthers and stigmas being prominent and pollination is by small insects such as dipteran flies. Fruit and seed formation proceeds as the plants are left higher above the water-line and the tiny seeds are eventually dispersed by the wind. The problem faced by the seeds is that they must become attached to a bare rock surface in a situation where they may eventually become exposed to water rushing past at high speed. If the seed lands on a moist area of rock, a mucilaginous 'corona' around the seed is formed which acts as an adhesive to cement the seed to the rock, especially if it is finally dried as it often is by the time the river reaches its lowest level. By this time, all that remains of the plants themselves are totally dried-out encrustations on the rocks which are often the most evident signs of the presence of the podostemaceans. Only a few species, such as *Hourera fluviatilis* from the Caroni River in Venezuela, which live in places not subject to great fluctuations in water level show vegetative growth throughout the year.

Through the evolution of this very intricate series of adaptations the Podostemacea have come to terms with the demanding conditions imposed upon macrophytes by seasonally fluctuating tropical river systems. In fact these

Fig. 4.20 Podostemaceans may show zonation down rocks depending on their response to light and water level. Here *Ledermaniella* occurs above a zone of *Tritchia*. (Photo: Dr P. J. Harris, Coventry Polytechnic)

plants can be found in some of the most difficult conditions for primary producers, for example in the black-water rivers of South America, where the pH can be as low as 4.8 and the conductivity little higher than that of distilled water at $9\,\mu S\,cm^{-1}$ (Grubert, 1975). They have also been able to colonize waterfalls and rapids.

4.9.5 Biomass

The recorded ranges of biomass, in dry-weight terms, of above-ground vegetation are $0.5–4.3\,kg\,m^{-2}$ for tropical areas with no dry season, $0.3–2.5\,kg\,m^{-2}$ where a dry season is usual, and $0.6–3.6\,kg\,m^{-2}$ in arid areas (Westlake, 1982).

Each of these ranges is rather wide although they are all on a similar scale, even for arid regions. Aquatic vegetation in arid areas generally relies upon water derived from some distance away and does not, therefore, reflect the otherwise dry nature of the immediate environment. There can be considerable variation in the estimation of emergent plant biomass at any single site owing to variations in the degree of clumping, and also because of the edge effect caused by the plants becoming less dense towards open water; sampling programmes need to take this into account. In the relatively few instances where these variables have been considered for a single cosmopolitan species, there has been a suggestion that standing crops of emergent plants in tropical regions can be rather higher than those from temperate regions. For example, *Phragmites australis* tends to occur in stands exceeding $3\,kg\,dry\,wt\,m^{-2}$ in suitable locations in the tropics compared to an average of around $1\,kg\,m^{-2}$ in temperate areas (Westlake, 1982).

The magnitude of the biomass or standing crop of plants found is generally very high compared to that of other elements in the community. In the case of the other major primary producers, the phytoplankton, their biomass can be smaller by one or two orders of magnitude (Westlake, 1981). In Lake George for example, which is a highly productive lake, the average phytoplankton standing crop was $46\,g\,dry\,wt\,m^{-2}$ whilst that of the marginal papyrus swamp was $2.5–4.5\,kg\,m^{-2}$. Phytoplankton standing crops in general rarely exceed $70\,g\,m^{-2}$. The reason for this difference is that the macrophytes need an elaborate three-dimensional canopy to maximize their light utilization which requires considerable supporting material. The material for this accumulates over a longer period of time than is possible amongst phytoplankton cells where rapid cycles of replication and mortality in a supportive medium is the rule. In Tasek Bera Swamp, Malaysia, biomass of both above- and below-ground parts of the reed *Lepironia* was some $2\,kg\,dry\,wt\,m^{-2}$, which was regarded as being slightly higher than the mean standing crop of terrestrial grasslands in temperate regions (Ikusima *et al.*, 1982).

Submerged macrophytes are, to some extent, an intermediate case, since on a wet-weight basis they can have a biomass similar to those shown by emergent types, although on a dry-weight basis they often weigh much less, since the buoyant effect of the surrounding water reduces the need for supporting tissue and the water content of the plant as a whole can be high. Nevertheless, even on a dry-weight basis, biomass of submerged plants can be substantial. For example, *Cryptocoryne griffithii*, attained a biomass of 364 g dry wt m^{-2} in the open waters of the Tasek Bera Swamp and even higher densities of 568 g dry wt m^{-2} along more shaded channels (Ikusima *et al.*, 1982).

Floating plants can also occur at very high densities and produce standing crops only slightly less than those of emergent vegetation. Banks of the water-cabbage, *Pistia* and the water-hyacinth, *Eichhornia* can completely block waterways to traffic and considerably reduce flow rates when they occur up to 1.5 kg m^{-2} (Westlake, 1981), and in the floating meadows of the Amazon the grass *Paspalum* achieves standing crops up to 1.14 kg m^{-2} (Junk, 1970).

One final particular difficulty in estimating the standing crops of aquatic macrophytes is that a substantial proportion may be in roots, rhizomes, or stolons which can be underground in the case of emergent plants or under water in the case of those which float. Many of these root systems accumulate over long periods of time and effectively constitute the energy reserves of the plant to produce new leaves following substantial die-back or loss of the shoots due to fire, which can be common in seasonal swamp areas. In addition, they often have a considerable influence on the physical structure of the community and the growth form of the plant colony. In this respect the links provided by the rhizomes, which enable papyrus mats to form over deeper water, and the shelter provided by the submerged roots of floating plants have already been mentioned. The stolon connecting individual *Eichhornia* plants can bind the plants together very effectively and makes them even more difficult to deal with when they reach nuisance proportions. In temperate regions, where the whole aerial shoots die back every winter, a considerable proportion of the reserves can be held in the underground parts of emergent plants and these may constitute five times the biomass of the above-ground shoots (Westlake, 1981). In many although by no means all of the tropical situations examined, however, the underground portions weigh considerably less than the above-ground, presumably because there is not the same regular complete die-back in the tropics and therefore less need for an energy source to regenerate the whole plants. For example, the below-ground biomass of *Typha augustata* in Uttar Pradesh, India was 40–50% of the aerial shoots (Sharma and Pradnan, 1983) and a similar ratio was found for stands of the reed *Lepironia* in Tasek Bera Swamp, Malaysia (Ikusima *et al.*, 1982). For the grass *Paspalum repens* in the floating meadows of the Amazon the proportion was as little as 18% (Junk, 1970), although in this case the colony may only last a few months. Some floating plants can have more substantial root systems, however, so that, for example, the water-hyacinth, can have between 30 and 65% of its biomass made up of the roots (Knippling *et al.*, 1970).

4.10 COMMUNITY STRUCTURE AND THE DISTRIBUTION OF BIOMASS

The information on the material structure, in terms of biomass, of tropical freshwater communities is far from comprehensive. However, an appraisal of the estimates dealt with in the preceding sections does give some impression of the relative orders of magnitude of the various components in these communities (Figs. 4.21 and 4.22). Since a variety of units were used originally, approximate conversions into dry weight equivalants per square metre are needed to determine this.

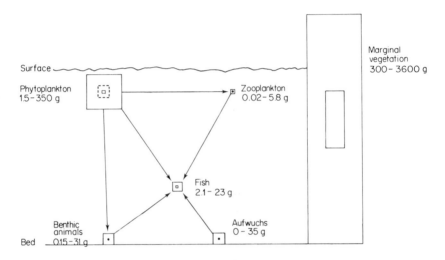

Fig. 4.21 Ranges of biomass found for the various components of tropical lake communities. Area of inner rectangle represents minimum and outer the maximum biomass. Phytoplankton maxima occur under the rather exceptional conditions of soda lakes and the maximum to be expected under more typically freshwater conditions is represented by the dotted line. Dry weight values are given

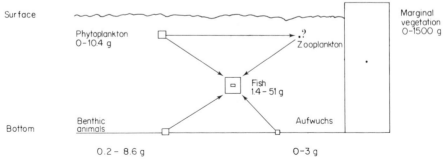

Fig. 4.22 Ranges of biomass found for the various components of tropical river communities. Area of inner rectangle represents minimum and outer the maximum dry weight biomass.

This approach emphasizes that the standing crop of phytoplankton in lakes tends to be rather higher than that in rivers and that the relative weight of zooplankton can be quite small. Indeed, although the abundance of zooplankton is known to be low in rivers, no indication of their actual biomass appears to be available at present. The fish standing crops in both lakes and rivers are of a similar order; in rivers this partly reflects the amount of organic material imported from upstream which plays a major role in sustaining the community since primary production can be limited. Through their feeding relationships, the fish play a central part in the redistribution of organic material within the community (Figs. 4.21 and 4.22).

The weight of organisms present at any one time, however, does not necessarily reflect the rate of production of new material or the rate at which energy is being processed. For example, the biomass of marginal vegetation can considerably exceed that of the phytoplankton and the biomass of the fishes tends to be greater than that of zooplankton. However, both the marginal vegetation and the fish have accumulated that material over years, whereas the phytoplankton and zooplankton have a very rapid turnover with generation times measured in weeks. Therefore, they have little tendency to accumulate biomass.

A more realistic measure of the contribution of each of the components to the energy and resources available within the community is, therefore, the rate of production generated by the biomass present. Dimensions are also needed for the pathways showing feeding relationships in Figs. 4.21 and 4.22 since the rate of consumption can influence both production and biomass. All these processes are considered in the next chapter.

GENERAL READING

Beadle, L. C. 1981. Tropical swamps. In *The Inland Waters of Tropical Africa*, Ch. 8, 2nd edn. Harlow: Longman.

Boney, A. D. 1975. *Phytoplankton*. Studies in Biology, No. 52. London: Edward Arnold, 116 pp.

Edmondson, W. T., and Winberg, G. G. 1971. A Manual on Methods for the Assessment of Secondary Production in Fresh Waters. IBP Handbook 17. Oxford: Blackwell.

Elliott, J. M. 1977. *Some Methods for the Statistical Analysis of Samples of Benthic Invertebrates*, 2nd edn. Scientific Publications of the Freshwater Biological Association, UK, Vol. 25, 160 pp.

Hawkes, H. A. 1975. River zonation and classification. In *River Ecology*, Whitton, B. A. (ed.), 312–374. Oxford: Blackwell.

Hynes, H. B. N. 1970. *The Ecology of Running Waters*. Liverpool: University Press.

Junk, W. J. 1970. Investigations on the ecology and production of the 'Floating Meadows' (*Paspalo-Echinochleotum*) on the middle Amazon. Part I: The floating vegetation and its ecology. *Amazoniana* 4: 449–495.

Junk, W. J. 1975. The bottom fauna and its distribution in Bung Borapet, a reservoir in Central Thailand. *Verh. Internat. Verein. Limnol.*, **19**: 1935–1946.

Lowe-McConnell, R. H. 1975. *Fish Communities in Tropical Freshwaters*, Chs. 2 and 8. London: Longman, 337 pp.

Reiss, F. 1977. Qualitative and quantitative investigations on the macrobenthic fauna of the Central Amazon lakes. I. Lago Tupé, a blackwater lake in the lower Rio Negro. *Amazoniana*, **6**: 203–235.

Round, A. E. 1972. *The Biology of the Algae*. New York: St Martins Press Inc.

Vollenweider, R. A. 1969. *A Manual on Methods for Measuring Primary Production in Aquatic Environments*. IBP Handbook, No. 12, Ch. 2. Oxford: Blackwell.

Welcomme, R. L. 1979. *Fisheries Ecology of Floodplain Rivers*, Ch. 3. London: Longman.

Westlake, D. F. 1975. Macrophytes. In *River Ecology*, Whitton, B. A. (ed.), pp. 106–129. Oxford: Blackwell.

Westlake, D. F. 1982. The primary productivity of water plants. In '*Studies on Aquatic Vascular Plants*' Symoens, J. J., Hooper, S. S., and Compere, P. (eds.), pp. 165–180. Brussels: Royal Botanical Society of Belgium.

Wiegert, R. G., and Owen, D. F. 1971. Trophic structure, available resources and population density in terrestrial vs. aquatic ecosystems. *J. Theoret. Biol.*, **30**: 69–81.

CHAPTER 5

COMMUNITY DYNAMICS

5.1 THE BASIS OF PRODUCTION

In the previous chapter, some indications were given of the biomass or standing crop of each component of tropical aquatic communities present at any one point in time. Plants and animals are, however, constantly assimilating new supplies of materials and energy whilst, at the same time, there is a continual loss from each population or trophic level through mortality, respiration, and excretion. The incorporation of new energy and materials into the organisms constituting the biomass is termed production and this is usually given per unit time. If the losses occur at the same rate as accumulation or production then there will be no net change in biomass. This also means that the size of the biomass is not necessarily related to the rate of production. For example, phytoplankters are microscopic and their biomass can be quite small; their life cycle, however, is very short, perhaps a few weeks, and the rapid rates of reproduction and mortality lead to a very high rate of turnover and production. By contrast, the biomass due to fish, whose life cycles are measured in years, results in the accumulation of energy over a much longer period of time whilst production may actually be quite low.

There is, of course, a difference in production accomplished by plants as opposed to that of the organisms which depend upon them. Plants transform light into chemical energy whilst heterotrophs only reassemble the chemical energy of their food into that of their own tissue. The distinction is therefore made between the primary production due to plants and secondary production from subsequent trophic levels.

Significant production can occur within a community although the biomass can remain stable, since losses may equal gain (Fig. 5.1). The greatest proportion of light energy striking the surface of the water is usually lost through reflection and absorption by the water itself, and only a few per cent may actually be available to the suspended phytoplankton for photosynthesis. Some of this is then converted to organic compounds, which represent the primary production. The percentage of energy from the incident light at the surface

124

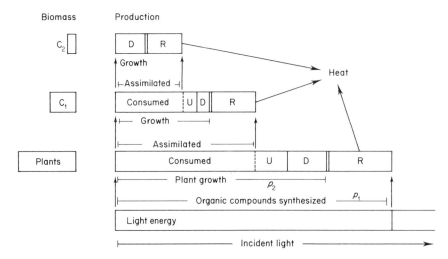

Fig. 5.1 The fate of the energy from gross primary production amongst the trophic levels of an aquatic community when the biomass is stable and shows neither gain nor loss: (P_1) total amount of energy incorporated in organic matter by photosynthesis (gross productivity); (P_2) energy available for plant growth or to the rest of the community after plant respiration and maintenance accounted for (net productivity); (R) respiration; (D) dead for decomposition; (U) unassimilated, i.e. faeces; (C_1) first stage consumer (herbivore); (C_2) second stage consumer (predator)

which is actually incorporated into new material in the phytoplankton is termed the photosynthetic efficiency. As indicated above, this is generally fairly low; a survey of lakes and rivers in India, for example, showed the range of photosynthetic efficiencies to be between 0.1 and 1.44% (Ganapati and Sreenivasan, 1972) and values are generally of this order.

All the time photosynthesis is going on, respiration is also taking place. A proportion of all new material synthesized will therefore be required to provide the energy for respiration and so a distinction must be made between gross productivity, which is the total production resulting from photosynthesis, and net production, which is that remaining after respiration has been accounted for (Fig. 5.1). This can be summarized:

Gross production = Net production + respiration

A small and often uncertain proportion of the gross primary production may also be lost as excretory or extracellular products. Under conditions of equilibrium, the net production will all be either grazed by the first-stage consumers or lost to the decomposers after death. Of the plant material ingested by the grazers, some passes through the gut to be voided as faeces which, therefore, represents resources unavailable to the animal although the faecal material can be used by the decomposers. The remainder is absorbed or assimilated in the gut by the animal and this can be expressed as a percentage of the original amount consumed to give the assimilation efficiency. Some of the assimilated energy is lost in the excretory products whilst a large proportion is dissipated as respiration (Fig. 5.1). The respiratory fraction is usually considerably larger in animals than plants, particularly microphytes such as the phytoplankton which have no supporting, non-photosynthesizing tissue to maintain, partly owing to the high levels of activity in animals. The energy and materials remaining after respiration has been accounted for can accumulate either as individual growth or increase in numbers through reproduction, which then represents the secondary production available for the next trophic level. That which is not consumed is lost through natural mortality to the decomposers.

The same sequence of events occurs in the subsequent levels occupied by the predators and it is evident (Fig. 5.1) that there can be considerable flux of energy and materials through the whole system without necessarily affecting the apparent density or biomass of organisms in the community. However, if, for example, consumption and death do not account for all the net production then some can accumulate in the form of growth of the individuals or an increase in population size, which is added to the biomass. Conversely, excessive consumption or loss will mean that some of the original standing crop will also be removed, which may impair the capacity for further production.

5.2 PRIMARY PRODUCTION

5.2.1 Light

For plants living under water, the problems of receiving sufficient light for effective photosynthesis are considerably greater than those of terrestrial plants owing to the much greater absorptive, reflecting, and refracting properties of water compared to air. Light intensity diminishes logarithmically with depth (Fig. 5.2; see also Fig. 4.2) and although theoretically it is never completely extinguished, an intensity of 1% of the incident surface light usually denotes the limit for effective production and the euphotic zone. The depth at which this occurs, however, is variable and depends upon suspended material in the water, including the algae themselves. Since the depth of the euphotic zone can vary from a few centimetres to 50 m or more, it is often convenient to rescale this region in optical terms to allow comparisons of photosynthetic events to be made in this productive region. Each unit of 'optical depth' is the point at which the intensity of the most deeply penetrating green light is halved compared to the previous unit, down through the water column. The extent to which the algae themselves cut down the light is difficult to determine, but one estimate for Lake Lanao in the Philippines showed that 30% of the variation in light penetration was due to changes in the algal standing crop (Lewis, 1974). Since a positive relationship is found between standing crop and the degree of light attenuation (Lewis, 1974; Talling, 1965), this 'self-shading' effect will be higher in lakes with higher densities of phytoplankton.

Direct measures of light energy and light penetration are obtained by lowering a selenium photocell down through the water column whilst taking readings

126

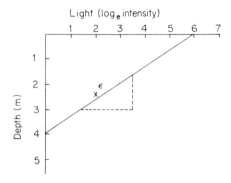

Fig. 5.2 A logarithmic transformation of the exponential attenuation of light through a uniform water column typically produces a linear relationship the gradient of which gives ε, the extinction coefficient

at appropriate depths. A measure of light penetration in the form of ε, the vertical extinction or attenuation coefficient, can be obtained from a plot of the natural logarithm of the percentage incident light against depth. The extinction coefficient ε is then the slope of this line and it will increase as the water becomes more turbid.

Not all of the light energy penetrating to a particular depth can actually be utilized by the algae for photosynthesis. The various photosynthetic pigments tend to absorb light of a particular wavelength so that, for example, chlorophyll *a* has its peak absorption at the red end of the spectrum at a wavelength of some 680 nm. The photosynthetically active irradiance in water is normally taken to be between 400 and 700 nm wavelength. The light climate of the phytoplankton is determined, therefore, not only by the absolute energy or quantum flux from the light but also by the wavelength. Yellow/green light usually penetrates the furthest whilst the blue light tends to be absorbed more rapidly (Fig. 5.3). Consequently, not only does the total light energy decrease with depth but its spectral composition changes until it is restricted to only those wavelengths which are least attenuated, usually around 500–600 nm. Since each wavelength has its own attenuation characteristics, the attenuation coefficient can be taken from determinations with an unselective photocell or, by using a series of colour filters, the coefficient for particular wavelengths can be determined. In many circumstances the coefficient of the wavelength of greater penetration and least attenuation, ε_{min}, is the most useful measure of light extinction. The presence of phytoplankton itself can alter the spectral composition of the light. Very clear waters tend to favour the blue end of the spectrum, but with generally increased attenuation due to the presence of phytoplankton or other compounds such as the 'yellow substance' related to humic materials, which are frequently present in natural waters, there is a shift to the longer wavelengths at the red end (Talling, 1982).

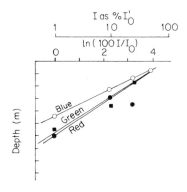

Fig. 5.3 Penetration of light of different wavelengths, blue (380–465 nm), green (485–565 nm), and red (615–670 nm) through the waters of the Lago Carioca, Brazil, plotted on a logarithmic scale. (*Reproduced by permission of Schweizerbart'sche Verlagsbuchhandlung from Reynolds et al., 1983.*) Intensity (I) is taken as a percentage of incident surface light (I_0)

5.2.2 The Measurement of Primary Production

There are three methods by which the rates of phytoplankton production are commonly estimated, whilst a fourth has recently become available. These are:
(a) Enclosure with measurement of oxygen flux: the light/dark bottle method.
(b) Enclosure with measurement of carbon dioxide flux: the ^{14}C method.
(c) Monitoring of oxygen flux in whole water bodies.
(d) Use of satellite imagery.
In the light/dark bottle techniques, samples of the water to be examined are exposed in glass bottles with the assumption that enclosing the plankton community in this way has no significant effect upon its performance.

If the basic activity of photosynthesis can be summarized as

$$6CO_2 + 6H_2O \rightarrow C_6H_{12}O_6 + 6O_2$$

then both carbon dioxide utilization and oxygen production must be related to the rate of synthesis of new organic compounds. In the oxygen light/dark bottle method, changes in oxygen are used to estimate productivity, whilst in the ^{14}C method the labelled carbon is used to follow carbon dioxide uptake.

(a) Light and dark bottle oxygen method

Two bottles of the same size are filled with water at the site and one, the dark bottle, is covered with foil or black paint to exclude light, whilst the other, the light bottle, remains uncovered. This pair of bottles can then be suspended in the water at the depth at which the sample was originally taken, and pairs of bottles at different depths will produce estimates of a dark production profile down through the water column. In the dark bottle, from which light is excluded, only respiration can take place, whilst in the light bottle both photosynthesis and respiration occur. Oxygen concentrations can be determined titrimetrically by the Winkler technique, so that by determining the initial concentration from a reference sample taken at the start of the experiment and obtaining final concentrations in the light bottle, any increase in oxygen over the time course of the exposure is equivalent to net production. However, significant bacterial production will not be taken into account by this procedure. To obtain the gross production, the oxygen being used concurrently for respiration must be accounted for and this can be obtained from the drop in oxygen due to respiration in the dark bottle (Vollenweider, 1969). To summarize therefore:

Light bottle $[O_2]$ −   initial $[O_2]$ → O_2 from net production
Light initial $[O_2]$ + (initial − dark $[O_2]$) → O_2 from gross production
    Net production   Respiration

The problems associated with this method are that prolonged exposure can lead to the accumulation of metabolites and that the glass surface of the bottles is an ideal substratum for bacterial growth and replication, which selectively increases the respiratory contribution and gives an underestimate of net production. These two factors can be minimized by exposing the bottles over a relatively short space of time. Lewis (1974), amongst others, wondered if the size of bottle might have an effect, with the larger ones providing a relatively smaller surface area to the water volume. However, no significant difference was found.

A rather more subtle problem is that of photorespiration. Some plants expend more energy in respiration when photosynthesis is occurring than in the dark when it is not, the difference being the phenomenon known as photorespiration. In which case, the respiration in the dark bottle is not directly equivalent to that in the light. As a consequence of this, estimates of gross production can be in error and net production may be the only reliable factor. Many workers have, however, used this method to produce estimates of both net and gross production. The respiratory rate obtained from the dark bottle does not, of course, represent only the rate of respiration of the phytoplankton, since bacteria and zooplankton are also enclosed along with the phytoplankton. This, in itself, does not invalidate the use of this estimate of respiration for obtaining gross production from net production and it is generally very difficult to separate the respiratory contribution of each component. However, in Lake Lanao, Lewis (1974) used abundance, body size, and metabolic rates to obtain such a separation and estimated that, of the total respiration of suspended organisms in the water, 80% could be attributed to the phytoplankton, 5% to the Crustacea, 12% to the Protozoa and 3% to the bacteria.

Since oxygen production is a function of the synthesis of organic compounds, there should be a proportional relationship between this and carbon fixed from carbon dioxide which would enable the conversion of production estimates from units of oxygen produced to units of carbon incorporated in new material. However, the ratio of carbon fixed to oxygen produced varies depending upon which type of compound is being synthesized. For example, when carbohydrates are the major product as in the equation above (p. 127), the ratio of $O_2 : CO_2$ is $1 : 1$ giving a photosynthetic quotient of 1, but this can range up to 3 when lipids alone are being produced. Normally, however, the plants are synthesizing a mixture of compounds and a quotient of around 1.25 is often taken (Stickland, 1960), which is very close to that determined for Lake Lanao of 1.26 (Lewis, 1974) and is equivalent to a conversion of 1 mg O_2 = 0.31 mg C.

(b) Light and dark ^{14}C method

The use of ^{14}C to label the carbon in CO_2 allows the amount of carbon synthesized into new organic compounds to be measured more directly. The ^{14}C is usually added to a sample enclosed in the bottle in the form of $H^{14}CO_3^-$ which acts as a source of carbon since it is in equilibrium with free CO_2 in the water (see Section 2.1). The amount of labelled ^{14}C added is known and the amount of unlabelled ^{12}C available can be obtained from a simple titration of alkalinity and a measure of the pH (Vollenweider, 1969). The quantity of ^{14}C taken up and fixed by the phytoplankton can be assessed by filtering a sample and counting the activity from the isotope in the plankton on the filter paper using a scintillation counter. The total amount of carbon fixed is then given by

$$\text{Total } ^{12}C \text{ assimilated} = \frac{^{14}C \text{ in phytoplankton}}{^{14}C \text{ introduced}} \times {}^{12}C \text{ available}$$

Carboxylation of bicarbonate can occur leading to incorporation of ^{14}C into the algae with scarcely any gain in chemical energy. This, however, can be estimated by keeping a bottle in the dark and measuring the ^{14}C uptake by the algae. The rate of uptake in the dark can then be subtracted from that obtained in the light to give a more precise estimate of photosynthetic carbon fixation alone.

No estimate of respiration can be obtained by this method since the total output of carbon dioxide cannot be measured. Following from this is the uncertainty as to whether the ^{14}C method is actually measuring net or gross productivity. In Lake Lanao, Lewis (1974) found that production measured by the ^{14}C method was very similar to net production derived from the oxygen method, whilst in Lake George, Ganf and Horne (1975) observed that production measured from ^{14}C was consistently less than oxygen-based estimates of gross production and was probably intermediate between this and net values. Work carried out in an oligotrophic area of the tropical Pacific Ocean, using extremely sensitive techniques of determining oxygen concentrations by microprocessor-controlled photometric Winkler titrations, showed, however, that ^{14}C estimated productivities were very close to gross production values from the oxygen technique (Williams et al., 1983). The situation, therefore, remains unclear. It is even possible that exposure of the ^{14}C bottle for a short time will give an approximation of total carbon fixation alone, i.e. gross productivity, whilst leaving it for a longer period, during which some of the labelled material may have been respired or secreted as some form of extracellular product, would be more likely to produce an estimate closer to net production (Vollenweider, 1969).

The ^{14}C method is nevertheless very sensitive and remains the only widely available method for use in oligotrophic waters. Ganf and Horne (1975) concluded that the oxygen and ^{14}C methods are complementary since the former allows reasonable estimates of net productivity which is important for considerations of the energy available to the next trophic level.

By measuring the ^{14}C accumulating in the water after driving off the inorganic ^{14}C by acidification, the quantity of organic extracellular products secreted by algae can be estimated (Soeder and Talling, 1969). These can be appreciable, particularly at high light intensities, and in Lake George, for example, on occasions accounted for between 6.2 and 58% of the gross carbon fixation.

(c) Oxygen flux in a whole water body

Photosynthesis and respiration can cause a lake to show the same effects over a 24-hour cycle as are found in a pair of light and dark bottles. During the day, both respiration and phytosynthesis are occurring, whilst at night only respiration takes place. If there is sufficient light for significant net production to occur during the day, there will be a tendency for the oxygen concentration to increase and, indeed, under eutrophic conditions the water may even become supersaturated by early evening. At night, respiration of all organisms in the water will cause the oxygen concentration to fall to a minimum just after dawn.

If the change in oxygen concentration is monitored then it is possible to calculate the net and gross productivity for the whole lake, if its physical dimensions are known, in a similar fashion to that for the light and dark bottles. There is, however, the added complication that oxygen in the water is in equilibrium with that in the air and therefore will tend to lose or gain the dissolved gas as the concentration deviates from the 100% saturation level. This needs to be compensated for by calculating the net flux of gas between the air and water, which can be accomplished by using a measured or assumed coefficient of gas transfer (Odum, 1956; Talling, 1957). One advantage of using oxygen changes throughout the whole water body is that the contribution of all plants, including submerged macrophytes, is included and, of course, there is no uncertainty over the effects of enclosure of small samples in bottles.

Talling (1957) used this method in three waters, the Gebel Aulia reservoir on the White Nile, a lagoon on the edge of the Sudd and in Pilkington Bay on Lake Victoria and found that the order of magnitude of productivity estimated in this way was similar to that determined concurrently by the light and dark bottle method. The estimates, however, did differ in detail, which probably reflects the different assumptions governing the two methods. Using diel oxygen fluctuations can be rather insensitive and is normally appropriate only to relatively productive waters where marked changes in oxygen can be found.

This method can be adopted for use in flowing waters by measuring the changes in oxygen between upstream and downstream sites in a river (Owens, 1969), although this appears not to have been used at tropical sites.

(d) The use of satellite photography

The availability of infra-red images of the earth's surface from the Landsat programme has opened up a further possibility of estimating lake productivity. This has been demonstrated in Lake Chad where the photographs were used to estimate radiance at the lake surface, a factor which is influenced by the quantity of suspended phytoplankton and other material (Lemoalle, 1980). Concurrent determination of some optical properties of the water are required to correct for reflection and other factors. The relationship between chlorophyll, production, and transparency also needs to have been determined for the lake beforehand (e.g. Lemoalle, 1979) but from all this information, the total lake production can be estimated. In the case of Lake Chad, this amounted to 5640 tonnes O_2 day^{-1} over a surface area of 1500 km^2, which is equivalent to 3.76 g O_2 m^{-2}.

5.2.3 Patterns of Productivity

(a) Lakes

If pairs of light and dark bottles suspended at every metre down a weighted line are lowered into a lake on a sunny day and the oxygen changes within them over 1 or 2 hours are noted as outlined in the previous section, then profiles of production and respiration similar to those represented in Fig. 5.4 would often be found.

Although light availability would be at a maximum at the surface (Fig. 5.2) the peak of production is commonly found at some intermediate depth. High incident radiation often causes inhibition of photosynthesis close to the surface whilst the optimum light climate occurs lower down the water column. Respiration may also reach a peak at a similar depth as the densities of organisms may be highest here (Fig. 5.4). Below the production maximum there is a progressive decline in both net and gross production as the light becomes attenuated, until a point is reached where the oxygen produced by photosynthesis no longer exceeds that utilized by respiration.

The depth at which oxygen production just equals oxygen utilization is the compensation point. Above this, energy incorporated in new materials through photosynthesis exceeds that used for respiration and consequently there can be accumulation as net production. Below the compensation point there may be sufficient light for some photosynthesis, but the pigment will be operating below full light saturation and synthesis will be less than breakdown for respiration (Fig. 5.4). The compensation point defines the depth of the euphotic zone and occurs where there is approximately 1% of the incident light intensity available. The photosynthetic pigments themselves cease to be fully light saturated when the light intensity is of the order of 10% of the incident.

A typical series of production profiles in relation to incident sunlight over a day for Lake Lanao in the Philippines is shown in Fig. 5.5 (Lewis, 1974). All

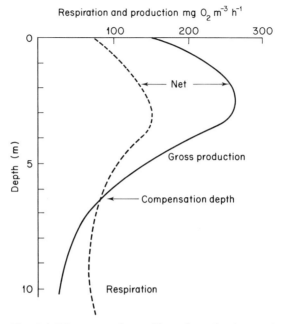

Fig. 5.4 Diagrammatic profiles of production and respiration with depth

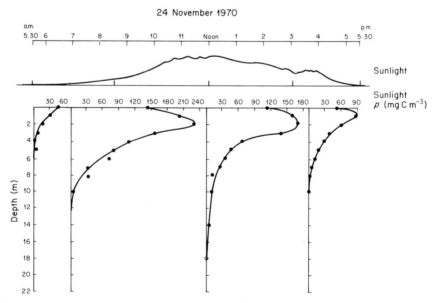

Fig. 5.5 Production profiles based on the ^{14}C method under four different light intensities during a day in Lake Lanao, Philippines. (*Reproduced by permission of the Ecological Society of America from Lewis, 1974*)

except the earliest show the surface inhibition of photosynthesis, and maximum production occurs around 2–3 m depth. Just after dawn, light intensity is normally low, but in this case heavy cloud also contributed to the low incident light with the result that the surface intensity was too low to cause inhibition and it was here, at this time, that production was greatest (Fig. 5.5). Under these circumstances the low levels of irradiation also mean that sufficient light for photosynthesis does not penetrate very far into the water and therefore effective production ceases quite close to the surface. As light increases during the day, the depth at which production can occur and the rate of maximum production both increase and subsequently decrease during the afternoon following the radiation pattern. In this particular instance the ^{14}C method was used and production estimated to be close to net productivity (Lewis, 1974).

The production maximum in Lake Lanao can occur at depths between 0 and 5 m, depending upon the light intensity at the euphotic zone, and can extend to some 12 m. By contrast, in a shallow lake such as Lake George, Uganda, photosynthesis can be limited to the first metre below the surface with a rapid decline below 15 cm (Ganf, 1975; Ganf and Horne, 1975). In this case the high density of phytoplankton causes considerable self-shading and the euphotic zone is restricted to some 75 cm, giving a considerable telescoping of the production system. Not surprisingly, the maximum rate of photosynthetic uptake of carbon was considerably higher in Lake George than in Lake Lanao with a peak at the optimum depth of some 1220 mg C $m^{-3} h^{-1}$ compared to maximum rates of less than 300 mg C $m^{-3} h^{-1}$. To take these results as an indication that Lake George was so much more productive than Lake Lanao would, however, be rather misleading. These values, taken from ^{14}C bottle experiments, are given as carbon fixed per unit volume of water and do not, therefore, take into account the difference in depth of the productive zone. To do this requires an integration of all the rates down through the water column to give the total production taking place beneath 1 m^2 of water. This can be accomplished either from the production profile by planimetry or from one of the equations produced by Talling (1965) or Fee (1969). That applied to East African lakes by Talling (1965) has the form

$$\Sigma A = \frac{A_{max}}{1.33\varepsilon_{min}} \cdot (\ln I_0 - \ln 0.5 I_k)$$

ΣA = integral photosynthesis as O_2 or C per unit area ($mg\ m^{-2}\ h^{-1}$);

A_{max} = rate of photosynthesis as O_2 or C at light saturation ($mg\ m^{-3}\ h^{-1}$);

ε_{min} = minimum value of attenuation coefficient over the spectrum 400–700nm;

I_0 = mean surface light intensity;

I_k = light intensity indicating light saturation of photosynthesis.

When integration of the profile is carried out for Lake George and Lake Lanao, their total production rates from ^{14}C experiments are 4.54 g C m² h⁻¹ and 1.7 g C m² h⁻¹, indicating that the differences are not quite so great as those indicated on a per unit volume basis. To obtain truly comparable rates of productivity, the production per unit area must be considered.

Another valuable comparative measure is the rate of production per unit biomass of the producer. An index commonly used for the phytoplankton is the oxygen production or carbon incorporated per milligram of chlorophyll a per unit time; when estimated from the light-saturated peak production in the water column, this is called the phytosynthetic capacity and designated P_{max}. From his observations of Lake Victoria, Talling (1965) noted how consistent this value was irrespective of the wide range of pigment densities observed throughout the year—the average was 25 mg O_2 (mg chl. a)⁻¹ h⁻¹ with a range of $\pm$ 6 mg which is two to three times the photosynthetic capacity found in temperate lakes. In Lake Tanganyika, another deep, clear lake, a similar P_{max} of 20 mg O_2 (mg chl. a)⁻¹ h⁻¹ was found (Hecky and Fee, 1981). The phytoplankton of the shallow Lake George, although much denser, still has an average photosynthetic capacity of 19.1 mg O_2 (mg. chl. a)⁻¹ h⁻¹ (Ganf, 1975) whilst in Lake Carioca, a small lake in Brazil, which has moderate phytoplankton densities, a value of 16.5 mg O_2 (mg chl. a)⁻¹ h⁻¹ was estimated (Reynolds et al., 1983).

In the very dense blooms of algae found in soda lakes, the photosynthetic capacity may be a little lower; for example in Lake Nakuru, P_{max} values at high algal densities ranged from 10 to 15 mg O_2 (mg chl. a)⁻¹ h⁻¹ (Vareschi, 1982), and in Lake Elmentitia the P_{max} changed from 15 mg O_2 (mg chl. a)⁻¹ h⁻¹ when blooms were dense, to 25.6 mg O_2 (mg chl. a)⁻¹ h⁻¹ during one of the periodic crashes in phytoplankton densities (from Melack, 1981). This may suggest that a unit of chlorophyll shows a diminished production rate under the crowded conditions found in these shallow soda lakes where the environment strongly favours the growth of certain algae.

Instead of relating the peak of production in the water column to the appropriate chlorophyll concentration, the ratio between the production per unit area (g O_2 m⁻² day⁻¹) in the euphotic zone and the equivalent chlorophyll total (mg chl. a m⁻²) can be used. From a review of this O_2/chl. a ratio, Lemoalle (1981) concluded that values around 0.14 were characteristic of many African freshwater lakes whilst their temperate North American counterparts tend to show a rather lower production per unit biomass of the order of 0.052. African soda lakes, particularly those with blooms mainly consisting of Spirulina, have relatively low values, suggesting that perhaps the severe self-shading at the very high algal densities reduces the O_2/chl. a ratio to values rather close to those found in temperate regions during the summer (Lemoalle, 1981; Vareschi, 1982). For this reason, Lake Nakuru in Kenya, where can be found one of the highest known standing crops amongst natural lakes, proportionally does not have such a high production rate even though nutrients and light energy are abundant and a constant daily circulation is maintained (Vareschi, 1982). Even so, purely in terms of biomass, daily production of new material is equivalent to 4–6% of the standing crop in this lake.

It appears, therefore, that the output per unit of photosynthetic pigment can be rather less in temperate phytoplankton than in tropical. This may be due to differences in such factors as temperature or light availability rather than in the relative effectiveness of the pigments concerned, but it does imply that if the concentration of pigment in temperate waters is no greater than that in the tropics, then the absolute rate of production will be lower.

Comparing productivity in tropical and temperate waters, Brylinsky and Mann (1978) and Brylinsky (1980) found there to be a significant negative correlation between gross primary production over a growing season and latitude, indicating that production does tend to be higher in the tropics (Fig. 5.6). From this relationship it appears, for example, that the seasonal production at 10° N. is, on average, some three times greater than at 60° N.

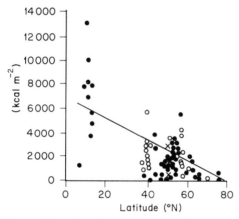

Fig. 5.6 Relationship of phytoplankton production during a growing season to latitude. (*Reproduced by permission of the American Society of Limnology and Oceanography Inc. from Brylinsky and Mann, 1978*)

In temperate lakes rates of primary production rarely exceed $3 \, g \, C \, m^{-2} \, day^{-1}$ (Talling, 1965) even in shallow eutrophic waters, but in the tropics the upper limit observed is around $11 \, g \, C \, m^{-2} \, day^{-1}$.

Lower productivity on a seasonal basis will be partly the result of a shorter growing season in temperate zones, but lower light intensities or temperatures may also contribute to inferior rates of production or diminished photosynthetic capacities. Light intensity at the surface in the tropics is generally higher during the day than in temperate regions but the effect of this on production is rather less than might be imagined owing to the photosynthetic pigments becoming light saturated at quite moderate intensities and therefore unresponsive to further increases. Higher incident radiation does increase the depth of the productive zone, but the response of total production to increased light intensity is rather weak (Talling, 1965). In Lake Lanao, Lewis (1974) examined the extent to which production per unit area increased with intensity of incident sunlight and found that the influence of light explained only 30% of the variability in production.

Temperature has been considered a major contributor to the higher rates of photosynthesis observed in the tropics (Talling, 1965). However, in Lake Lanao, no correlation was found between temperature and productivity (Lewis, 1974) although, as in most tropical lakes, the range of temperatures involved was very limited. In Lake Chad, the relative photosynthetic rate occurring between 20 and 30 °C (i.e. the Q_{10}) at the same locality only increased by factors of 1.15–1.21 (Lemoalle, 1983). The algae, therefore, appeared relatively unresponsive to changes in temperature within their normal range.

A significant correlation can be demonstrated between mean annual temperature and gross production over the temperate/tropical range to the effect that the rate at 30 °C is, on average, almost twice that at 10 °C (from Brylinsky and Mann, 1978).

There are cold lakes in highland regions of the tropics, but some of these can have quite appreciable rates of production. For example, Lake Titicaca in the altiplano of the Andes proved to have a mean rate of $1.45 \, g \, C \, m^{-2} \, day^{-1}$ (Widmer et al., 1975) which is not dissimilar to some lakes of the lowland tropics. In contrast, however, Lake Mucabaji in the Venezuelan Andes had a very meagre rate of only $0.168 \, g \, C \, m^{-2} \, day^{-1}$ (Lewis and Weibezahn, 1976) and the neighbouring Laguna Negra, $0.186 \, g \, C \, m^{-2} \, day^{-1}$ (Gessner and Hammer, 1967). The principal difference between these lakes is not one of temperature but of nutrient content. Lake Titicaca has a high ionic concentration of some $1170 \, \mu S \, cm^{-1}$ which is typical of many lakes of the dry Puna regions (Section 3.1.3) whilst the Venezuelan lakes had the demineralized waters, characteristic of the Paramos, with concentrations equivalent to only $12 \, \mu S \, cm^{-1}$. This gives some indication of the importance of nutrient content irrespective of light or temperature.

In the lowland tropics by no means all lakes are highly productive. The relatively few lakes found in Malaysia appear to have quite low productivities of the order of $0.4 \, g \, C \, m^{-2} \, day^{-1}$ even though temperature and light intensities are high. This is probably due to their siting on old, much weathered soils which allow only a poor nutrient input, as indicated by the low conductivities in the waters of $13–28 \, \mu S \, cm^{-1}$ (Richardson and Jin, 1975).

Nutrient availability can therefore be of particular significance in the tropics where other conditions for photosynthesis are often very favourable. Even with high light intensities and large algal standing crops, the rate and efficiency of photosynthesis will be inhibited if nutrients are in short supply. The availability of nutrients is dependent upon two factors—firstly the rate of recycling of these between the phytoplankton and other components of the system, and secondly the effectiveness and frequency of mixing processes within the water body (Section 5.2.4).

Within the tropics, if nutrient supply is good, then conditions can be ideal for photosynthesis. This is particularly true of the Class III soda and volcanic lakes (Section 3.4) which have high ionic concentrations frequently including a substantial carbonate and bicarbonate content and a pH above 9. In such lakes, the carbonates and bicarbonates ensure that carbon for photosynthesis is not a limiting factor. Phosphate concentrations may also be elevated and the waters can have a high buffering capacity, which prevents the pH of the water from becoming too alkaline and therefore inhibitory when rapid photosynthesis is occurring. Many of these lakes are also quite shallow, which encourages circulation of the entire water column and also means that the euphotic zone includes or is close to the mud surface where some mineralization of organic material can take place. If the lake bed receives sufficient light, a significant contribution to total production will be added by the benthic algae. Often, however, the phytoplankton blooms in these lakes are very dense and may restrict the euphotic zone to a few centimetres.

Production in soda lakes can be very substantial; for example hourly rates in Lake Aranguadi, Ethiopia, reached $30 \, g \, O_2 \, m^{-3} \, h^{-1}$ (Talling *et al.*, 1973), whilst maxima from other soda lakes vary from 1.2 to $10 \, g \, O_2 \, m^{-3} \, h^{-1}$ (Melack, 1981). The daily areal value for Lake Nakuru, Kenya, amounted to $36 \, g \, O_2 \, m^{-2} \, day^{-1}$ (Melack and Kilham, 1974) more than twice that of Lake George. However, rates are not uniformly maintained since such lakes are prone to periodic crashes in phytoplankton. So, for example, over a 3-year period, in Lake Nakuru, the highest daily rate was $21.2 \, g \, O_2 \, m^{-2} \, day^{-1}$ compared to the lowest, during a trough in phytoplankton activity, of $7.2 \, g \, O_2 \, m^{-2} \, day^{-1}$ (Vareschi, 1982).

The rapid rates of production in soda lakes are usually associated with very great phytoplankton densities. These are commonly between 300 and 600 mg chl. *a* m^{-3}, most often due to blue-green algae such as *Spirulina* and *Anabaenopsis* spp., whilst, in Lake Aranguadi, the mean phytoplankton density attained an astonishing 2500 mg chl. *a* m^{-3} (Talling *et al.*, 1973). In dry-weight terms, the standing crop in such lakes may achieve 500 g m^{-2} (Vareschi, 1982).

In contrast, the larger deeper Class I or Class II lakes contain very low phytoplankton pigment densities. Lake Victoria, for example, can have a concentration of 5 ± 0.5 mg chl. *a* m^{-3}, which gives rise to a net production rate of around $100 \, mg \, C \, m^{-3} \, h^{-1}$ (Talling, 1965). Both of these values are approximately 100 times less than those for soda lakes. In deep lakes, however, the euphotic zone can extend to a considerable depth but again, in Lake Victoria, the chlorophyll within the euphotic zone was only 44 mg m^{-2}. The production from the whole water column, however, amounted to $4 \, g \, C \, m^{-2} \, day^{-1}$ which, in spite of the low chlorophyll values, is similar to that of the shallow, more obviously productive, Lake George.

Several of the larger, deeper tropical lakes, such as Lake Victoria, Lake Malawi (Talling, 1965), Lake Lanao (Lewis, 1974), and Lake Tanganyika (Melack, 1980; Hecky and Fee, 1981), are similar in that they have a high transparency, low nutrient concentrations, and a low phytoplankton density, features which, in temperate regions, would suggest an oligotrophic nature. Their productivity, however, is much greater than that of oligotrophic temperate lakes and disproportionately larger than is indicated by the small phytoplankton biomass. In Lake Tanganyika the algal population of the phytoplankton had one of the most rapid growth rates recorded under natural conditions, even though the standing crop was very low (Hecky and Fee, 1981). This lake evidently contains a small, very rapidly growing population in contrast to Lake George, for example, where the dense population proved to grow much more slowly. The algal populations of two other deep lakes, Lake Lanao and Lake Titicaca, also had grown some ten times more quickly than that of Lake George (Hecky and Fee, 1981).

To sustain such algal growth and such high rates of production per unit biomass must require an efficient recycling of nutrients. It is also surprising that, given such rapid growth, the biomass itself does not accumulate to a greater extent. In Lake Tanganyika, for example, from the seasonal minimum in October–November when the phytoplankton biomass was $0.09 \, g \, C \, m^{-2}$, at the observed growth rate of $3.2 \, day^{-1}$, it should take less than 2 days to attain the maximum seasonal biomass observed in the lake which patently did not happen at that time (Hecky and Fee, 1981). It must, therefore, require equivalent losses to offset these rapid rates of growth.

The deep epilimnion of these lakes would tend to reduce or slow down losses from sedimentation. Other sources of loss could include extensive grazing, disproportionately elevated levels of respiration, or high cell mortality. All of these would increase the turnover of nutrients in the epilimnion, which in turn, would support rapid growth. A deep epilimnion, therefore, also allows more nutrient regeneration and generally retards nutrient loss whilst the low biomass permits a high transparency to the water and minimizes self-shading (Hecky and Fee, 1981). The very rapid recycling of essential nutrients in the water column is particularly significant in maintaining the substantial photosynthesis often observed in the tropics, particularly in the deeper lakes (Sections 5.2.4 and 5.3).

Measurement of photosynthesis during the daylight hours is a useful indication of lake productivity, but for there to be a net gain in organic material, synthesis during the day must exceed loss due to respiration over the full 24-hour period. In Lake George, total daytime oxygen production was estimated at $15.5 \, g \, O_2 \, m^{-2}$ in 12 hours, whilst respiration amounted to $9.1 \, g \, O_2 \, m^{-2}$, leaving a considerable net production (Ganf and Horne, 1975). However, nocturnal respiration required a further $6.4 \, g \, O_2 \, m^{-2}$, giving a total respiratory uptake in 24 hours of $15.4 \, g \, O_2 \, m^{-2}$; consequently there was a net gain of only $0.1 \, g \, O_2 \, m^{-2}$ over the full 24-hour cycle. The greater respiratory demand during the day began to build up above the temporary thermocline as the lake became stratified. Respiration below the thermocline, however, remained close to nocturnal rates which were re-established throughout the lake when stratification broke down at the end of the day.

In Lake Nakuru, the mean net production over a 12-hour period was estimated as $16.7 \, g \, O_2 \, m^{-2}$, whilst over the full 24 hours it was found to be

7.6 g O_2 m^{-2}, rather greater than in Lake George. At this time, the circulation of the lake meant that individual algal cells were only in the euphotic zone for 2 hours per day, which indicates how rapid production must be to compensate for long periods of respiration without photosynthesis (Vareschi, 1982).

Within a day there can be considerable variation in the level of production as light and other factors change (Fig. 5.5). For example in Lake George, the greater part of production takes place in the morning, although the highest insolation occurs during the afternoon (Fig. 5.7). Not only does the rate of production fall during the day but various aspects of the efficiency of the process also decline. In Lake George photosynthetic capacity (P_{max}) fell from 28.0 to 18.2 mg O_2 (mg chl. a)$^{-1}$ h^{-1} between 10.00 and 15.30 hours (Ganf, 1975), whilst in Lake Nakuru the efficiency with which light energy was converted to net production showed a progressive reduction from 1.76% between 06.30 and 10.30 hours to 1.02% between 14.00 and 18.30 hours (Vareschi, 1982). A similar pattern has also been found for Lake Chad (Lemoalle, 1983).

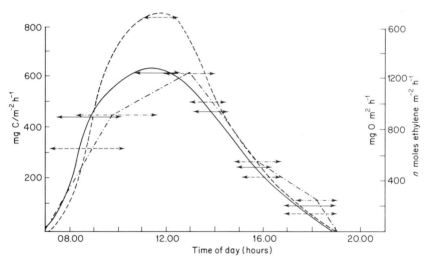

Fig. 5.7 Diel oxygen production, carbon and nitrogen fixation in Lake George, Uganda. (*Reproduced by permission of Blackwell Scientific Publications Ltd. from Ganf and Horne, 1975.*) The horizontal lines indicate the time span of each experiment; – – – O_2 production, - - - - C fixation, -.-.- N_2 fixation

The decline in production during the afternoon could be due to sinking of algal cells below the euphotic zone, which would then be recirculated at night. In Lake George, this is promoted by the regular nocturnal winds. Other reasons for the decline include a possible physiological depression of phytosynthesis later in the day, due to such factors as high light intensities or a reduction in the availability of the essential inorganic nutrients for photosynthesis, particularly nitrogen and phosphorus, as stratification builds up during the day. In Lake George, there is little or no detectable inorganic phosphorus or nitrogen in the water during the day; some progressive nutrient stress as the day passes may contribute to the reduced photosynthesis and diminished efficiency during the afternoon (Ganf and Horne, 1975). In Parakrama Samudra, Sri Lanka, the efficiency of photosynthesis was greatest in the afternoon, although the rate of production was highest before noon (Dokulil *et al.*, 1983). This was due to increased phosphorus availability at this time caused by phosphatases released by the algae themselves which split phosphorus from otherwise unavailable organic forms in the water (Gunatilaka, 1983).

(b) Rivers

Conditions for photosynthesis by phytoplankton in rivers may not be quite so suitable as they are in lakes and the relatively low densities of standing crops and photosynthetic pigments have been mentioned previously (Section 4.5.3). Many rivers carry a heavy load of suspended material which cuts down light considerably and condenses any productive zone into a narrow layer close to the surface. The white-water rivers of the Amazon basin exemplify this well, for although they are quite well supplied with inorganic nutrients from the volcanic rocks of the Andes (Section 2.3.1), the euphotic zone can be restricted to 1 m or less (Fisher and Parsley, 1979). The constant turbulence of the river gives a good circulation of nutrients with no stratification or local depletions building up, but the same circulation may also mean that individual algal cells do not stay in the euphotic zone for very long, an additional feature which will impair productivity. Consequently, in a white-water river such as the middle Amazon above the junction with the Rio Negro, production can be quite low with a mean of $0.063 \, \mathrm{g \, C \, m^{-2} \, day^{-1}}$, as estimated by the ^{14}C method (Fisher and Parsley, 1979), which is 10–100 times less than most of the tropical lakes mentioned in Section 5.2.3(a).

Many rivers are similar to this with a high nutrient content but low light availability whilst others, the 'black-waters' which drain from podzol areas (Section 2.3.1), not only have low nutrients and very low pH but also low light penetration. In the black-water Rio Negro of the Amazon, primary production declined with depth and there was no surface inhibition of photosynthesis owing to the rapid attenuation of light down through the water column. The black waters contain highly coloured material derived from the incomplete breakdown of humic substances which will also give a spectral bias to light penetration. Although poor in nutrients as well as light, the production from the Rio Negro is, on average, $0.063 \, \mathrm{g \, C \, m^{-2} \, day^{-1}}$ (Schmidt, 1976), and is generally at a similar level to that of white waters. The highest recorded rates are also very similar with $0.14 \, \mathrm{g \, C \, m^{-2} \, day^{-1}}$ for the middle Amazon and $0.10 \, \mathrm{g \, C \, m^{-2} \, day^{-1}}$ for the Rio Negro.

In contrast to both of these cases, the 'clear-water' rivers, although possessing lower concentrations of nutrients than white-water types, nevertheless have much better optical characteristics and allow considerably more light penetration. An investigation of the clear-water Rio Tapajoz (Schmidt, 1982) in the

Amazon basin showed that the euphotic zone extended down to between 5 and 10 m and also that the production profile consistently showed surface inhibition of photosynthesis by light, having a production peak around 1.5 m as in many lakes (Fig. 5.8). There are many similarities between these clear-water rivers and the deep tropical lakes such as Lake Lanao and Lake Victoria (Section 5.2.3(a)).

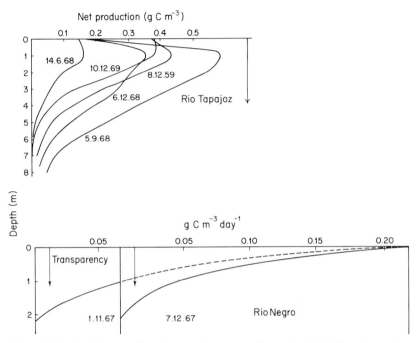

Fig. 5.8 Production profiles for a clear-water river, the Rio Tapajoz and a black-water river, the River Negro from the Amazon basin. *Reproduced with permission from Schmidt, 1976, 1982).* The secchi disc transparency is indicated by the vertical arrows

In particular, the phytoplankton themselves are the major component of the suspended material and the productivity is very high in comparison with the standing crop. In fact, the time of year at which the standing crop is least and transparency is highest—at low water in September—is the time when production is at a maximum ($2.41 \, g \, C \, m^{-2} \, day^{-1}$ net production). As in the lakes, this indicates a high efficiency of the photosynthetic process, which can be expressed in terms of carbon turnover. Schmidt (1982) found that this occurred at between 0.3 and 2.2 days. Similar calculations for Lake Lanao showed rates for carbon turnover within the phytoplankton as taking between 0.35 and 4.9 days to occur, with a mean of 1.24 days which was considerably faster than for turnover times in temperate lakes (Lewis, 1974). The average net production in the Rio Tapajoz was $1.37 \, g \, C \, m^{-2} \, day^{-1}$ which is rather higher than for the white- and black-water parts of the river system and is similar to that found for the deep, clear lakes. It also exceeds estimates for the relatively shallow varzea lakes which are fed by Amazon water. Here, production can obtain values of 3.9 tonnes ha^{-1} yr^{-1}, but the Rio Tapajoz can produce some 5 tonnes whilst the Rio Negro only gives 239 kg ha^{-1} $year^{-1}$ (Schmidt, 1976, 1982).

In both clear-water rivers and transparent lakes, the fast turnover of materials indicates an extremely rapid recycling of dead material, presumably through the bacterial microflora. In these situations, the material does not accumulate as detritus, hence the high transparency of the water and lack of suspended material other than the phytoplankton, as the organic material is mineralized very rapidly, perhaps almost in a matter of hours. The nutrient recycling (Section 5.2.4) and decomposition (Section 5.3) within the water column itself would appear to be of major importance here.

Not all rivers with appreciable silt contents are unproductive, however. In the White Nile above Khartoum, even though the euphotic zone was confined to the uppermost metre of water, largely due to the load of inorganic particles the rate of production achieved an average of 2.4 g C m^{-2} day^{-1} which exceeds that of any of the Amazon waters and is equivalent to rates for many tropical lakes. The Nile, however, is particularly rich in nutrients and the site examined was downstream of the Gebel Aulia reservoir where gross phytoplankton productivity reaches a similar level (Prowse and Talling, 1958). No doubt the reservoir contributes quite significantly to the phytoplankton population of the river below it. As in most rivers, the seasonal fluctuation can cause tremendous changes in productivity (Section 6.2). In the Nile, the production is 100 times greater at the maximum as the water level falls than at the minimum during the floods (Sinada and Karim, 1984).

Although conditions may not be particularly propitious for phytoplankton development in shallow, faster-flowing streams and rivers, some production may be detectable. In the Rio Santo Domingo in the Andes of Venezuela, at altitudes between 2540 and 200 m, the net primary production ranged from 0.352 to 0.230 mg C m^{-2} day^{-1} (Lewis and Weibezahn, 1976). As with all waters of such low productivity, the ^{14}C method had to be used as there was insufficient oxygen flux to be detectable over a limited period. A very sparse phytoplankton comprising diatoms and nannoplankton types could be found in the river. In this river which was almost completely unshaded, the ecological efficiency, that is, the proportion of incident light used, is extremely low, $6-9 \times 10^{-5}$ %. One contributory factor in such shallow rivers—the Rio Santo Domingo is around 50 cm deep—may be photoinhibition throughout the water body. Shading of the river to halve the incident light at midday in a single experiment almost doubled the production rate, indicating that photoinhibition is a real possibility.

At a slightly lower altitude the Rio Limon, which empties into Lake Valencia, had phytoplankton with a rather higher range of net production, having a mean of 2.46 mg C m^{-2} day^{-1}. However, although this is approximately twenty times higher than that of the Santo Domingo, it is still very low (Lewis and Weibezahn, 1976). The Rio Limon is almost entirely shaded by trees, and on a sunny day only 6.2% of the incident light penetrates down through the tree canopy to the surface of the water, but this is evidently sufficient to sustain primary production in the stream. Moreover, in addition to phytoplankton, there were also mats of algae, largely diatoms and blue-greens, on the bed and encrustations of a red alga, *Hildenbrantia,* on the rocks. Enclosure of samples of these in light and dark bottles gave net production values of 1189 mg C m^{-2} day^{-1} for the algal mats and $13.5 \text{ mg C m}^{-2} \text{ day}^{-1}$ for *Hildenbrantia.* In this stream, therefore, the most productive element amongst primary producers are the benthic algae. This may not be an untypical situation in streams. The total primary production from the Rio Limon at this particular time was, therefore, some $0.73 \text{ g C m}^{-2} \text{ day}^{-1}$ which is well below the productivity of many tropical lakes but still within the lower limit of the range.

The primary production of phytoplankton may be of only minor importance as an energy supply to many tropical river systems for, as we saw earlier, an important additional source is that of particulate organic material, some autochthonous and some allochthonous, which is washed down with the current. The relative importance of this is only beginning to receive attention, but some impression of the magnitude can be gained from the fact that 60 million tonnes of particulate organic carbon is carried out by the Amazon each year of which 16% comes directly down the main river and 85% from the tributaries (Richey, 1981). In addition, 22 million tonnes of dissolved organic material is also carried down. At any single stretch on the river it appears that only half the imported material is exported downstream with the remainder being utilized as a substrate for respiration (Richey, 1981) thus further emphasizing the extensive mineralization which is occurring in the water. Some of the organic material will have originated from the phytoplankton but some will also have come from the marginal vegetation. Like many rivers, the Amazon has little immediate vegetation in its littoral area, but on the varzea lakes lie the dense 'floating meadows' (Section 4.9.3) which produce much material in a season and some of the resulting detritus is exchanged with the river as the water levels drop (Junk, 1970, 1982). Rivers with extensive flood-plains which are regularly inundated must also be expected to carry consider-able quantities of suspended organic materials. The relative contribution of the terrestrial system to organic carbon in rivers is unknown but contributions within forested river basins are likely to be higher than from those originating in grassland.

(c) Swamps and marginal vegetation

The macrophytes, which are the most prominent primary producers in many shallow-water systems, are more difficult to measure with regard to production than are the algae of the phytoplankton. It is not easy, for example, to enclose these large plants in order to measure oxygen or carbon dioxide fluxes as is com-monly done for the phytoplankton (Section 5.2.2). Instead, the usual method for determining productivity of emergent, submerged, or floating vegetation is by estimating changes in biomass over a given length of time.

Where growth is seasonal, as it is in temperate regions and as it may also be in some tropical situations such as floodplains, the increase in biomass between the seasonal minimum and maximum during the growing season will give an index of accumulated production in dry-weight terms (Westlake, 1969). With this approach, it is evident that losses due to leaf fall and consumption by herbivores must be accounted for to achieve a complete estimate of net production, whilst losses due to respiration must be estimated to ascertain gross production. A convenient summary of this is given by (Kira *et al.*, 1967; Ikusima *et al.*, 1982):

$$\Delta P_g = \Delta Y + \Delta L + \Delta G + \Delta R$$

where ΔP_g, ΔY, ΔL, ΔG, and ΔR are rates of change in gross production, biomass, leaf loss, grazing by herbivores, and respiration respectively.

For complete estimates of production, changes in the underground portions of the plants must also be taken into account. This can present difficulties because above- and below-ground parts can grow and die at different rates. They also exchange material by translocation which could lead to the inclusion of the same material twice in an estimate. It has been shown, for example, that 40% of the growth of papyrus culms may be translocated from senescent shoots (Thompson, 1976). Errors of this type could also arise in floating plants, like *Eichhornia crassipes,* which are connected by stolons.

Much aquatic vegetation in the tropics, however, grows with some consistency throughout the year and does not produce substantial seasonal changes in biomass. The accent here, therefore, has to be put upon determining the growth, concurrent losses, and turnover rates of the shoots (Westlake, 1969; Thompson *et al.*, 1979; Ikusima, 1978).

Westlake (1982) collected together a wide selection of production rates from aquatic plants from which only the top 10% of estimates were considered for each category to indicate upper levels and to facilitate comparison under 'good' conditions (Table 5.1). Although the data are limited, there are some indications from this information that tropical emergent and submerged plants can have higher rates of production than their temperate counterparts. The highest recorded belong to those such as papyrus, possessing the C_4 photosynthetic pathway (Table 5.1), which enables them to make better use of high-intensity sunlight than plants with the C_3 process which tends to dissipate more energy in photorespiration at higher light intensities.

Both emergent and floating vegetation can show higher rates of production than phytoplankton (Table 5.1). It is quite striking, however, that the phytoplankton production is similar to that of emergent vegetation, bearing in mind the very large differences there are in the standing crop (Section 4.10; Fig. 4). The phytoplankton, therefore, have a very high rate of production per unit biomass because the algae are small and do not need to support a large non-photosynthesizing superstructure. Their total production is limited, however, through an inefficient canopy leading to inevitable shelf-shading and an unstable biomass which is a result of the short generation time (Westlake, 1982).

Table 5.1 Upper ranges of productivity of aquatic plants (top 10% of estimates; reproduced with permission from Westlake, 1982).

	Annual production m.t dry wt ha^{-1} yr^{-1}
Emergent C$_4$ plants, tropical freshwaters* (esp. *Cyperus papyrus*)	60–90
Emergent C$_3$ plants, freshwaters (esp. *P. australis, Typha* spp.)	50–70
Emergent, floating C$_3$ plants, sub-tropical freshwaters (esp. *Eichhornia crassipes*)	40–60
Terrestrial plants (forests, grasslands, crops)	20–85
Phytoplankton	15–30 (60)
Submerged plants, marine*	40–60
tropical freshwaters*	20
temperate freshwaters	5–10

Total light energy available to both phytoplankton and submerged macrophytes in the field will be more limited owing to the absorption of light down through the water column.

The difference between macrophyte and phytoplankton production at a given location may not be so great. In one of the varzea lakes, Lago do Castanho, phytoplankton production was equivalent to 6 tonnes dry wt ha^{-1} year^{-1} (Schmidt, 1973a,b) whilst one of the principal components of the Amazonian floating meadows, the grass *Paspalum repens*, produced 6–8 tonnes dry wt ha^{-1} year^{-1} (Junk, 1970). Nevertheless, whilst other phytoplankton populations such as those in some of the clear-water rivers can produce twice as much as that of the Lago do Castanho (Section 5.2.3(b)), within Amazonia, floating meadows can produce up to 30 tonnes dry wt ha^{-1} year^{-1} (Junk, 1982).

The values given in Table 5.1, whilst useful for comparative purposes, do very much represent the higher rates of productivity. The sedge, *Lepironia articulata* growing in the Tasek Bera swamp, Malaysia, produced between 7.7 and 17.8 tonnes ha^{-1} year^{-1} (2.1–4.9 g m^{-2} day^{-1}) of which some 40% remained as net production (Ikusima *et al.*, 1982). This was only about one-tenth of the rate found for *Cyperus papyrus*, another sedge, in equatorial Africa (Turner, 1981). It could be that *Lepironia*, unlike *Cyperus*, is only a C$_3$ plant, but it also showed a low emergence rate of new culms and a low standing crop (Section 4.9.2). The efficiency of solar energy conversion to gross production in this case was 0.5% which is less than half that of many other macrophyte stands (Ikusima *et al.*, 1982).

The ratio of prodution which takes place above as opposed to underground is variable. For example, in *Lepironia*, underground production is only 12–14% of that above ground (Ikusima, 1978) whilst in *Cyperus*, it was equivalent to 33% (Thompson *et al.*, 1979). It is impossible, therefore, to make generalizations with the limited information currently available.

One consequence of the rapid turnover in stands of tropical aquatic vegetation is the continued litter of dead leaves and organic material shed by the plants on to the bottom mud.

Although possessing a small biomass, a stand of *Lepironia* supplied $2.24 \, g \, m^{-2} \, day^{-1}$ or 8.2 tonnes $ha^{-1} \, year^{-1}$ of dead organic matter to the swamp bed, which is almost equivalent to the rate derived from a temperate forest (Ikusima *et al.*, 1982). In the case of *Lepironia* in the Tasek Bera swamp, much of the organic matter accumulated as peat. This is also commonly observed beneath papyrus swamps in East Africa where layers of peat 10 m thick are found. In contrast, the layer of organic debris beneath a dense *Typha* swamp around Lake Chilwa, Malawi, was scarcely 10 cm deep (Howard-Williams and Lenton, 1975) and the lack of accumulation of litter has also been commented upon in the Amazonian floating meadows (Junk, 1982), and the marginal vegetation of lakes in Ecuador (Steinitz-Kannan *et al.*, 1983). The extent of accumulation would appear to depend upon the suitability of conditions for decomposition and the nature of the material to be decomposed (Section 5.3). The quantities of energy entering the ecosystem from this source, however, would appear to be very great and probably represent the major route since grazing directly on the plants is often insignificant. Many swamp areas can be swept by fire in the dry season, which dissipates the energy but considerably enhances the rate of mineralization and nutrient turnover.

5.2.4 Nutrient Recycling

(a) Which nutrients limit photosynthesis?

The elements generally considered to be the most essential for plant production are phosphorus and nitrogen, the first because of its role in the formation of high-energy compounds in the cell and the latter because it is a major constituent of the proteins which can be formed by photosynthesis. In fresh waters both are commonly found in very low concentrations and their concentrations are very variable from water to water as outlined in Sections 2 and 3.

In their analysis of the factors influencing lake production, Brylinsky and Mann (1978) showed significant correlations between phytoplankton production during the growing season and both total nitrogen and phosphorus contents in waters. Total phosphorus plus light energy availability together could explain 77% of the variation from production estimates over a wide range of latitudes. One other feature of this analysis, however, was that an equally high proportion of the variation can be attributed to the influence of light energy availability combined with conductivity.

We have seen previously (Section 3.4) that conductivity is a measure of total ionic concentration, a high proportion of which is due to bicarbonates and carbonates since these are most often the dominant anions. Their link with the carbon dioxide equilibrium in the water (Section 2.1) means that they constitute the major carbon source for photosynthesis. Unlike the sea, where bicarbonate concentrations are quite constant, or the air, which also possesses a constant percentage of carbon dioxide, fresh waters show considerable variability in bicarbonate concentration and therefore in the availability of carbon. In fresh waters, therefore, carbon itself can be a limiting factor to primary production. However, against the framework of potential productivity set by incident light and carbon reserves in the water, the availability of phosphorus, nitrogen, and micro-nutrients can place substantial constraints on the actual level of production attained.

In most temperate waters, phosphorus has proved most frequently to be the principal limiting factor of production. In tropical waters, Talling and Talling (1965) suggested from their observations on East African lakes that nitrogen was probably more critical than phosphorus. Subsequent experimental work on the enrichment of samples of natural plankton with nitrogen and phosphate compounds showed that in Lake Chilwa (Moss, 1969) and Lake George (Viner, 1973), for example, inorganic nitrogen was a more potent stimulus to phytoplankton production and growth than phosphorus. In the experiments on phytoplankton from Lake George, ammonia compounds were used as a source of nitrogen rather than nitrate since it was in this form that most uptake appeared to take place within the lake itself. The relative importance of nitrogen has been demonstrated in a similar fashion for reservoirs in Brazil (Henry and Tundisi, 1982), the varzea Lago Jacartinga (Zaret et al., 1981), Lake Titicaca (Wurtsbaugh et al., 1985) and the more turbid waters of the Amazon system (Grobbelaar, 1983). In the black waters of the Amazon, however, phosphorus enrichment appeared to have the greatest effect (Zaret et al., 1981) as was also found in Lake McIlwaine, Lake Kariba, and three other man-made reservoirs in Zimbabwe (Robarts and Southall, 1977). With regard to Lake Kariba, however, the use of the aquatic fern, *Salvinia*, rather than planktonic algae for the enrichment assay had indicated nitrogen to be the major limiting factor so there remains some question here.

Enrichment experiments, therefore, do suggest that nitrogen is commonly a limiting factor in tropical waters and that phosphate is not quite so important as in temperate areas. However, enrichment of small samples enclosed in flasks are rather far removed from the conditions to which the community as a whole are exposed. Enrichment by adding fertilizers to whole water bodies can overcome this problem to some extent. For example, the addition of superphosphate to a fishpond fed from a small reservoir in Tanzania produced little effect on the phytoplankton (Fig. 5.9), whilst a subsequent introduction of ammonium sulphate rapidly stimulated a bloom of algae. Ammonium as a form of nitrogen, in this case, shows a similar effectiveness to the experiments on the samples from Lake George, a lake where NH_4^+ is a major source of nitrogen for the phytoplankton and is rapidly utilized. Nitrogen rather than phosphorus, therefore, appeared to be the principal limiting factor of phytoplankton growth in the pond and the beneficial effects could be detected at the next trophic level by an increase a month later in the growth rate of the phytoplankton-feeding tilapia, *Oreochromis esculentus*, in the same pond (Payne, 1971).

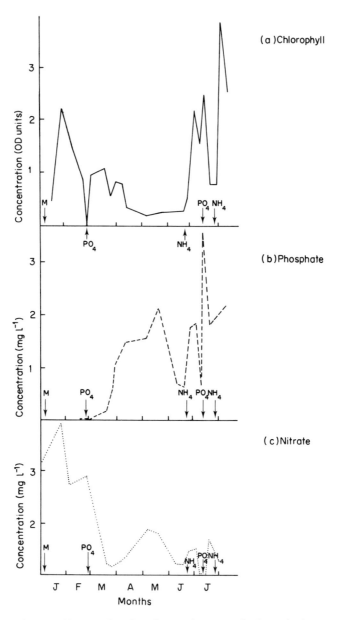

Fig. 5.9 Changes in phosphate, nitrate, and phytoplankton pigment density in a fertilized fishpond at Malya, Tanzania. The bed of the pond was initially prepared with 390 kg ha^{-1} cattle manure (M), and the treatment of the pond with 250 kg ha^{-1} superphosphate (PO$_4$) and 500 kg ha^{-1} in June and 250 kg ha^{-1} in July of ammonium sulphate (NH$_4$) is also indicated by arrows. (*Reproduced by permission of Academic Press Inc. (London) Ltd. from Payne, 1971*)

This approach confirms the importance of nitrogen indicated by the laboratory enrichment experiments but also demonstrates the more complex conditions found in whole water bodies. For example, following the introduction of the phosphate fertilizer in February there was no immediate rise in phosphate concentration in the pond water, presumably because it had immediately been incorporated directly into the phytoplankton during the small peak which occurred shortly after the addition or, more probably, it had been adsorbed with the mud surface to give the slow release noted later (Fig. 5.9). Furthermore, the two peaks of phosphate in July coincided with the peak of phytoplankton density which produced very low oxygen concentrations in the morning, causing the release of ions from the mud/water interface (Section 3.5.1(b)). The total amount of an element within the ecosystem does not therefore dictate how much is actually available to the producing organisms.

(b) The distribution of nutrients

Inorganic nitrogen and phosphorus are often present in negligible concentrations in tropical waters even when considerable production is taking place. In Lake George, for example, which, as we have seen, is a very productive lake with a high standing crop of phytoplankton, nitrate was rarely detectable and ammonia was present in low concentrations, $40-70 \mu g \, N \, l^{-1}$, only at night and never during the day (Viner, 1973). The lack of inorganic nutrients free in the water does not, however, mean that they are absent from the system. These compounds can be envisaged as being distributed between a number of components of the ecosystem in a similar fashion to the summary provided by Golterman (1975) for phosphate (Fig. 5.10).

In productive water, a substantial proportion of the nutrients may actually be incorporated into the algae themselves. This can be observed happening as nutrient-rich white water, in which productivity is normally low owing to restricted light penetration (Section 5.2.3(a)), flows from the Amazon into one of the varzea lakes along the connecting channel (Fisher and Parsley, 1979). The appreciable concentrations of phosphate and nitrate are progressively reduced to virtually undetectable levels as the river water moves along the several kilometres of channel whilst particulate organic nitrogen, which includes that in the phytoplankton, increases from 154 to $490 \mu g \, l^{-1}$.

In temperate waters, the rapidly growing phytoplankton populations deplete inorganic nutrients in the water during the spring and, since stratification occurs in the summer, the nutrients are not replenished. During the summer, therefore, rates of production can be very low through lack of nutrients even though temperature and illumination are at their maximum. In the tropics, a high sustained production whilst inorganic nutrient concentrations appear very small or non-existent is a common phenomenon.

Although there would appear to be a perpetual nutrient depletion in Lake George, the high algal biomass is remarkably constant. Similarly, the stripping of the nutrients from the inflowing Amazon water is virtually complete, yet

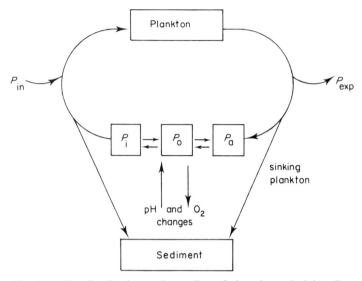

Fig. 5.10 The distribution and recycling of phosphorus in lakes. P_{in}, imported inorganic phosphorus; P_i, dissolved inorganic phosphorus; P_o, phosphorus incorporated into organic detritus; P_a, phosphorus adsorbed onto the surfaces of suspended particles; P_{exp}, phosphorus exported

biomass and production in the varzea lakes are increased and sustained for several months.

High sustained production in the presence of nutrient depletion is a typical feature of tropical fresh waters, although other seasonal factors can interrupt this (Section 6.2). What seems to make this possible is the very high turnover rate of these elements between the phytoplankton and the water. We have already seen (Section 5.2.3) in Lake Lanao and Amazon clear-water rivers, how the turnover of carbon, a major element in the process of synthesis, may occur within a few hours and the same can also be shown to be true for phosphates. In temperate waters phosphate turnover may occur approximately twenty times a year (Golterman, 1975) whilst in the Lago Janauaca and several East African lakes it can occur as frequently as every 1–3 hours (Fisher and Parsley, 1979; Peters and MacIntyre, 1976). To maintain such high turnover rates, the decomposition and mineralization of dead material and organic compounds in the water must proceed very rapidly. This suggests high activity from the microbial populations (Section 5.3), but the zooplankton and other herbivores also have a role to play (Section 5.4.1).

In Lake George, for example, the phytoplankton is grazed by copepods and by two cichlid fish species. Determination of the excretory rates showed that the zooplankton excreted 3300 tonnes of nitrogen and the fish population 220 tonnes which together exceeded the total annual input of 2800 tonnes (Ganf and Viner, 1973). The phytoplankton feeders themselves, therefore, recycle rather more nitrogen within the lake than enters it in a year and, with regard to phosphate, they proved to recycle two to three times the annual input. There must be an intimate relationship between the producers and the mineralization process to ensure that the inorganic nutrients are utilized immediately they are released and not allowed to become locked up in an inert component of the system and thereby become unavailable for long periods of time.

Dissolved nutrients are not the only source for phytoplankton, however. Phosphate, nitrate, and other ions exist adsorbed or otherwise bound to suspended particles in the water, and although they are not detected by the conventional analytical tests for dissolved compounds, they can be utilized by the algae. Golterman (1975) demonstrated that populations of the green algae, *Scenedesmus*, could remove as much phosphate from the suspended clay sediments of the Kainji River, Uganda, amounting to some 10% of the bound fraction, as could be removed artificially using a chelating agent. Similarly, in the turbid waters of the Amazon, up to 95% of the adsorbed NO_3 nitrogen and 54% of the PO_4 phosphorus adsorbed on to suspended sediment can be utilized by algae. Even so, the total amount of nitrogen adsorbed is small and, since nitrate in the free water is also very low, nitrogen remains the principal limiting factor of algal growth in these waters (Grobbelaar, 1983).

In a river in Papua New Guinea, half the phosphorus was dissolved in the water and half was available from the suspended sediments; closer to the estuary, however, by far the largest proportion was obtainable from the sediment (Viner, 1982). In this case the most effective particle size, judging from its ability to support the alga *Stichococcus* with phosphorus, was that of fine clays, less than 4 µm diameter, probably owing to their relatively high surface area to volume ratio.

The interplay between the pool of nutrients incorporated into the phytoplankton and those dissolved in the water or associated with particulate matter, inorganic or organic, is obviously of considerable importance in determining the rate of production, but an additional factor is the role of the sediment surface at the mud/water interface (Fig. 5.10). The mud surface has a considerable affinity for ions under aerobic conditions (Section 3.5.1(b)) and can absorb them readily as seems to have been the case in the fertilizer experiments mentioned earlier (Fig. 5.9). They are not totally unavailable in this form since they can be released under conditions of low oxygen or pH or when turbulence stirs up the sediment. In shallow lakes surface turbulence may well stir up the mud surface and, in Lake George, one of the few tropical waters in which the role of the mud has been examined, the quantities of ammonia and phosphate in the near-surface mud are very low and it seems unlikely that much decomposition and mineralization actually take place here (Ganf and Viner, 1973). The concentrations of inorganic nitrogen and phosphate increase below depths of 5 cm in the mud and mineralization was probably at a maximum at around 15 cm depth where oxygen consumption was greatest (Ganf and Viner, 1973). However, release of ammonium and phosphate from cores was very slow and therefore the contribution of the sediment to nutrient recycling is probably relatively small (Viner, 1975a,b). This may not be the case for all shallow lakes since Gaudet and Melack (1981), for example, found that the mud was a major factor in the distribution of nutrients in Lake Naivasha, Kenya.

In stratified lakes, particularly where deoxygenation occurs below the thermocline, substantial release of nutrients from the mud surface can occur (Section 3.5.1(b)). These, together with inorganic nutrients derived from organic debris which has sunk below the thermocline into the hypolimnion, form a potential pool for enrichment but are dependent upon the mixing process within the water body to make them available to the phytoplankton (Section 3.5.1(d)). This may occur when stratification breaks down seasonally (Section 6.2), but even when this is permanent other mixing processes may occur such as the tilting of the thermocline under the influence of a strong wind, as observed in Lake Malawi and Lake Tanganyika, or the effect of cool marginal water sinking, as in Lake Victoria (Section 3.5.1(d)). In the deepest lakes, thermal discontinuities may occur within the epilimnion as this warms up. Production tends to be confined to the most superficial layer whilst nutrients can accumulate below this due to decomposition of sinking plankton (Lewis, 1974). Strong winds may then cause mixing within the epilimnion with the consequent increase in nutrients; even though stratification may remain, the driving of the epilimnion deeper by the wind can result in some exchange of nutrients.

Although, with respect to stratification and low nutrient concentrations, tropical lakes can resemble temperate lakes in summer, nevertheless they can sustain seasonal rates of production several times higher than their temperate equivalents (Section 5.2.3(a)). This is a contrast to tropical oceans which are, for the most part, less productive than those in temperate seas. This is largely due to the presence of a permanent thermocline which does not break down and consequently leads to a consistent drain on nutrients, whilst in tropical lakes, even when the thermocline occurs through much of the year, there is usually some mixing process which gives some replenishment of nutrients to the euphotic zone. In tropical lakes, therefore, production benefits from the high light intensities and long growing season combined with a reasonable nutrient turnover in a way not seen in the oceans.

There is one further aspect of nutrient availability which has received little attention and this is nitrogen fixation. As described in Section 4.5.2, many, although by no means all, blue-green algae are able to fix nitrogen as N_2 into other chemical compounds within the organelles known as heterocysts. The extent to which they can do this is demonstrated by a series of measurements in Lake George which showed that the blue-green algae, which dominate the phytoplankton, fix 1280 tonnes annually. This is almost equal to the total of 1452 tonnes of nitrogen input from all other sources such as rivers and direct rainfall (Ganf and Viner, 1973; Ganf and Horne, 1975). The contribution from this source will obviously depend upon the suitability of the water for blue-greens. They tend to be found in the largest quantities in alkaline waters and can reach huge numbers in soda lakes, for example, whilst in more neutral or acidic waters, groups such as diatoms and desmids are more prevalent.

Nevertheless it is evident that nitrogen fixation can play an important part in the nutrient budget of tropical waters.

5.3 MICROBIAL ACTIVITY AND DECOMPOSITION

The utilization of organic matter as a respiratory substrate of bacteria and fungi in the water, with the consequent reduction of this to its inorganic components, is of evident value in maintaining sustained primary production. In addition, however, part of the metabolized energy from the organic compounds goes into growth of the micro-organisms themselves and in this way becomes part of the trophic structure of the community. The organic material itself can be either dissolved or particulate, i.e. detritus. The extent to which the particulate organic material needs to be converted to bacterial substance before it can be used by animals within the community presents a difficult problem. The benthic suspension feeder, *Simulium*, for example, will consume and digest bacteria but grows poorly on these alone; in natural circumstances bacterial consumption can be well below that required to maintain the animal. Consequently, a large proportion of assimilated load can be non-living detrital material (Ladle and Hansford, 1981; Baker and Bradnam, 1976). A similar situation has been found for the midge larva *Chironomus*, but bacteria do not appear to be utilized at all by the aufwuchs-feeding ephemeropterans such as *Ephemerella*. Some plankton-feeding fishes, including the tilapia *Oreochromis mossambicus*, are able to digest and assimilate bacteria (Bowen, 1976), but the importance of micro-organisms and particulate organic matter and their relative contribution to the energy supply of other trophic levels is probably one of the largest problems in tropical limnology. Rivers, in particular, have often considerably more energy in the form of particulate organic material than from primary production, but how much is actually available remains to be seen. The particles themselves may have bacteria attached to them which makes identification of the roles even more complex.

From the little work that has been done on aquatic bacteria in the tropics it appears that total numbers can be very high. Temperate oligotrophic lakes often have numbers less than $5 \times 10^8 \, 1^{-1}$ whilst in eutrophic waters they can exceed $20 \times 10^8 \, 1^{-1}$; total bacterial counts from a selection of varzea lakes in Central Amazonia showed densities between 4.2 and $15.6 \times 10^9 \, 1^{-1}$ in the dry season (Rai, 1979). Only in highly enriched, polluted temperate waters do bacterial numbers exceed $10^9 \, 1^{-1}$ (Rai and Hill, 1980). This would put the varzea lakes amongst eutrophic waters by temperate standards even though their chemical features would not indicate this. It was evident that bacterial activity tended to follow changes in primary production since this activity, measured as the rate of glucose uptake (V_{max}), was between eight and thirty-two times greater at low water than at high when primary production was also very low. In addition, there was a highly significant correlation between V_{max} and chlorophyll density indicating that bacterial activity is high when the phytoplankton is abundant (Rai, 1979).

At low water when the lakes are shallow, allowing rapid nutrient recycling, primary production becomes very high and the bacteria will be using either the organic material from the dead plankton or the organic compounds, including glycolic acid and glucose, which are secreted by photosynthesizing algae as extracellular products. These can attain quite appreciable levels; for example, in Lake George, the extracellular products ranged from 6.5 to 58% of the gross carbon fixation (Ganf and Horne, 1975). The background levels of naturally occurring compounds which can be used as substrate by bacteria were found to be much higher in the varzea lakes than in any temperate lakes (Rai, 1979; Rai and Hill, 1980).

Within the varzea lakes the greatest bacterial activities were associated with the most productive waters whilst black-water lakes such as the Lago Tupé, where pH and light penetration are low, show greatly reduced activity (Rai and Hill, 1980). Nevertheless, bacterial numbers in these black-water lakes still fell within the range for eutrophic temperate waters and the activity in the most productive lakes, measured as the rate of glucose uptake, was higher than for almost all temperate examples, including those enriched by organic pollution. In these Amazonian lakes, therefore, bacterial activity is apparently very intense and commensurate with the rapid nutrient recycling deduced from primary production studies.

The bacteria are distributed down through the water column but their peak of activity, at least in the varzea lakes, may occur close to the surface or at some intermediate depth (Rai, 1979). It has also been found that peaks in pheopigments, which are the breakdown products of chlorophyll, and also in zooplankton numbers coincide at about 1 m depth which is the euphotic zone, above the thermocline in the stratified lakes. The pheopigments represent the degradation products of chlorophyll and therefore indicate a layer of detritus derived from the phytoplankton. The coincidence of the zooplankton peaks at this depth suggests that the pigments result from the grazing activity of the zooplankton since passage through their gut results in the chlorophyll being converted to these compounds. This detritus, consisting of zooplankton faecal debris produced as a result of grazing, would form a good substrate for bacterial activity and therefore promote the rapid nutrient turnovers which have been observed (Rai, 1978). The zooplankton, therefore, play an important role in the rapid recycling of nutrients in these systems (Sections 5.4.1, 5.4.2).

This procedure of rapid mineralization and reuse of inorganic compounds by the primary producers, with little accumulation of these nutrients outside the living organisms themselves, is very similar to that found in the tropical rain forests where an intimate relationship between the plant roots and the decomposers ensures that all nutrients are immediately taken up and reincorporated into the biomass of the plant rather than allowed to accumulate in the soil.

Such a rapid transfer system in tropical waters would explain the lack of detritus which accompanies high productivity in deeper lakes and clear-water rivers, and also why shallow lakes with restricted nutrient inflows can sustain very high productivity rates over much of the year (Section 5.2.3). In other situations there may also be a notable lack of accumulated organic material. For example, a number of coastal lakes in Ecuador had considerable marginal vegetation and swamps which would be expected to produce much debris although very little was actually found (Steinitz-Kannan *et al.*, 1983). Similarly, even in the region of the dense floating meadows of the central Amazon lakes, there is relatively little accumulation of organic material (Junk, 1970). The implication in these cases is that the dead plant material is being broken down very rapidly and, indeed, up to 50% of dry plant material placed in wide-mesh litter bags in the Amazon lakes was lost due to decomposition within 2 weeks and, after 6 months, only 15% remained (Howard-Williams and Junk, 1976). The decomposition process is further speeded up if the lake margins or flooded areas become exposed to air during the dry season. Alternate wetting and drying organic material also appears to accelerate the rate of decomposition. For example, mineralization of organic material in Lake Naivasha, Kenya, during the drawdown period contributed 21% of the phosphorus, 39% of the nitrogen and 4% of the sulphur to the annual nutrient budget and it was evident that the grazing of terrestrial herbivores through conversion of plant material to dung generally speeded up nutrient release (Gaudet and Mutmuri, 1981). Such processes are particularly important in man-made lakes and reservoirs where large drawdowns often occur.

The quantities of organic material in the sediments of a number of Brazilian reservoirs proved to be very low, between 3.7 and 16.9% dry weight, compared to those from temperate lakes where levels of around 45% were considered more common (Esteves, 1983). This again suggests a rapid mineralization of organic material in the sediment as well as in the water body.

The higher temperatures will no doubt contribute to the rapid rate of bacterial decomposition in the tropics which could be four to nine times faster than in temperate waters (Ruttner, 1940).

The bacteria responsible for decomposition in rivers occur in at least two phases—those which are free living in the water and those which are attached to suspended particles. In the middle Parana River, Argentina, there was a very strong correlation between the attached or episammic bacteria and the density of suspended solids as well as river discharge (Emiliani, 1984). These, in turn, depended upon the rainfall, which was heaviest in the spring and summer (November to March), and this suggested that the attached bacteria were associated with the suspended solids washed out from the land by the rains.

In contrast, the free-living bacteria showed no such correlation and were principally associated with the nutritional status of the water. The implication is that these free-living types are mainly inhabitants of the river water, i.e. they are autochthonous, whilst the episammic group consists basically of soil bacteria which are washed into the rivers by rainfall (Emiliani, 1984). Bacteria are found in lake waters in both attached and free-living forms, but it is unlikely that this indicates two different sources as it may do in rivers. The relative activity of these two states to the general energy flow and decomposition in rivers remains to be seen.

The average numbers of bacteria found in the middle Parana were greater than those found in the Amazon but less than those determined for the Nile and the temperate Mississippi (Rai and Hill, 1978) whilst the relative activities of bacteria in tropical and temperate rivers have yet to be determined.

5.4 SECONDARY PRODUCTION

5.4.1 Feeding Relationships

The food consumed by animals provides the energy for production and respiration just as light does for plants. Moreover, just as some light energy impinges upon the plant but is not actually used for photosynthesis, so some of the energy and materials consumed by animals passes through the gut and is voided as faeces without being absorbed or assimilated (see Fig. 5.1). The extent to which assimilation occurs varies with the nature of the food, with plant material being rather less digestible than animal owing to the high proportion of compounds such as cellulose and lignins for which most animals do not possess the necessary cellulase enzymes. The percentage of food actually absorbed from the gut and therefore available for the metabolism of the individual is called the assimilation efficiency.

The few studies done on grazing animals in tropical waters indicate that their ability to assimilate plant materials is similar to that of animals from temperate areas. For example, young and adults of the copepod *Thermocyclops hyalinus* from Lake George assimilated 35% of ingested carbon from the blue-green alga *Microcystis* whilst the nauplius larvae could extract 58% from similar food (Moriarty *et al.*, 1973). The phytoplankton-feeding cichlid fishes in the same lake, *Oreochromis niloticus* and *Haplochromis nigripinnis*, using a similar diet, had mean efficiencies of 43 and 66% respectively which is considerably less than the average for fish in general of 85% given by Winberg (1956) in his review. This is because most of the examples available to Winberg at that time were from temperate fishes which are largely omnivores or predators with few, if any, truly plant-feeding types. The long growing season in the tropics means that there is sufficient phytoplankton growth over long periods to sustain planktivorous fishes. The low assimilation efficiencies of the cichlids reflect the plant nature of their diet as opposed to those feeding wholly or in part on animals. For example, the predatory Indian murrel, *Channa striatus*, can absorb up to 90% of the material in the prey which it consumes (Pandian, 1967).

Many benthic animals, like plankton feeders, consume a mixture of algae and detritus and consequently have similarly low assimilation efficiencies. The atyid shrimp, *Caridina nilotica*, which is found in littoral areas throughout East and Central Africa, feeds on periphytic algae and detritus and was found to have an efficiency ranging from 20 to 45% (Hart and Allanson, 1981). Detrital particles, which may in fact partly consist of faecal debris from other animals such as copepods, tend to have a high proportion of materials resistant to digestion and this is why enrichment from attached bacteria may be valuable to animals which utilize this food source. Nevertheless, as seen in Section 5.3, although there may be some utilization of bacteria by benthic types such as *Simulium* and *Chironomus*, detrital particles alone must also form a part of the food assimilated (Baker and Bradnam, 1976). However, *Simulium* species kept on mixed diatoms, absorbed between 53 and 66% of the material consumed whilst, on natural detritus alone, the proportion was only 1.7–1.9% (Ladle and Hansford, 1981). So the percentage of digestible material in detritus can be very low, but the usual strategy of dealing with this is to process large quantities in the gut very rapidly to obtain sufficient for the total energy requirements. Food can pass through the gut of *Simulium* within 20 minutes to 1 hour (Ladle, 1972).

In general, the proportion of food actually assimilated is more directly related to the diet than to geographical locality or the type of animal involved.

The assimilation efficiency tells us what proportion of the food intake is actually utilized by the animal, but the total amount consumed by a population is a product of the total numbers of individuals and their rates of feeding. In addition, of course, the actual rate of production made available from the previous trophic level will influence or even determine this. In the few instances where it has been estimated, the amount of this production which can be consumed is quite high. In Lake George, daily estimates of phytoplankton consumption amounted to $34\,mg\,C\,m^{-2}$ for the cichlid fishes and $504\,mg\,C\,m^{-2}$ for the dominant zooplankter, *Thermocyclops hyalinus* (Moriarty *et al.*, 1973). In addition to this $538\,mg\,C\,m^{-2}$ lost to the grazers, a further $1\,g\,C\,m^{-2}$ is taken by the benthic organisms although a third of this could be supplied from the incompletely digested algae in the faeces of the grazers. The total daily requirements are, therefore, $1240\,mg\,C\,m^{-2}\,day^{-1}$, which is inconsiderable beside the phytoplankton standing crop of $30\,g\,C\,m^{-2}$, although the estimated net primary production was only $600–800\,mg\,C\,m^{-2}\,day^{-1}$. This is rather less than the daily requirement and does not allow for losses from sedimentation or phytoplankton washed away down the outflow. This will be due to currently unavoidable discrepancies in the methodologies but it does indicate that a substantial proportion of the daily net phytoplankton production can be taken by other trophic levels.

In some ways Lake George may not be typical of shallow tropical waters in that 95% of the total biomass is due to phytoplankton alone and the density of grazers is relatively low. This could be because few types of zooplankton can deal with the large colonial algae such as *Microcystis* (Burgis *et al.*, 1973). In one of the Amazonian varzea lakes, however, it was estimated that the very dense zooplankton could graze sufficient of the phytoplankton biomass to cause a 100–300% turnover each day (Fisher and Parsley, 1979). Lewis (1974) also thought that the zooplankton could take a substantial proportion of algal biomass during productive periods but, since biomass and production are not necessarily directly linked, concluded that the effect on production was unclear.

5.4.2 Some Effects of Grazing

Consumption of phytoplankton by the grazers can be a controlling factor of primary production although the interactions may be subtle. Since zooplankton numbers vary on a longer time-scale than those of phytoplankton, they provide a cropping effect which is more stable than the short-term variations in primary production caused by daily changes in light and nutrient availability, for example. This, therefore, tends to damp down oscillations in the phytoplankton (Lewis, 1974). A statistical analysis of the relationship of the maximum rate of photosynthetic production in Gatun and Madden Lakes in Panama to a number of biotic and abiotic factors showed this to be only significantly correlated with two of them—chlorophyll a density and grazing intensity (Gliwicz, 1975). High production was a result of lower chlorophyll densities which in turn was caused by intense grazing. Apparently the considerable production from the lower biomass was due not to reduction in self-shading amongst the algae but to the lower proportion of nutrients locked up at any one time in the phytoplankton. Once again, as suggested by Rai (1978) for the Amazon varzea lakes, the indications are that the grazing of the zooplankton, whilst reducing biomass, actually stimulates production by speeding up the nutrient turnover.

In the varzea lakes, the peak in zooplankton abundance commonly coincides with the peak in chlorophyll degradation productions, probably from faecal material, at about 1 m depth presumably as a result of the zooplankton grazing on the lake algae. This may represent the zone of maximum rate of mineralization (Rai, 1978). A second peak in chlorophyll degradation products can be found close to the lake bed owing to accumulation of dead algal cells. From the ratio of degradation products to photosynthetic pigments it was suggested that 30% of organic matter was due to living material in the varzea lakes at low water level and only 10% at high water (Rai, 1978).

Grazing by the zooplankton may also influence the composition of the phytoplankton. Experiments with small enclosures of phytoplankton from Gatun Lake, Panama, in which numbers of the dominant zooplankton, the small cladoceran *Ceriodaphnia cornuta* and the larger calonoid copepod *Diaptomus gatunenis*, were artificially increased, showed that nine of the sixteen phytoplankton species, including *Scenedesmus* and *Cyclotella*, decreased as a result of this whilst three species showed no change and four were more plentiful (Weers and Zaret, 1975). Those which declined were evidently readily taken whilst the remainder were either not taken or were not digested. Most striking were those types which became more common as a result of grazing. These tend to be green algae with a gelatinous covering, such as *Ankistrodesmus*. There are indications from parallel observations on temperate species that the gelatinous covering gives some protection in the passage through the crustacean gut but, in addition, some factor in this passage actually enhances the probability of cell division leading to the increase in numbers observed. One apparent result of this is that the gelatinous types form 75–90% of the phytoplankton in Gatun Lake (Weers and Zaret, 1975).

5.4.3 Rates of Production

(a) Zooplankton

There are no simple or convenient ways of measuring animal production in natural communities. Its determination requires estimates of numbers, biomass, age structure, growth and reproductive rates (Edmondson and Winberg, 1971), and the complexity of the information needed explains why there are very few estimates of production for any types of aquatic animals in the tropics. It is necessary to measure all production of new material either in growth of the individual or production of young, over a given period of time. For crustacea such as copepods the situation is perhaps made simple by the fact that once the adult size is reached the animals cease to grow and any further production can be estimated in terms of eggs or young. Moreover, the growth from larva to adult is punctuated by a series of steps ending in a moult, which produce a subsequently identifiable stage. The time taken to reach each stage can be estimated in the laboratory and then applied to the numbers encountered in the field situation.

In small crustaceans, the time taken for each stage can be quite short; for example in *Thermocyclops hyalinus* from Lake George the egg stage lasts for 1.5 days, the nauplius larva 6 days and each copepodite stage 11 days (Burgis, 1974). In this particular case, the final production estimate for the population of this species in the lake was 44 mg dry wt m^{-2} day^{-1} which is equivalent to 19.6 mg C m^{-2} day^{-1}. This is no greater than zooplankton production rates which might be expected from temperate waters. A preliminary estimate for the calanoid copepod, *Pseudodiaptomus hessei*, in Lake Sibaya, however, proved to be 140 times lower than that of *Thermocyclops hyalinus* in Lake George (Hart and Allanson, 1975). This may be due to the location of this lake in the sub-tropical zone of Southern Africa, particularly since seasonal fluctuation in the production by the copepod was correlated more strongly with temperature than with phytoplankton biomass. However, it could also indicate that this lake, which lies in an open sandy, coastal region, may be truly oligotrophic. Even so, production amounted to 11.3% of the standing crop each day whilst the equivalent in Lake George was 7.9%.

In contrast, the zooplankton of the saline Lake Nakuru where primary production is extremely high (Section 5.2.3(a)), shows equally substantial secondary production. The rotifers, largely *Brachionus dimidatus* and *B. placatilas*, show production equivalent to 90 mg dry wt m^{-3} day^{-1} or 40.5 mg C m^{-3} day^{-1} (Vareschi and Jacobs, 1984). This gives an approximate areal value of 180 mg dry wt m^{-2} day^{-1} which is considerably more than that in Lake George and is generally in accordance with their respective rates of primary production (Section 4.6.4), but the high production is due to a fast reproductive rate of the rotifers incorporating a very short juvenile phase of only 2 days and rapid egg production. At their peak, copepods showed production similar to that of the rotifers but their contribution was far more erratic (Vareschi and Jacobs, 1984).

The mean annual production of the cyclopoid *Lovenula africana* in Lake Nakuru ranges from 30 to 80 mg dry wt m^{-2} day^{-1}. A combined estimate during 1973 for rotifers and copepods gave a total zooplankton production of 190 mg dry wt m^{-3} day^{-1}.

In Lake Chad the zooplankton contains a higher number of common crustacean species than in Lake George or Lake Nakuru, with eight species of Cladocera and four species of copepods contributing significantly to the biomass and production (Lévêque and Saint-Jean, 1983). The biomass of rotifers in this lake was small but, as shown in Lake Nakuru, this does not mean that their production is necessarily insignificant. In Lake Chad, the mean daily rate of crustacean zooplankton production ranged from 9.6 to 98.8 mg dry wt m^{-3} day^{-1} over the whole lake with a mean of 41.9 mg dry wt m^{-3} day^{-1}. Per unit area, the average production rate is some 129 mg dry wt m^{-2} day^{-1} which is equivalent to 57.5 mg C m^{-2} day^{-1}. The greatest single contribution to this was most often from the Cladocera, particularly *Moina micrura, Bosmina longirostris*, and *Diaphanosoma excisum*, which provided between 40 and 66% of the total crustacean production, with the cyclopoid copepods being the next most important.

The zooplankton production rate is less than that of Lake Nakuru but greater than that of Lake George. The primary production of both these lakes is greater than that in Lake Chad. The lower zooplankton production in Lake George is most probably due to the unsuitable nature of much of the phytoplankton for zooplankton grazing. The mean daily production amounted to between 10 and 19% of the biomass (i.e. the P/B ratio was 0.1–0.19) in Lake Chad compared to some 8% in Lake George.

Amongst the major groups in the zooplankton there appear to be considerable differences with regard to their specific growth rates. Some of the small representatives, such as the rotifer *Brachionus dimidatus* in Lake Nakuru (Vareschi and Jacobs, 1984) or the small cladoceran, *Moina micrura* in Lake Chad (Lévêque and Saint-Jean, 1983), can have P/B ratios of 40–55% day^{-1} and 27–87% day^{-1} respectively. Calanoid copepods may have lower values, being 1–7% in Lake Chad, although in Lake Turkana a rather wider range of 4–56% was found for *Tropodiaptomus banforamus* (Ferguson, 1975). Cyclopoid copepods which have been examined have similar P/B ratios, 4–5% for *Lovenula africana* in Lake Nakuru and 10–27% for cyclopoids in Lake Chad.

(b) Benthic animals

Similar types of information to that needed for zooplankton are required to estimate production from benthic organisms. The atyid shrimp, *Caridina nilotica*, had an estimated production rate of 24 mg dry wt m^{-2} day^{-1} in the subtropical Lake Sibaya where it is one of the dominant littoral species with a mean standing crop of 2.7 g m^{-2} (Hart, 1981). These animals take periphyton and detritus (Section 4.7.2) and the efficiency with which this food was used for growth, a factor known as the production or net growth efficiency, varied between 21 and 42% in the females and 18 and 32% in the males (Hart and Allanson, 1981). Breeding females, in which egg production was taking place, were actually able to use the food for growth more efficiently than non-breeding females.

The average net growth efficiency for the natural population of *Caridina nilotica* in Lake Sibaya was 52%, which is much higher than that for populations of equivalent benthic animals such as *Asellus aquaticus* and *Mysis relicta* in temperate areas where values for populations tend to fall between 20 and 30%. Young animals always have higher efficiencies than older ones in using food for growth, and in the youngest *Caridina nilotica* over 80% of the food assimilated and 40% of the food consumed can be used for growth. The high value for the population as a whole reflects the large proportion of young individuals, owing to round-the-year spawning of this species in Lake Sibaya. Extended spawning seasons are often a feature of the life cycles of tropical animals and, therefore, this demographic feature for growth may be a common characteristic. It may also provide an explanation as to why production rates in tropical aquatic animals can be quite high whilst standing crops are relatively modest (Hart and Allanson, 1981).

The total energy requirements of individual *Caridina* in Lake Sibaya show an increase with temperature, mainly through elevated respiratory rates. This effect, however, was much less marked above 24 °C which may suggest some regulatory mechanism in tropical animals for minimizing the effects of high temperatures upon their overall energy demand (Hart and Allanson, 1981).

Amongst the very few strictly tropical lakes for which the production levels of benthic animals have been estimated, Lake Chad proved to have a rate close to 20 g dry wt m^{-2} $year^{-1}$, or approximately 54.8 mg dry wt m^{-2} day^{-1} (Lévêque and Saint-Marie, 1983), whilst the mean daily rate for Lake Nakuru was 40.1 mg dry wt m^{-2} day^{-1} (Vareschi and Jacobs, 1984).

These are only slightly higher than the production of *Caridina* in the subtropical, oligotrophic Lake Sibaya and are also within the upper limits of production estimates from temperate lakes.

The large biomass of benthic animals in Lake Chad is heavily dominated by prosobranch snails and other molluscs, which contribute 75% of the total organic production. In addition, their shell production is considerable, amounting to about 100 g m^{-2} $year^{-1}$, or 1 883 000 tonnes for the whole lake annually. The shells contain 37% calcium in the form of argonite and the molluscs require 700 000 tonnes of calcium per year for this purpose, which is four times the annual supply of this element to the lake or half the dissolved stock in the lake water. The molluscs must, therefore, play a very significant part in the recycling of calcium within the ecosystem (Carmouze, 1976).

In Lake Nakuru, the benthic fauna consists of two chironomid species, *Leptochironomus deribae* and *Tanitarsus horni*. The rate of production of these animals is quite variable. Years with high productivities, 56 and 79 mg dry wt m^{-2} day^{-1}, were observed in years of low salinities and low algal abundance. This production is sustained by a relatively small biomass of organisms, on average 0.43 g m^{-2}, when compared to the considerable biomass of molluscs, around 3.3 g m^{-2}, in Lake Chad. The percentage production or P/B ratio for the chironomids of Lake Nakuru was 9%, whilst that of the molluscs from Lake Chad was only 1.2%. To this extent, chironomids are more efficient converters and producers than molluscs.

(c) Fish

Fish are, in some ways, even more difficult to deal with from the point of view of production estimates than are the zooplankton and benthic invertebrates. As they are large, active animals, it is more difficult to obtain measures of biomass (Sections 4.8.2, 4.8.3) than for small invertebrates, and all estimates of age and growth must be made in the field. In addition, fish continue growing throughout their lives and therefore adult growth must be taken into account. Fishes from temperate regions lay down rings on their scales annually at the transition from the cold winter to the warmer spring and these can be used to determine age and growth rate. Seasonal temperature differences are not marked in the tropics, but rivers in particular do have a pronounced cycle based upon the wet and dry seasons. This often imposes a seasonal cycle upon the fish which results in ring formation on the scale (Payne, 1976a), and even fish in equatorial lakes such as Lake Victoria will form rings at certain times of the year (Garrod and Newell, 1958). In both of these cases, ring formation may be related to calcium resorption from the scales during maturation and spawning (Garrod and Newell, 1958; Payne 1976a,b) but in other cases a correlation with lack of food during the dry season has been suggested. Such factors require critical observation before they can be used in determining the age of fish from tropical waters, but nevertheless it has been possible to use marks on scales, otoliths, or other hard parts to estimate age and growth rates within populations (Fig. 5.11). In some lakes, however, more than one ring appears every year and so the time period between them must be established. Reliable periodic rings are by no means always present, however, and in the equatorial Lake George, their absence hindered the estimate of fish production in an otherwise very thorough study. Alternatives to the use of scale rings to estimate age include the monitoring of changes in the mean size of individuals making up discernible population peaks with time, often known as the Petersen method (Tesch, 1966). The reliability of this method, particularly in tropical fish communities, has been enhanced by the use of computer techniques (Pauly and David, 1981).

Amongst the few estimates of fish production which have been made in the tropics, the Kafué River in Zambia produced 63–87 g fresh wt m^{-2} year^{-1} (Kapetsky, 1974), whilst Lake Sabonilla and Lake Luisa in Cuba produced 22 and 27.6 g m^{-2} year^{-1} (Holcik, 1970). The highest of the Kafué estimates, when converted to dry weight on a daily basis, gives a value in the region of 50 mg m^{-2} day^{-1} which is a similar order of magnitude to the few studies made on zooplankton and benthic invertebrates. The range of production rates for the small tilapia, *Oreochromis alcalium grahami* (Vareschi and Jacobs, 1984) in the fertile Lake Nakuru was 41–$72 \text{ mg dry wt m}^{-2}$ day^{-1}.

163

Fig. 5.11 A scale from *Oreochromis niloticus* of 17.3 cm total length from Lake Manzala, Egypt, showing three checks on the scale

The principal fishes in the area of the Kafue under survey were *Tilapia rendalli* and *Oreochromis andersonii* which both feed on plant material (Kapetsky, 1974). In both species a high proportion of production occurred during the first 2 years of life. In *Tilapia rendalli*, for example, 53% had been produced by the first year and 98% by the second year. So, just as with the shrimp *Caradina*, the young make a disproportionate contribution to the total production. In addition many tropical fishes have a relatively short life of perhaps 2–3 years, unlike their temperate counterparts, and the population structure tends to be dominated by younger age-groups.

5.5   ENERGY FLOW WITHIN THE ECOSYSTEM

Some details of the few examples available on production at each level of tropical aquatic ecosystems have been given in this chapter. However, even when production rates are known, this does not tell us what proportion will be available to other components of the ecosystem; for example, how much net primary production goes to the grazers and how much to the decomposers? Energy flows through the ecosystem to be ultimately dissipated as heat of respiration at one of the trophic levels. Once the distribution of energy within the system has been charted, its participation and transfer amongst the functional components can be represented as an energy flow diagram which is effectively putting dimensions to the relationships described in Figs. 4.1 and 4.2. There are virtually no tropical examples in which our current level of understanding permits the construction of an energy flow diagram to the extent which is possible for several temperate ecosystems. One such temperate aquatic example, the River Thames, is included here as an illustration of the aim of this approach (Fig. 5.12). In the energy flow diagram, the size of each compartment represents the biomass expressed in energy units whilst the width of the pathways represents the magnitude of the energy transfer. In the case of the River Thames, the largest single energy input can be seen to come from upstream import of allochthonous organic production. Primary production, whilst significant, provides a little less than this, and this pattern is common to many rivers. One further general point demonstrated is that animals often accumulate a much higher biomass than plants, although their relative production is much smaller.

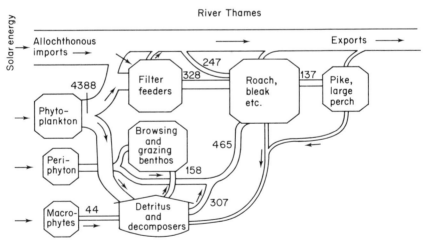

Fig. 5.12 Energy flow through the River Thames ecosystem. (*Reproduced by permission of Blackwell Scientific Publications Ltd. from Mann, 1975*)

One tropical investigation of a river, which has led to the establishment of a preliminary energy budget and energy flow diagram (Fig. 5.13), is that carried out on the Rio Limon which flows into Lake Valencia, Venezuela (Lewis and

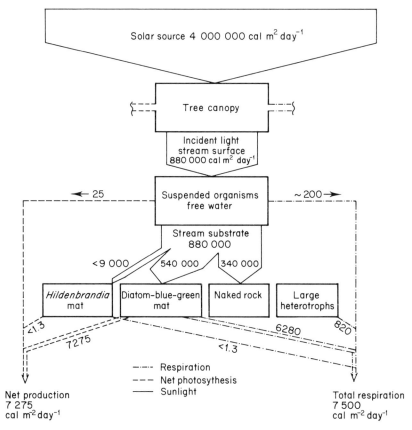

Fig. 5.13 Preliminary energy flow diagram for the Rio Limon, Venezuela. (*Reproduced by permission of Schweizerbart'sche Verlagsbuchhandlung from Lewis and Weibezahn, 1976*)

Weibezahn, 1976). In this, however, the biomass was not determined and only the dynamics of the system were considered. In addition all the heterotrophs, both invertebrates and fish, were taken together and their respiration was not measured but deduced from the literature. What the diagram demonstrates is that only 6% of incident solar energy actually filters through the tree canopy to the river surface. A fraction of this energy is assimilated by the suspended algae to produce a small but measurable contribution to production whilst the remainder is either used by the benthic algae, which make the greatest contribution to primary production, or impinges unprofitably on the bare rock of the river bed. The heterotrophic vertebrates and invertebrates largely feed on suspended material imported from upstream. This was not measured in the Rio Limon study although the situation on the Thames (Fig. 5.12) shows how important this can be.

The most notable feature of energy utilization in the Rio Limon is that there is a significant net primary production and that community energy production as photosynthesis exceeds community energy utilization as respiration. The P/R ratio is around 1.8. The river can, therefore, be described as autotrophic as opposed to heterotrophic as appears to be the case in several temperate, shaded forest rivers (Mann, 1975) where community respiration exceeds synthesis with the difference being made up through imported allochthonous energy.

One of the reasons why production appeared to be so high in the Rio Limon was the efficiency with which the producers used the relatively small amount of light which filtered through the overhanging tree canopy. The proportion of light reaching the water surface which was transferred into actual production —that is, the utilization efficiency—was 1.46% compared to 0.44% in a moderately productive tropical lake such as Lake Lanao (Lewis, 1974). This prompted the suggestion that tropical streams may be rather efficient at using light, certainly more so than their temperate counterparts, and therefore more prone to being autotrophic in nature (Lewis and Weibezahn, 1976).

It is also possible to derive a picture of energy flow through a tropical lacustrine system based upon information from Lake Chad (from Carmouze et al., 1983). It is remarkable (Fig. 5.14) how small the biomass of the phytoplankton is compared to its gross production, that is, the combined respiration and net production. The biomass is even smaller than that of the dependent zooplankton, and the phytoplankton only represents about 30% of the energy content of the total plankton biomass. By contrast, the standing crop of phytoplankton in Lake George was equivalent to 1254 kJ m^{-2} and constituted 98% of the plankton standing crop (Burgis and Dunn, 1978). However, the standing crop of algae is high in Lake George and that of the zooplankton unusually low for reasons previously mentioned. The pattern shown by Lake Chad is probably more typical of tropical lakes.

In Lake Chad, only the gross production was determined directly; the net production was estimated to be 10% of this with the remainder being dissipated in respiration (Carmouze et al., 1983). However, this is an unusually low proportion and, even in Lake George, 28% of the gross was available as net production (Burgis and Dunn, 1978). Moreover, experimental work on copepods in Lake Nakuru showed that they required an energy consumption six times that of their eventual production (Vareschi and Jacobs, 1984). Applying this factor to the Lake Chad case shows that zooplankton consumption alone would exceed the net production from the phytoplankton, which in this case, therefore, could be something of an underestimate. The gross production, however, is high and is probably similar to that of the macrophytes, largely papyrus, despite the enormous disparity in biomass. The phytoplankton production in Lake Chad is equivalent to 0.25% of the incident radiation which is within the usual range for tropical lakes. For example, Lewis and Weibezahn (1976) found a range of 0.016–0.54% for a series of lowland lakes in Venezuela whilst the equivalent value for Lake George is 0.79%. The phytoplankton of Lake Chad is not, therefore, as efficient as some in converting the incident solar energy into organic material.

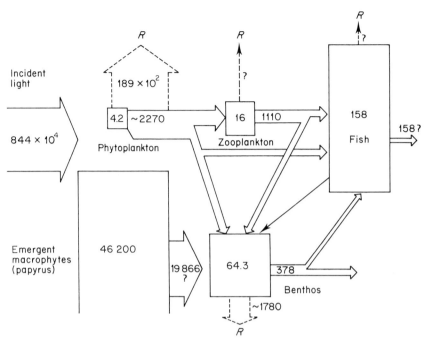

Fig. 5.14 Preliminary energy flow diagram for Lake Chad, West Africa (from data in Carmouze *et al*, 1983a). Biomass in $kJ m^{-2}$ and rates of production and respiration (R) in $kJ\, m^{-2}\, y^{-1}$

The rapid production of the phytoplankton in Lake Chad is due to a very fast turnover and short generation time. Conversely, the much larger biomass of some of the other components, such as the macrophytes, benthos and fishes, is due to a longer-term accumulation of material. Animal production is relatively small and the gross production of macrophytes, represented in Lake Chad by extensive papyrus beds, is probably similar to that of the phytoplankton, even though there is a tremendous disparity in their standing crops (Section 5.2.3(a)).

The varied roles played by fishes within the tropical ecosystem are emphasized by the energy flow diagram from Lake Chad (Fig. 5.14). In this lake, there are primary consumers such as *Sarotherodon galilaeus*, feeding on the algae, although the majority are secondary or tertiary consumers feeding upon zooplankton, other fishes, or upon benthic organisms. This is in contrast to Lake George where 60% of the fish biomass is due to phytoplankton-feeding types, which reflects the huge predominance of phytoplankton in the standing crop and its rapid rate of production.

The energy-flow approach allows comparisons to be made between whole systems. However, the energy-flow diagram remains a purely descriptive device. It does not tell us what causes more energy to go down one pathway than another or how the partition of energy between the components of the com- munity is regulated. As a snapshot, the diagram provides a framework, but the identification of factors regulating energy flow and their interaction to provide stability, or at least continuity, within the community remains one of the greatest problems of synecology.

GENERAL READING

Berrie, A. D. 1976. Detritus, micro-organisms and animals in fresh water. In *The Role of Aquatic and Terrestrial Organisms in Decomposition Process*, Anderson, J. M. and Macfadyen, R. (eds.), pp. 323–338.

Brylinsky, M., and Mann, K. H. 1973. An analysis of the factors governing productivity in lakes and reservoirs. *Limnol. Oceanogr.*, **18**: 1–14.

Goltermann, H. L. (ed.) (1977). *Interactions between Sediments and Fresh Water*. The Hague: Junk, 473 pp.

Greenwood, P. H. 1976. Lake George, Uganda. *Phil. Trans. R. Soc. Lond.*, **B274**: 375–391. A useful summary of IBP investigations.

Lewis, W. M. 1974. Primary production in the plankton community of a tropical lake. *Ecological Monographs*, **44**: 377–409.

Mann, K. H. 1975. Patterns of energy flow. In *River Ecology*, Whitton, B. A. (ed.), pp. 248–263. Oxford: Blackwell.

Moriarty, D. J. W., Darlington, J. P. E. C., Dunn, I. G., Moriarty, C. M., and Tevlin, M. P. 1973. Feeding and grazing in Lake George, Uganda. *Proc. R. Soc. Lond.*, **B184**: 299–319.

Rai, H., and Hill, G. 1980. Classification of central Amazon lakes on the basis of their microbiological and physico-chemical characteristics. *Hydrobiologia*, **72**: 85–99.

Ricker, W. E. 1968. *Methods for Assessment of Fish Production in Fresh Waters*. IBP Handbook No. 3. Oxford: Blackwell, 326 pp.

Schmidt, G. W. 1982. Primary production of phytoplankton in three types of Amazonian waters. V. Some investigations on the phytoplankton and its primary productivity in the clear water of the Lower Rio Tapajóz (Pará, Brazil). *Amazoniana*, 7: 335–348.

Sioli, H. 1975. Tropical river: the Amazon. In *River Ecology*, Whitton, B. A. (ed.) pp. 461–488. Oxford: Blackwell.

Talling, J. F. 1965. The photosynthetic activity of phytoplankton in East African lakes. *Int. Rev. ges. Hydrobiol.*, **50**: 1–32.

Vollenweider, R. A. 1969. *A Manual on Methods for Measuring Primary Production in Aquatic Environments*. Oxford: Blackwell, IBP Handbook No. 12, 211 pp.

CHAPTER 6

SEASONALITY

6.1 ARE THERE SEASONS IN THE TROPICS?

It is not uncommon to read in textbooks that conditions in the tropics are relatively constant and therefore seasonal changes are rather small or absent. Temperature variations are, of course, rather restricted but, nevertheless, most of the tropics do have at least one rainy season. In lakes this may have the effect of increasing inflowing waters from rivers and streams with their associated suspended material and dissolved nutrients (Section 3.3), whilst in rivers the rains can lead to increased discharge rates and water levels which may lead to flooding of the terrestrial zone (Section 2.4). The rains are also often associated with a change in the wind pattern which, as we have seen in the previous chapter, is an extremely important factor in the mixing processes and nutrient distribution of water bodies.

In addition to carrying suspended material and dissolved substances, rivers and inflows carry one other thing—information. If animals and plants are to have a seasonal cycle in reproduction for example, they require some environmental trigger, the 'proximate factor', which will initiate the appropriate behaviour or growth pattern. In temperate areas changes in temperature and day length are common proximate factors, but these are rarely sufficiently variable to act as stimuli in this way in the tropics. Instead, potential triggers supplied by the rains are changes in current speed, water volume, or dissolved salt concentration or there may be something less tangible such as the 'taste' of the water. When the rains follow a dry season many compounds, both organic and inorganic, will be flushed from the soil into the watercourses and can no doubt be detected by the chemosensory organs of the animals. Where seasonality is found amongst animals of large lakes in which the influence of inflows is limited, however, it is more difficult to envisage exactly what the proximate factors might be.

The most common objective of seasonal cycles in animals and plants is to maximize the utilization of resources in favourable periods and to minimize the influence of unfavourable times. Essential requirements of the organisms which can change under the influence of seasonal environmental fluctuations, such as food supply, water availability, spawning sites, or stones to hide under, are known as the 'ultimate factors'. Proximate factors often prepare a population for changes in ultimate factors. For example, some aspect of the rising water levels, the proximate factor, induces many fish species to start gonad maturation which results in spawning in the early rains; the young are then produced as the surrounding land is flooded, yielding a large increase in an ultimate factor, the food supply, which the recently hatched young can exploit. The other aspect of seasonality is well illustrated by the small rock-bound plants of the Podostemacea dealt with earlier (Section 4.9.4). Here, the exposure of the plant above the water as the level falls triggers flowering which rapidly produces large numbers of tiny seeds as the plants dry out. The resistant seeds can then spend the dry season on the exposed rock following dispersal to germinate under more favourable conditions when water again becomes available with the rising river level. Some small cyprinodont fishes which live in seasonal streams have a very similar approach to this in that they spawn in the late rains as the stream is beginning to dry up. The eggs, which are resistant to desiccation, an unusual feature in fish, remain on the dry stream bed until the stream fills again with the next rains and the young hatch out.

Although some seasonal events are more obvious than others and certain species have marked cycles, nevertheless the whole community is often influenced by the changes. The periodicity of changes is also not necessarily geared to the annual cycle and daily changes may be more important. The study of these changes and their underlying mechanisms has now become sufficiently distinct as to be grouped into the discipline of phenology.

6.2   THE PHENOLOGY OF PRODUCTION

Melack (1979) has suggested that there are three patterns of periodic changes in production shown by tropical lakes; firstly those with pronounced fluctuations which constitute the majority, those possessing a low degree of seasonal fluctuation and finally a small group of lakes which have irregular but abrupt changes. Those relatively few lakes showing very limited changes can occur anywhere in the tropical zone and there is no trend from the equator to higher latitudes in this respect.

In lakes where seasonal fluctuations do occur the differences between the peaks and troughs of production are rarely as great as those seen in temperate waters but they can be quite distinct. Whilst variation in light intensity may play a minor role in causing these seasonal changes—for example only 30% of the variation in primary production in Lake Lanao could be explained by differences of light intensities—the major causes tend to be changes in mixing patterns and water inflow into a lake. A thorough examination of the pattern of production in Lake Lanao showing that the lowest rates of production were found just after thorough mixing of the lake occurred in December, whilst the peaks appeared just at the onset of stratification in March–April (Lewis, 1974).

Before December the lake had been strongly stratified and, therefore, since the mixing under the influence of the monsoon winds should involve the redistribution of nutrients from the hypolimnion, an increase in production might have been anticipated. However, one other aspect of vertical water circulations is that not only does it bring nutrients to the surface but it also carries phytoplankton downwards. In a deep lake such as Lanao this means that many algae are transported downwards out of the euphotic zone into regions where there is insufficient light for photosynthesis.

This situation persists whilst the isothermal, fully mixed conditions exist, but as the turbulent period subsides, around March, the more stable water column allows plankton populations to build up again within the euphotic zone as they take advantage of the improved nutrient levels. This can also be observed when calm periods of a few days occur during the isothermal season when mixing normally takes place. High levels of production are therefore re-established just as stratification begins to occur. Even when the thermocline has been fully formed, production does not crash to the very low levels often seen in temperate systems during the summer. This is due to the rapid recycling of nutrients within the epilimnion dealt with earlier (Section 5.2.4) and also to superficial mixing processes. Occasional windy periods during protracted stratification on Lake Lanao tend to force the thermocline deeper and are followed by peaks in phytoplankton production and increased efficiency of production. This is probably the result of the thorough mixing of the epilimnion where several secondary thermoclines may have formed, trapping nutrients below the uppermost thermocline which coincides with the euphotic zone. A similar peak in the phytoplankton of Lake Victoria was found in January and March following windy periods which did not actually cause the thermocline to break down completely but did drive it considerably deeper (Talling, 1966).

Lake Victoria, like Lake Lanao, does have a season of deep vertical mixing when in fact the lake becomes isothermal. During June and July the established thermocline breaks down under the seasonal onset of the south-east trade winds and for a brief period at the end of July the main body of the lake becomes isothermal with respect to depth (Talling, 1966). This mixing also causes the phytoplankton to become more homogeneously distributed in the water column, but it does not lead to a drop in overall production as it does in Lake Lanao. However, the mixing period is characterized by an abrupt increase in the numbers of the diatom *Melosira nyassensis*. This species is prone to rapid sinking and it seems probable that the mixing processes, as well as redistributing algal cells from close to the surface, may also cause the resuspension of types such as *Melosira* from the sediments or lower depths and bring them closer to the surface (Talling, 1966, 1969) which presents yet another aspect of the mixing process.

In Lake Victoria a peak in production and biomass appears in the more stable water column as stratification begins to become re-established in August. The cycle is, therefore, similar to that in Lake Lanao although rather more condensed. Unlike Lake Lanao, however, these peaks do decline quite rapidly to much lower levels presumably as nutrient availability declines in the stable conditions of the dry season. The seasonal range of production is some $0.4–1.0\,g\,m^2\,h^{-1}$, a threefold variation. In East Africa there are fewer of the irregular storms and cyclones experienced by the Philippines throughout the year and therefore fewer occasions for incidental mixing.

Small lakes may also show seasonal patterns in productivity. Lake Carioca, a shallow rather sheltered lake in eastern Brazil, showed a higher productivity during the cooler season in July compared to the warmer period in January. The temperature differences between the seasons were sufficient to cause the lake to become isothermal in the winter. This allowed daily mixing of nutrients which promoted the rather higher rate of primary production during the cooler season when the water was some 5 °C lower than in January (Barbosa and Tundisi, 1980).

In lakes at higher altitudes seasonal temperature changes can cause the lakes to become isothermal with consequent mixing. In Lake Titicaca, which lies at 3800 m above sea-level in the Andes, the winter mixing does promote a small peak in production but this is rapidly depressed and the main production peak occurs in the rainy season in December, presumably when the runoff increases nutrient levels significantly (Widmer *et al.*, 1975). Here the main mixing period coincides with the winter when temperatures are lower and conditions less suitable for production. The peak of production occurs in the spring/summer, even though stratification has set in, owing to the coincidence of the rains. Even so, the differences are relatively small, with the highest rates only being some twice that of the lowest, and zooplankton abundance can remain relatively constant. One other feature of the altitude of the lake is the severe surface inhibition caused during the most transparent conditions by the high levels of UV light experienced at these altitudes.

Since the mixing processes caused by the winds are linked to their influence on nutrient availability, it is curious that many large lakes do not show a seasonal response to other influences upon nutrients, such as inputs from rainfall and fluctuations in inflow volume. There are, however, some waters which are strongly governed by changes in the hydrological regime. In rivers, of course, this is of paramount importance. During the dry season current speeds tend to be low, providing more suitable conditions for phytoplankton growth, whilst with the more rapid flow at high water levels there is a much greater probability of the plankton being washed away. The sediment load carried by a river can also vary between high and low waters, and this can affect the optical properties of the water. In both the clear-water Rio Tapajoz and black-water Rio Negro of the Amazon, rates of production were three and ten times greater respectively between low water, when there was little water movement, and high water (Schmidt, 1976, 1982). Similarly in the Niger, the White Floods of July and August and the Black Floods of February to April both resulted in an almost tenfold dilution of chlorophyll *a* in the river water (Imevbore, 1970).

The reduced production is not only due to a lower standing crop of phytoplankton but often also to reduced light availability as a result of increased turbidity in the floodwaters. For example, in the Nile before the closure of the Aswan High Dam, the turbidity could be elevated from negligible levels to more than 300 mg l^{-1} over June and July as the waters rose, giving the water a greatly decreased transparency (Talling, 1976b). In the White Nile, during August to October when more than 80% of the annual flow occurs, secchi disc transparency can be reduced to less than 5 cm and primary production is undetectable at this time (Sinada and Kerin, 1984). Peak production occurs in November after the flood has subsided when rates of 3.9 g C m^{-3} h^{-1} have been recorded.

In most rivers this is the usual cycle of events with a substantial increase in suspended material coinciding with the earliest part of the rains or rising water levels and clearing towards the height of the flood. In the Sokoto in northern Nigeria and also the Niger, the period of maximum transparency was found to coincide with the end of the rains, when the river was still high but no more silt was being washed off the land, whilst the phytoplankton density itself had not begun to increase (Holden and Green, 1960).

A related aspect of the productivity of rivers is the supply of organic material brought downstream, upon which many components of the riverine communities depend. As we have seen, much of this suspended material is allochthonous and is washed or falls into the river from the land, although a proportion will be autochthonous and may have originated from phytoplankton or marginal swamps if these are present (Fig. 4.4). Where the seasonal effect on suspended organic material has been examined it appears to be similar to that of the total suspended load. In the River Taia, Sierra Leone, for example, a sharp peak in oxidizable organic material appeared during the early rains as the river rose (Fig. 6.1) but this rapidly subsided as the river level reached its height (Wright, 1982). The peak was three times the dry-season level and much of the organic material was apparently derived from water coming down seasonal streams carrying with it material which had accumulated on the land during the dry season. The extent to which this seasonal spike of organic material is retained within the river system, rather than being washed into the sea, remains to be seen. It coincides with the most turbulent periods of the river's cycle when rates of sedimentation will be at their lowest. Up to 50% has been estimated as being broken down in any particular stretch of the Amazon due to bacterial activity and therefore enters the biotic community in this way (Grobbelaar, 1983). However, when, where, and how much becomes available to benthic invertebrates and fishes is not clear.

Lakes which are intimately associated with rivers such as reservoirs and the varzea lakes of the Amazon are considerably affected by the hydrological regime although stratification may also play a part in the seasonal cycle. Impoundment of a river generally encourages phytoplankton growth partly due to more stable conditions with enhanced nutrient concentrations and partly due to increased light availability following sedimentation of the river's suspended load. Where the river fluctuates seasonally, however, the increased quantity of suspended solids often associated with the floods causes a deterioration of the

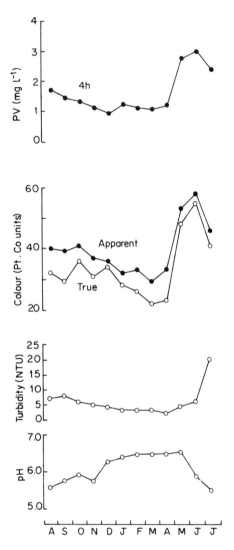

Fig. 6.1 Seasonal variation in oxidizable organic material as indicated by the permanganate value (PV) in relation to water colour, turbidity, and pH in the River Taia, Sierra Leone (Wright, 1982)

light climate, and the accelerated water flow can produce an increased flushing rate of the reservoir. At this time it becomes an 'expanded river' like the Volta Lake where consistent depression of phytoplankton occurs during the floods (Biswas, 1975). This effect is particularly pronounced in small reservoirs with relatively large inflows where the turnover time during the rains may only be a matter of days and presents virtually no opportunity for phytoplankton to become established. In Malawi, Moss (1969) observed ponds in which so much inorganic turbid material was washed in during the rains that effective light penetration was reduced to as little as 4 mm and the dominant green algae and euglenoid phytoplankton were restricted to a surface skin. In large man-made lakes it can take some time for the turbid floodwaters to penetrate the whole length of the water body. It used to take 1 month for the Nile flood to reach Aswan from Khartoum but now, with the formation of Lake Nubia/Nasser, it can take between 5 and 12 months (Entz, 1976). The arrival of the flood at the southern end can have a dramatic effect upon transparency with secchi disc readings diminishing from 70–140 cm to 20–30 cm or less in a matter of hours. As the floodwaters proceed slowly down the lake sedimentation takes place, so that by the time the dam itself is approached the water is relatively transparent with the only turbidity being due to the phytoplankton.

Many reservoirs and man-made lakes do become stratified in the non-flood period except in those cases where the flushing rate remains high throughout the year. The stratification is usually broken down by the flood but can also be dependent upon the pattern of water abstraction since manipulation of the water flow can create its own quasi-seasonal effects. For example, phytoplankton productivity in the Gebel Aulia reservoir on the White Nile is enhanced following full closure of the dam to catch the floods in August and substantial pulses of algae can be detected downstream in the Nile itself at Khartoum in the following months (Prowse and Talling, 1958). In shallower reservoirs other factors may also influence stratification, for example in Subang reservoir in Malaysia destratification occurred spasmodically following heavy rainstorms in most months (Arumugam and Furtado, 1980) as it also did in Queensland (Finlayson et al., 1980). In general, however, although stratification may take place in these impoundments, the production cycle tends to be similar to that of the river with lower rates during the rains or floods and more elevated levels during the falling or low-water period.

In some ways the incursion of Amazon river water into the natural varzea lakes of the river's floodplain has similar effects to those observed in reservoirs. Differences stem from the fact that communications between the lakes and the river are rather more indirect, usually by means of channels or *furos* several kilometres in length. As the river rises from November onwards water passes down these channels which, in the case of the middle Amazon (River Solimoes), is nutrient-rich white water. The advent of this water with a high sediment content initially suppresses phytoplankton blooms, but in the still conditions of the lakes the particles sediment out. However, by this time, considerable amounts of water have flooded into the varzea raising the level by up to 10 m. Even though the inflowing waters are relatively nutrient rich, the overall effect is one of dilution of phytoplankton and this is the low point of production during the year. An additional feature is the direct effect of rainfall and the 'shore factor', the runoff into the lake from the surrounding land which also dilutes the ionic concentration and nutrient content of the water itself. Once the floods are past their peak in May, the rate of primary production builds up and, as the water subsides, much of the flocculent material, both inorganic and detritus which sedimented out initially, is churned up in the shallow waters. This makes the conditions rather turbid, but by this time the water is too shallow for the long-term stratification that could be found earlier when the water level was higher. The mixing of the whole water volume when so little water remains puts the productive zone into intimate contact with the mineralization process and dense phytoplankton blooms are the result. Over the annual cycle, therefore, the varzea lakes move from distinctly oligotrophic conditions at high-water level to eutrophic or hypertrophic conditions at low water. This prompted Rai (1978) to suggest the term 'allotrophic' for such systems which oscillate between the extremes of production under the influence of the hydrological regime.

As the water level falls, in fact, organically enriched water leaves the varzea for the main river. Part of this enrichment is due to the phytoplankton, but the major portion is probably due to the floating meadows which grow luxuriantly along the decanted white-water lakes, although not on those fed by black-water rivers where the chemical composition is too extreme. These meadows of the grasses *Paspalum* and *Echinocloa* are effectively annual and they may be stranded as the water level drops; also their decomposition in the inundated zone may provide nutrients for the following year. However, during the productive time of year much organic material in the form of dead leaves and other parts will be released into the water to add to the detritus and the browsing of large animals such as the manatee, *Trichechus inunguis*, now unfortunately much reduced in numbers, can convert a proportion of the living plant material into a more finely divided form which would contribute significantly to the fertility of the water. Much of this material is, therefore, ultimately siphoned out into the main river in the form of suspended organic material to enrich its potential productivity. The seasonal relationship between the Amazon and its floodplain lakes is, therefore, a two-way one with the river replenishing the inorganic nutrients of the lakes directly and the lakes making a substantial contribution to the organic detritus in the rivers.

6.3 SEASONAL VERSUS DAILY CHANGES IN PRODUCTION

Many tropical waters do, therefore, show a distinct seasonal cycle of production usually in response to either a periodic mixing cycle often associated with the rainy season or to changes in the inflow. By no means all waters show such periodicity, however, particularly amongst the lakes (Melack, 1979). In Lake George, for example, the seasonal variation in phytoplankton and zooplankton abundance rarely exceeds a twofold difference over the year and no regular peaks or troughs are to be found (Ganf and Viner, 1973). In fact, diel variation in production rates exceeds annual. This stability is not interrupted by seasonally increased nutrient-bearing inflows into the lake (Viner and Smith, 1973) and in this it is similar to the nearby, much larger, Lake Victoria. The lake is, in fact, dominated not by an annual rhythm but by a daily one. As the day proceeds the waters become warmed and stratified which causes a redistribution of the phytoplankton (Section 5.2.3) and an asymmetrical production (Fig. 5.6), with the peak occurring in the morning and declining rates in the afternoon probably as nutrients are used which cannot immediately be replenished. There are windy periods during most evenings which break down the stratification built up during the day and cause the water column to become homogeneous once more and ready for the cycle to repeat itself the following day. The limits to this system are determined by the rate of nutrient recycling rather than new nutrient input.

As we have seen (Section 5.3), the amounts of nitrogen and phosphorus processed by the zooplankton and fish alone exceed the annual input and it seems probable that these herbivores can cause a complete turnover of the phytoplankton six to seven times a year (Ganf and Viner, 1973). The system could only remain stable, however, as long as frequent mixing occurs and to this extent is dependent upon the constant daily wind pattern and the shallow depth. In these respects it is similar to the varzea lakes at low-water level and both systems are highly productive (Section 5.2.3). Occasionally prolonged calm periods do occur which produce a persistent thermocline and the consequent deoxygenation of the hypolimnion which frequently attends this. Eventual breakdown may be associated with fish kills. The shallow depth of Lake George means that prolonged strong winds could stir up the anoxic bottom sediments which again could severely affect the whole system periodically. Therefore, although the lake shows remarkable annual stability, this stability is very fragile and depends upon a combination of biological and climatic factors which are not necessarily widely found in the tropics. Other shallow lakes also lack a seasonal rhythm although none are so well understood as Lake George. Lake Naivasha in Kenya also appears to be seasonally constant although in this case it may be the dense surrounding swamps which buffer the lake water from change (Melack, 1979). One other feature of the 24-hour cycle found in Lake George is that this is less than the generation time of the producers, therefore, since the maximum environmental changes are to be experienced within a day, individual algal species must be able to tolerate the widest range of conditions (Ganf and Horne, 1975). There can be no sequential change in species composition as the community adapts to progressive seasonal changes of the type found in temperate or seasonal tropical lakes. Consequently the species composition of the Lake George phytoplankton remains relatively constant throughout the year and consists of tolerant species in contrast to Lake Victoria, for example, where sequential changes are found (Talling, 1966).

In addition to the alternative seasonally variable and seasonally constant patterns of production found in tropical waters, a third state can also be identified (Melack, 1979). Some shallow soda lakes, including Lakes Nakuru and Elmentitia in Kenya, show unchanging high levels of production for several years, in what is virtually a single-species culture, often of the blue-green, *Spirulina*. Within a short space of time a sharp decline can occur corresponding with the decline of blue-greens and replacement with green algae. This may persist for a year or two but the original situation usually becomes re-established. These, therefore, constitute rather irregular changes. They are often associated with increases in salinity but the underlying mechanisms are unknown (Melack and Kilham, 1974; Melack, 1981; Vareschi, 1982).

Although seasonal changes are found in many tropical waters, the extent of the differences is rather variable. In some cases the changes, whilst regular and predictable, are rather muted. In Lake Victoria and Lake Titicaca, for example, extremes only differ by a factor of two to three. Extremes can vary by as much as twenty times in Lake Lanao, but on average oscillations are much more damped down than in temperate systems (Lewis, 1974). When nutrients are available in tropical lakes the more consistently high light intensities and temperatures generally ensure that the rate of production will be high. In temperate regions, by contrast, the peak of light availability and period of highest temperatures often coincide with the reduced nutrient circulation caused by stratification of the summer period which consequently severely inhibits potential productivity. It is these differences in the phasing of the seasonal cycle which makes tropical lakes several times more productive than temperature equivalents (Section 5.2.3).

Seasonal differences tend to be much more pronounced in systems mediated by rivers so that, for example, production rates in the Amazon varzea lakes, or in the more sheltered parts of the river itself, can regularly show tenfold differences. Within rivers, the nutrient distribution is rarely affected by stratification or other seasonal phenomena and the mechanisms are not too dissimilar between temperate and tropical rivers. Thus, although there are differences in temperature and light regimes, these do not act so out of phase with nutrient availability in temperate rivers as in temperate lakes. The productivity of temperate and tropical rivers appears to be of a similar order of magnitude (Rai and Hill, 1980) and the not dissimilar seasonal events in these may provide some explanation for this.

6.4 ANIMAL PERIODICITY

6.4.1 Seasonal Factors in Flowing Waters

With the arrival of the rains or a flood from a distant part of the catchment area, there is a gradual increase in the current speed of rivers and the water level begins to rise, giving a greater volume of water surrounding the animals. As the seasonal streams and the smaller tributaries begin to fill up so silt, organic, and inorganic material accumulated in the soil during the dry season are washed into these streams and into the main channel of the river. In small forest streams this accumulated material from the land can appear as a distinctive chemical 'spike' in the water following the first heavy rain (Payne, 1975). The animals in the river become surrounded by more turbulent, cloudy conditions which are detectably different chemically from those which they experienced in the dry low-water season. Since the rains often begin with a series of storms, the transition can be quite abrupt as the first violent rainstorm arrives, but progress is intermittent and depends upon the frequency of subsequent storms until the more sustained rains arrive. Amongst this sequence are a variety of potential information sources which animals can use as cues to co-ordinate behavioural and physiological patterns. It is not only the aquatic climate of the animals which alters, however, for those such as the zooplankton their whole way of life becomes interrupted as they are physically dispersed. A similar effect can be experienced by the benthic animals inhabiting turbulent streams. For example, the number of invertebrates from a pool in a forest stream in Sierra Leone following the scouring effects of the rains constituted only 5% of the total present in the dry season (Payne, 1975), and changes of this order observed in small Andean streams have been held to indicate that such spates are the main factors regulating the densities of benthic invertebrates, particularly the smaller ones such as mites and oligochaetes (Turcotte and Harper, 1982a). The rising river waters do, in fact, not only provide information but also alter the ultimate factors including a place to live, as well as the food supply.

6.4.2 Feeding and Food Availability

Early in the rains an increase in fine suspended organic material washed from the land is available for those animals able to use this (Section 6.2); in addition, the early rains often produce a large increase in terrestrial insect numbers, a proportion of which will fall into the water to enhance the aquatic food resources. Moreover, the faster current speed will increase the rate of delivery of food particles although it may also render them more difficult to catch. In rivers with large floodplains, however, the major increase in resources occurs when the river spills over its banks to cover large areas of land. Mobile animals such as fish can then move into this increased living space and exploit the newly available resources in the form of decaying vegetable material, seeds, fruit, and insects. They will often gear their reproductive cycle to this event. Rivers without floodplains will not show such a dramatic increase in resources and therefore the consequences may not be quite so great. Once the river water begins to fall these processes will go into reverse as the river level decreases and the current speed slows down until, during the dry season, the river area is so contracted that flow between stretches may be negligible or even cease in some cases.

Amongst the animals influenced by these changes are the fishes. As the river contracts after the flood there is a retreat of large numbers of fish, swollen by juveniles recently spawned, into a smaller area with a progressively dwindling supply of resources. This can be the chief period of mortality except amongst the predators when it can be the prime feeding period. The effect is particularly significant for rivers which virtually dry up into a series of pools during the dry season, such as the Rupununi River in the savanna region of Guyana, where food consumption of the survivors was much reduced (Lowe-McConnell, 1964).

During this period many species cease to feed and several South American species have modified alimentary tracts which act as respiratory organs at this time (Section 4.8). The transition from a time of plenty in the floods to one of scarcity as the water level falls may be sufficient to produce a growth check which appears on the scales in some species and which may be of value in estimating age and growth of the fish, owing to its seasonal nature (Daget, 1952; Lowe-McConnell, 1964). The situation may not be quite so severe in forest streams where there is a perpetual rain of debris from the overhanging forest, or in equatorial rivers with two flood periods per year where the contraction of the river is not quite so extreme between the flood periods. Even so, as observed in a small forest stream in Sierra Leone, the food consumption of the fish population may still be reduced during the dry season as indicated by the higher proportion with empty stomachs (Payne, 1975). This may reflect the seasonality of the terrestrial system in terms of a decreased litter fall from the trees during the dry season as well as a decreased rate of delivery as the water flow itself is reduced. Similarly, in a small forest stream in Panama numbers of both terrestrial and aquatic insects taken by a number of the fish species were found to be almost always greater in the wet than the dry season (Zaret and Rand, 1971). However, in these situations the reduction is only relatively small; for example in the Sierra Leone stream 75% of the fish had food in their stomachs even in the dry season and the situation is less marked than for some of the savanna rivers.

There is little evidence on the variation of food intake with season in lakes. Some species, particularly those which brood the young in the mouth, may cease feeding during the reproductive period which itself may be seasonal but precise data are scarce. The tiger-fish, *Hydrocynus vittatus*, was shown to consume more fish during the summer in Lake McIlwaine, Zimbabwe, than later in the year (Munro, 1967) which could have been due to the seasonally higher temperature in this high-veld location.

Food intake of fishes, however, may commonly vary with time over a 24-hour cycle in both lakes and rivers and in a few instances this has been demonstrated. For example, daily consumption and digestion cycles have been shown for the cichlids, *Oreochromis niloticus* and *Haplochromis nigripinnis* in Lake George (Moriarty, 1973) and *Oreochromis niloticus* in Lake Turkana (Harbott, 1976), which begin at sunrise and end at sunset. Conversely, many fish are known to be nocturnal in their habits and therefore tend to feed during the hours of darkness. The division of the community into day- and night-time feeders is one way in which the resources have become shared out between different species. No doubt some species will feed throughout the 24 hours with perhaps peaks at dawn and dusk as happens amongst temperate fishes but few species from the tropics are known in this detail.

One further feature which can vary with time in both lakes and rivers is the quality of the diet as well as the quantity. Often this reflects the effects of time on the food source itself rather than upon the feeder, but again most information is available for the fishes. In rivers with wide amplitude fluctuations in level and discharge, especially where extensive flooding occurs, the cycle is one of abundance in the wet season and scarcity in the dry-season pools. This is often linked to the spawning season since fish like the enormous characin, the tambaqui, *Colossoma macropomum*, spawn in the early floods and then move into the inundated forests where their large molar-like teeth can be used to crack hard nuts and seeds as the fish exploit the fruits as well as insects falling into water from the forest trees (Goulding, 1981). As the waters recede food of this type is no longer available in such quantities and the fish must either stop feeding or survive on what they can get, supplemented by energy from the large fat stores laid down in the body cavity during the prime feeding season. This trend from relatively specialized feeding to taking whatever is available as the river changes from flood to low water has been noted in the Rapununi River, Guyana (Lowe-McConnell, 1964) and in the Zaïre River (Matthes, 1964). In clear-water forest streams of the Amazon (Knoppell, 1972), central America (Zaret and Rand, 1971), West Africa (Payne, 1975), and Sri Lanka (Kortmulder, 1982a,b), although there may be some fluctuations in availability, the nature of the food, mainly terrestrial and aquatic insects with allochthonous detritus, remains relatively constant. This is presumably because the fish are always close to their food source, the evergreen forest canopy.

In tropical lakes seasonal changes in diet are not usually so marked as in some rivers, although examples of annual variation are found. In Lake McIlwaine, Zimbabwe, the catfish *Clarias gariepinnis* takes a high proportion of chironomid larvae and pupae in the summer, when these are available, but takes proportionally more zooplankton in the spring and cool season (Munro, 1967). Other changes, however, can occur over a shorter time period. In the equatorial Lake Victoria cyclical changes in the proportion of chironomids and the ephemeropteran, *Povilla adusta*, occur in the diet of the mormyrid fish *Mormyrus kannume* but the periodicity of this was found to be determined by the lunar cycle (Corbet, 1961). Many tropical aquatic insect larvae synchronize their emergence as adults using the moon as a proximate factor, and during this process they are more vulnerable to fish predation. Some coincide with the new moon itself whilst others tend to appear at a predictable number of days after this event. Consequently the proportion of certain insect larvae in the diet of fish such as *M. kannume* and *C. gariepinnis* can be a function of the phase of the moon as well as season.

The moon may also affect the diet and distribution of zooplankton feeders. In Gatun Lake, Panama, the single most important predator of the major zooplankter, the copepod *Diaptomus gatunensis*, is an atherinid silver side, *Melaniris*. However, because the zooplankton approach the surface at night as part of their diurnal migration and are close to the bottom during the day (Section 4.6.5), this appears to offer some protection to the copepods since they do not figure to any great extent in the stomach contents of this fish (Zaret and Suffern, 1976). Only on moonlit nights do the silver sides take appreciable quantities. This, however, only reflects the increased visibility under these conditions and not an increase in abundance of the type associated with the emergence of aquatic insect larvae.

The diurnal and seasonal changes of the zooplankton can, however, influence the distribution of their predators. In Lake Kariba the daily migration of the cladoceran, *Bosmina longirostris*, as it approaches the surface at dawn and dusk, is almost directly paralleled by its predator, the freshwater sardine *Limnothrissa miodon* (Begg, 1976). In this way the zooplankton feeder keeps in touch with its prey. Moreover, when the lake is isothermal, both groups of animals can migrate more than 60 m in depth during this cycle, whereas during the season of stratification when the hypolimnion is anaerobic, all movements are confined to the upper few metres which constitutes the epilimnion. Begg (1976) considered that the sardine was not actively following the cladoceran as such, but was actually responding in the same way to the same proximate stimulus of light which brought the two populations into constant contact. The movement of another copepod, *Mesocyclops leukartii*, was identical to the sardine but this species did not appear to be taken by the fish. This cyclopoid copepod is itself a predator of the smaller rotifers which remain perpetually close to the surface, but its journey up to the surface at dusk allows it to prey upon these at a time when it is least vulnerable itself.

Generally, therefore, seasonal changes in feeding patterns tend to be rather muted in lakes although periodic variation in types of food taken and locality of feeding does occur. The more equitable supply of food throughout the year has allowed considerable specialization in some of the larger lakes as, for example, has occurred amongst the mbuna cichlid species flock which feed in and around rocks in Lake Malawi. Even here, however, amongst a group of species renowned for their fine degree of feeding specialization switches in diet are known. For example, two of the specialized algal scrapers will change to feeding on zooplankton, phytoplankton, detritus, and eggs when these are available and really only resort to their specialized food source when resources are limited (McKaye and Marsh, 1983). Even such a highly specialized type as the recently discovered cleaner fish which removes external parasites from the catfish, *Bagrus*, will switch to taking the eggs of the same species during the breeding season. These situations emphasize the maxim attributable to McKaye that 'fish are jacks of all trades and masters of one' (vide McKaye and Marsh, 1983). Many fish at least are capable of taking a wide range of foods, and if one type becomes available seasonally then they will take this whilst reserving their more specialized habit for the appropriate occasion.

6.4.3 Reproduction

(a) Rivers

Since the hydrological regime of many rivers controls the quality and abundance of food, as we have seen, it is not surprising that many riverine animals tend to have a seasonal reproductive cycle which ensures that their young are produced under conditions within which they can survive and at a time when their food is most plentiful. The changing current speed and water levels can also impose a seasonal cycle upon some plants of which a particularly good example is that of the small, rock-encrusting Podostemacea mentioned earlier (Section 4.9.4).

Both zooplankton and fish can show marked seasonal reproductive cycles. Zooplankton in the Sokoto River, for example, produce their nauplius larvae from March to July during the dry season and early rains (Holden and Green, 1960). None could be found from August to January at the height of the floods nor as the floods receded. The timing of this cycle will enable the larvae and young stages to profit from phytoplankton growth during the dry season and from the early wash-out of detritus normally associated with the first rains.

Many fish also begin spawning in the early rains, although by no means all. A proportion of fish spawn during the dry season, particularly those such as cichlids, which show some degree of parental care. Such fishes often produce relatively few large eggs which contain sufficient yolk to enable the young to pass through the early stages of development without feeding. They are independent of external food availability during the earliest phases of development and therefore can be spawned in the dry season. Often some form of nest is constructed and amongst *Tilapia*, for example, large numbers of their saucer-shaped nests may be seen along the sandy river edges during the dry season (Fig. 6.2). Although there may be a distinct reproductive period, individual fish may spawn more than once, and consequently their ovaries can contain batches of eggs at different stages of development which will be laid in sequence.

Fig. 6.2 Nests of *Tilapia* along the edge of a river during the dry season

A large number of species do, however, spawn in the early rains and often this can be associated with a migratory phase, particularly amongst characins, catfishes, or some cyprinoids, in order that the eggs can be laid in the most beneficial place for the young and to co-ordinate the meeting of the sexes. The fish can move upstream or downstream depending upon the circumstances, often in order to bring them to some suitable flooded area with a rich food supply which the young can exploit after spawning and hatching are completed. These fish often produce a very large number of small eggs. No parental care is exercised so that mortality amongst the young is high which necessitates the large numbers of eggs to offset the very high mortality rates. The small eggs also contain little yolk, therefore the young must have rapid access to food.

The migratory response itself must result from the interaction of proximate factors, such as changes in current speed or water level, with internal hormonally co-ordinated rhythms as the floods begin. However, not all migrations take place in the rains or floods. Two families of catfishes, the Pimelodidae and the Doradidae, show upstream low-water migrations in the Rio Madeira section of the Amazon system (Goulding, 1981). Normally catfishes travel close to the bottom, but in that particular river there are two sets of cataracts which the fishes have to battle up in a fashion which makes them obvious and allows details of their movements to be recorded. The reason for the low-water movement is unclear, although it could be for the purposes of spawning far upstream and that the timing of the migration is purely a function of the distance which must be travelled.

Some species can indeed travel long distances. In the Niger, Daget (1957) followed a shoal of *Alestes leuciscus* for 125 km during their upstream migration and deduced that some must move between 215 and 400 km. The movement of this particular species also seemed to depend upon the phase of the moon and the whole shoal was accompanied by a series of predators which used them as prey.

The annual migration may simply be a movement to the spawning grounds but in other cases a more complex strategy may be involved. In the Rio Madeira branch of the Amazon the major characin types such as *Colossoma* and *Brycon* show two migrations a year. The first coincides with the early floods when the shoals move down the tributaries, which are largely clear waters, into the main turbid river following which they move into the flooded forests after spawning at some point. The larger fish move once more into tributaries probably because the inundation of the forests lasts longer here. Following this, the fish migrate back upstream and re-enter the tributaries but usually one further up the system than the point from which they started out. It has been suggested that the main nursery grounds for the young are in the lower region of the main Rio Madeira as this, with its rich floating meadows, is a much richer source of food for the young than the less productive clear-water tributaries (Goulding, 1981). Moreover, since there are no young in the tributaries to be recruited into the adult populations, losses are replaced by the migration of fish from lower down in the second phase of the cycle. Migration is not therefore a simple annual event but a movement from one tributary to another further upstream over the course of several years.

The stimuli which trigger spawning migrations are not necessarily those which initiate gonad maturation or those which stimulate the act of spawning. The production of gonads requires a considerable investment of energy and may take some time. In *Barbus liberiensis*, for example, gonad maturation begins early in the dry season at a time when food availability is showing some reduction. In the males the initiation of testes development is associated with a loss in body weight which is only recouped slowly and the final burst to complete maturation coincides with the increased availability of food in the early rains (Fig. 6.3). The fall in weight seen in July is the result of spent males which have completed spawning appearing in the population. The females manage to maintain or even increase their weight slightly, but as development of the ovaries proceeds the weight of the somatic part of the body, that is, the weight without the ovaries, drops as there is evidently a shift of energy and resources from the body to this large organ. Once again, however, the most rapid phase of gonad growth, from March to May, coincides with the start of the rains when food abundance increases (Payne, 1975). The relative difference in gonad size between male and female is also evident in Fig. 6.3 where the ripe ovaries constitute some 9% of the body weight compared to 2.3% for the testes. *Barbus liberiensis* is typical of those types producing large numbers of small eggs which are shed in one single spawning in the early rains and it is evident that this represents a considerable annual loss of energy for the individual. This loss will be made up by feeding in the rains and the plentiful food supply often enables the fish to lay down a considerable fat store around the gut. This will enable the fish to get through the lean period of the dry season and also provide energy for the initiation of gonad development.

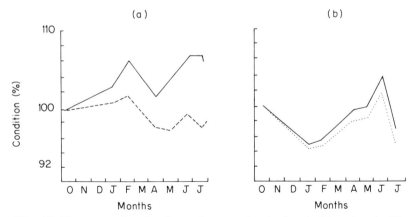

Fig. 6.3 The seasonal cycle of gonad maturation in female (a) and male (b) *Barbus liberiensis* as indicated by changes in total weight condition and somatic weight condition (i.e. excluding gonad weight) in relation to the post-spawning state in October. (*Reproduced by permission of the Zoological Society of London from Payne, 1975.*) Total condition ———, somatic condition – – – –

In the case of *Barbus liberiensis*, the start of gonad development precedes migration quite considerably. Gonad maturation is under endocrine control from the pituitary gland and the pituitary itself normally requires a trigger. In temperate fish this is often day length, but in the tropics variation in photoperiod is very small and whether this is sufficient remains to be seen. In any case, as with spawning, the timing of gonad maturation must generally coincide with a period of reasonable food supply or, alternatively, the animal must lay down so much fat over the feeding phase to sustain both itself and the whole of gonad development.

(b) Lakes

Seasonal changes in lakes are not quite so marked as in rivers and consequently seasonal cycles in reproduction may not be so apparent. No seasonal trends were found in zooplankton reproduction in Lake George (Burgis, 1971, 1974) nor even in the subtropical Lake Sibaya where nine generations per year were produced (Hart and Allanson, 1975). In the same locality, also, the benthic shrimp *Caridina nilotica* bred throughout the year. Amongst the fishes there are some types which do show reproductive seasons and some which do not. In the former category are a number of groups which inhabit lakes but retain a riverine reproductive biology. In Lake Victoria many of the non-cichlid fishes are of this kind, and maturing fish migrate up the inflowing rivers to spawn in the floodwater pools during the early rains (Whitehead, 1959). The rivers here, however, have two flood periods quite close together and the ascent of ripe fish on each occasion suggests that different individuals from the lake populations are involved with each peak. In a similar way, amongst populations which come to inhabit man-made lakes a proportion, often characins, cyprinoids, and catfishes, move out of the lake during the flood period to spawn in their traditional habitats. Some, however, such as the cichlids, have found their reproductive methods appropriate to the stable conditions of the lake and so remain. The artificially controlled seasonality of such lakes can present difficulties for the fishes. In Lake Kainji on the Niger, for example, the main period of drawdown coincided with the main period of tilapia reproduction and consequently many nests were left stranded. It therefore took several seasons for tilapia numbers to build up within the lake. There can also be considerable consequences for downstream fish populations. The Volta Lake, for example, can effectively impound the whole flood of the river and then allows it to pass through the hydroelectric turbines at a more or less constant flow rate all through the year. The flow in the river below the dam is therefore more or less constant (People and Rogoyska, 1969) with no information to tell the animals what season it is.

Even amongst fishes which are permanent residents in lakes, some do have a marked seasonality in spawning cycles. In Lake Victoria the phytoplankton-feeding tilapia *Oreochromis esculentus* has two breeding peaks in the north of the lake coinciding with the two phytoplankton peaks which follow isothermy, whilst the population at the south, where there is only one period of isothermy, has only one peak (Lowe-McConnell, 1956). On the other hand, *O. variabilis*, *O. niloticus*, and *O. leucostictus* which possess a similar diet are continuous spawners and have no peaks (Lowe-McConnell, 1979). In the same lake a whole assemblage of haplochromine cichlids found in Mwanza Gulf was divided into phytoplankton/detritus or zooplankton feeders which lived over mud and did show a clear reproductive peak, and those species living over sand which were largely insect or mollusc feeders which did not (Witte, 1981). The peak in planktivore reproduction occurred in May to June at the end of the rains. A peak in phytoplankton coincides with the rains following the breakdown of stratification together with the influence of runoff from the land. The rich feeding at this time may be used to promote gonad maturation and to sustain the mouth-brooding females through the reproductive season at the end of the rains during which they cannot feed. The food sources of the non-seasonal groups, insects and molluscs, may not show such discrete seasonal peaks.

Food availability is not the only factor which is associated with seasonal spawning cycles. In Lake Malawi are four species of closely related tilapias which have distinct breeding seasons although they breed in different places and also at different times of year. This may avoid competition for nesting sites and also reinforces the reproductive isolation of the species. A further illustration of the influence of biological pressures on the timing of breeding seasons is provided by the cichlids of Lake Jiloa in Nicaragua. In this lake there are two breeding peaks, one in the wet season and one in the dry season (McKaye, 1977). The wet-season peak is dominated by *Cichlasoma citrinellum* which has profited from an increase in its preferred food of insects at the beginning of the rains. This species aggressively excludes all other cichlids, apart from two small species, from the limited number of suitable nest sites available. Although these other cichlids have been shown to have the capacity to breed all the year round they in fact tend to breed in the dry season when the sites have been vacated by *C. citrinellum*. Even so, they themselves are segregated within this season by both depth and timing of reproduction.

The proximate factors which might trigger these reproductive cycles are even less obvious in lakes than in rivers. They could include some aspect of runoff from the rains, an increase in water level or possibly the slight changes in day length over the year may provide sufficient information to act as a cue. Socially oriented cycles may use interaction with other animals or lack of it. A comparison of the influence of sunlight, temperature, and rainfall upon reproduction of *Oreochromis leucosticta* in Lake Naivasha, Kenya, indicated that temperature and hours of maximum solar irradiance were the major factors initiating gonad developments, whilst the onset of rainfall appeared to stimulate peak breeding activity (Hyder, 1970). It could be that the physiological control mechanisms of tropical animals are more finely tuned than their temperate counterparts and will respond to smaller degrees of change. However, the seasonality of ing of *O. leucosticta* in Lake Naivasha is disputed by Siddiqui (1977) so that even in this lake the nature of the triggering mechanism is uncertain.

CHAPTER 7

DIVERSITY AND EVOLUTION

7.1 HOW DIVERSE ARE COMMUNITIES IN TROPICAL FRESH WATERS?

7.1.1 General Considerations

It has largely become axiomatic that tropical communities contain a high diversity of organisms; nevertheless it is not a point which should be taken for granted. In terrestrial systems there are many examples of communities or taxonomic groups which show increased numbers of species in equatorial as opposed to temperate areas; insects, birds, trees all become more profuse in the tropics. The most frequently used example is the mature tropical rain forest in which not only can the number of species, that is, the species richness, be high but also there is relatively little difference in the abundance of each species. The rain forest can therefore be said to have a high 'equitability' with resources being shared more or less equally between a large number of species and few, if any, being recognizably dominant.

Considering that diversity is regarded as one of the prime attributes of the tropics, it has received little analytical attention in aquatic systems and the approach has remained descriptive. Nevertheless, simply in terms of the number of recorded species, there are some spectacular examples of high diversity in tropical fresh waters. Amongst the Great Lakes of Africa, Lake Tanganyika contains 247 fish species, Lake Malawi 242, and Lake Victoria 238, with more almost certainly awaiting discovery, compared to the 172 fish species from their temperate equivalent, the North American Great Lakes (Lowe-McConnell, 1969; Greenwood, 1984a,b). A tropical region such as Thailand can contain 546 species of fish and Central America 456 compared to the European list of 192, whilst the major river systems such as the Zaïre and the Amazon show considerable species richness with 690 and 1000 being recorded to date and more still being discovered.

188

The fishes, therefore, evidently conform to the common pattern of high diversity in the tropics and, moreover, in many of the waters just referred to the proportion of the species which have evolved within the catchment areas, those referred to as being 'endemic', is also high. For example, in Lake Tanganyika 80% of all the fish species and 38% of genera are endemic and in the Zaïre basin 83% of the species are similarly endemic to the basin (Poll, 1959; Beadle, 1981). Some tropical situations, therefore, appear to promote the evolution of populations into new species, that is speciation, and thereby contribute to a high diversity, providing the rate of extinction is slower than the rate of accumulation of new species.

7.1.2 The Producers

The relatively high number of species found amongst the fishes of tropical lakes and rivers does not appear to extend to all other components of the communities, although studies which provide anything other than species lists are rather rare. The number of algal species found in Lake Victoria is around 100 and is of the same order as that from the North American Great Lakes (Talling, personal communication), whilst that from the considerably smaller and shallower Lake George is really rather modest (Ganf, 1974), particularly considering that small, shallow British lakes often contain 90–100 types of algae. A consideration of Lake Lanao in the Philippines and of the 'Sunda' lakes of Java, Sumatra, and Bali studied by Ruttner (1952), led Lewis (1978a) to conclude that, in South-east Asia, lakes ordinarily contain between 50 and 100 planktonic autotroph species and that half or slightly more of these could be found in a 1 ml sample at any time of year. The species richness of tropical lakes was suggested by Lewis to be rather less than for temperate lakes and, as a measure of this, it was demonstrated that the average number of genera found in temperate lakes was 36.3, whilst for tropical lakes in the Philippines, Indonesia, and Venezuela the average was 21. Some 79% of the genera were found in all tropical lakes, indicating rather uniform phytoplankton assemblages in the tropics. In addition, 45% of the genera were shared with the temperate waters, showing the relatively large proportion of cosmopolitan types present in tropical phytoplankton. There is commonly a low number of endemic species which is in sharp contrast to the situation amongst the fishes where large numbers of species, genera, or even families are restricted to the tropics (Section 4.8.1) and give the fauna very distinctive characteristics compared to temperate fish assemblages.

Naturally, when relying upon species lists alone as a measure of diversity, much depends upon the diligence of the observer in establishing the length of the list. Most of the planktonic studies used by Lewis (1978a), for example, were for ecological rather than taxonomic purposes so that large amounts of time were not spent looking for rarities; nevertheless, since they were all carried out with similar intent, they are probably comparable. An additional complicating factor is that open water or limnetic assemblages are often different from those of more marginal areas where periphyton and benthic algae may also be included, so that the relative time spent sampling marginal and pelagic areas can influence the impression of species richness. Even with these constraints in mind, however, there is no proven or apparent trend towards increased algal diversity in the phytoplankton as the equator is approached, although the data are rather patchy and the most complete information (Lewis, 1978a) does, in fact, suggest that phytoplankton in the tropics may actually contain fewer species than temperate communities although others would dispute this (Serruya and Pollingher, 1983).

In rivers the picture is even more fragmentary. Whilst phytoplankton in the varzea lakes of the Amazon (Uherkovich, 1976) are not particularly species rich and quite closely coincided with South-east Asian lakes (Lewis, 1978a) with regard to overall composition, the Rio Tapajoz and Rio Negro contained a great profusion of species. The clear-water Rio Tapajoz showed a considerable numerical dominance by a few cosmopolitan types such as the blue-greens *Anabaena hassalii*, *Microcystis aerouginosa*, and the diatom *Melosira granulata*. However, the greatest number of species was found to be amongst the desmids which numerically constituted less than 1% of the total numbers. A high species richness was also characteristic of the Rio Negro although total numbers of algal cells in this inhospitable environment were very low. The acidic waters of the Amazon, therefore, appear to be characterized by a large number of desmid species (Uherkovich, 1981).

The desmids, a family of green algae (Section 4.5.2(b)), similarly produce a species-rich flora in other South American waters such as those of Guiana where they constituted 78% of the 240 species collected from two lakes and a marshy area near Cayenne. Even here, only 47% could be regarded as typically tropical or subtropical taxa and only 29% were confined to the Americas and so, again, a large cosmopolitan element was present (Bourrelly and Couté, 1982). Forest rivers of West Africa, such as the Bandama in the Ivory Coast, also possess similar attributes in that up to 68% of the algal species from this river were chlorophytes with the majority being desmids, whilst the total algal numerical abundance itself was low (Iltis, 1982). The desmids appear to be particularly associated with neutral to acidic waters as they are in the temperate regions.

In comparison with lakes, the diversity of plankton in rivers is particularly difficult to assess as the algae in the water can have originated from so many different sources. As well as the truly planktonic types, the turbulence and abrasion will cause sedentary species from the aufwuchs on stones and marginal plants to be detached and suspended in the water. Pools, lakes, or marshes which drain into a river will also contribute their own characteristic species. The variability in flow patterns down a river itself will considerably influence the nature of the phytoplankton. For example, the Rio Tapajoz where, in the lower course, the river expands to form deep, very slow-flowing 'mouth bays' or 'river lakes' (Sioli, 1968) has phytoplankton blooms dominated numerically by typically lake-dwelling blue-green and diatom species which are superimposed upon a high underlying diversity contributed by the desmids, as mentioned above.

In headwater streams there is little, if any, algae suspended in the water and most are attached to the stones and rocks, although in a rather uneven distribution since some surfaces are more suitable than others. This makes assessment of species diversity even more difficult. In the Gombak forest stream in Malaysia, regular scrapings from rocks and submerged vegetation, together with examination of sediment and species cultured from the water at five stations down the stream, showed that along its course 194 species could be found of which between 58 and 98 occurred at only one station. The diatoms contributed the largest number of species and the assemblage was generally considered to be species rich (Bishop, 1973).

Position and nature of the river course together with additions from inflows, therefore, make the autotrophs of rivers a fairly heterogeneous assemblage. The information available on species diversity is rather scanty although the underlying impression is that, at least under some circumstances, the algal diversity might be relatively higher than that in tropical lakes. As in the lakes, there are a large number of cosmopolitan species which are shared both with temperate and tropical areas.

7.1.3 The Zooplankton

A consideration of the zooplankton from tropical waters indicates a similar pattern in the variety of species to that shown in lake phytoplankton. Using as an index collections from the most uniform habitat, the open-water limnetic region, in this case from a large number of standing waters in Sri Lanka, the total number of rotifer species found was eighty-nine (Fernando, 1980a), which is similar to the numbers found in some temperate lakes (Green, 1972). Indeed a systematic analysis of species number against latitude showed no significant correlation (Green, 1972). The number of types of rotifers in limnetic zooplankton, therefore, appears to be similar in tropical and temperate waters; only in the subpolar regions, such as Spitsbergen, is a reduced number apparent.

With regard to the Cladocera, the total numbers of species found in Sri Lanka was 58 and over the whole of Malaysia 64, Indonesia 65, and the Philippines between 49 and 65, which are rather less than for temperate regions such as Britain where up to 87 species can be found and Ontario, Canada, where 93 species occur (Fernando, 1980a). The low number of cladoceran species in South-east Asia and other parts of the tropics is partly due to the particular scarcity of the genus *Daphnia* which is so prominent in temperate areas (Fernando, 1980b). The paucity of cladoceran species is particularly evident amongst the true limnetic species where only 12 are found in South-east Asia compared to 30 in the Holarctic temperate region. A similar situation exists amongst the copepods where the cyclopoids (see Section 4.6.1; Fig. 4.6) show a diminished range of species compared to those found in temperate lakes. The same is probably true of the calanoids although their systematics needs to be clarified in some cases (Fernando, 1980a). There are commonly only two or three species of cyclopoid copepods per lake in the tropics and these often include the predatory *Mesocyclops leukarti* and the herbivorous *Thermocyclops* (Table 7.1).

As with the phytoplankton, therefore, the zooplankton assemblages are no more species rich than their temperate counterparts and may, indeed, be rather less diverse largely owing to a lack of cladocerans and, to some extent also, of copepods. Again, as with the phytoplankton, there is a significant proportion of cosmopolitan species and a large pantropical component of types such as *Keratella tropica* and *Brachionus* species (Fig. 4.6) amongst the rotifers, *Ceriodaphnia cornuta*, *Moina micrura*, and *Diaphanosoma* species (Fig. 4.6) from the cladocerans in addition to the two cyclopoid copepods mentioned above, which are widely distributed throughout the tropics. Although on a world-wide basis 48% of rotifer species do show a limited distribution, most of these are benthic or sedentary types, and many of the planktonic species are widespread within which a warm-water pantropical series can be identified (Ridder, 1981). The extensive distribution of so many species suggests that the zooplankton, like the phytoplankton, possess a very effective dispersal system, most probably wind and wildfowl, that renders the evolution of new species unlikely (Section 7.5).

One additional common feature of tropical zooplankton species is their small size with cyclopoid copepods and cladocerans generally less than 1.3 mm in length, which is rather less than many temperate species (Fernando, 1980a).

The diversity of the zooplankton does vary on a more local scale; for example, rivers and streams in Sri Lanka carried half the number of species of nearby ponds and reservoirs (Fernando, 1980a). Ponds tended to have the highest number of species, followed by small reservoirs, since mixtures of both littoral and limnetic species are to be found in these, unlike larger reservoirs (> 300 ha) where the limnetic component is predominant. However, although there were differences in the number of species found in the various types of water, the same common species tended to recur, indicating that many of the tropical zooplankters can live under a variety of conditions (Table 7.1).

Many of the locations sampled by Fernando (1980a,b) in his survey of Sri Lankan and South-east Asian zooplankton were reservoirs and, since these are relatively new waters, more species might possibly accumulate with time. However, a comparison with established natural lakes presents a similar picture. For example, the zooplankton of the Indonesian lake, Ranu Lamongan (Green *et al.*, 1976), contained 8 rotifer species whilst in East Africa, Lakes Naivasha and Oloiden, samples commonly contained up to 14 and 18 rotifer species respectively whilst in Winan Gulf, Lake Victoria, a similar score of 18 was found (Nogrady, 1983). Small temperate lakes in Britain commonly have 10–20 rotifer species in the zooplankton which is essentially similar to the tropical pattern, and most examples from other tropical lakes substantiate the South-east Asian survey. Zooplankton assemblages often rapidly become established and this again testifies to the efficiency of their dispersal strategies.

7.1.4 Benthic Organisms

Few indications of the diversity of benthic organisms for tropical waters have been obtained. In Lake George, the bottom fauna over most of the lake bed was

Table 7.1 Distribution of common zooplankton species from lakes throughout the tropics.

	Ranu Lamongan[a] (Indonesia)	Lake George[b] (Uganda)	Pakrama Samudra[c] (Sri Lanka)	Lake Izabal[d] (Guatemala)
Copepoda (cyclopoid)				
Mesocyclops leukarti	+	+	+	
Mesocyclops edax				+
Mesocyclops inversus				+
Thermocyclops hyalinus	+	+		
Thermocyclops crassus			+	
Macrocyclops albidus	+			
Eucyclops agiloides	+			
Copepoda (calanoid)				
Phyllodiaptomus annae			+	
Diaptomus dorsalis				+
Pseudodiaptomus culebrensis				+
Cladocera				
Bosmina longirostris				+
Ceriodaphnia cornuta	+	+	+	+
Moina dubia		+		
Moina micrura			+	+
Daphnia barbuta		+		
Alona sp.		+		
Eubosmina tubiccu				+
Diaphanosoma excisum			+	
Diaphanosoma fluviatile				+
Rotifera				
Keratella tropica	+	+	+	
Keratella cochlearis				+
Branchionus calcyflorus	+		+	
Branchionus caudatus	+	+	+	
Branchionus havannaensis				+
Branchionus falcatus	+		+	+
Lecane sp.	+		+	
Other species	3	14	10	6
Insecta				
Chaoborus	+	+	+	+
Ostracoda				
Cypria javana	+		+	

[a] Green *et al.*, 1976; [b] Burgiss *et al.*, 1973; [c] Fernando, 1980b; [d] Brinson and Nordlie, 1975.

poor in species, largely consisting of midge larvae *Chaoborus* and *Procladius*, which are only temporary residents of the lake bed (see Section 4.7.1), together with a further chironomid species, an ostracod and some oligochaetes (Burgis *et al.*, 1973; Darlington, 1976). No doubt the unstable and deoxygenated nature of the mud contributed to the lack of species diversity. Similar situations have been found in central areas of the varzea lakes on the Amazon (Reiss, 1977) and in the Bung Borapet reservoir in Thailand (Junk, 1975), although in this last case areas of firmer sediment also contained several species of bivalve.

Communities from tropical marine sediments, also, do not appear to be particularly diverse and showed no real trend from temperate to tropical seas in contrast to those animals which lived on the surface of the sediments, the epifauna, where a positive trend is evident (Thorson, 1957). The lack of diversity amongst animals living in the sediment is probably due to the physical and chemical instability of the environment. In Lake George, for example, the top 15 cm of sediment is regularly disturbed by turbulence.

In the littoral areas and amongst marginal vegetation, however, the diversity increases quite dramatically as a wide variety of insects appear including the nymphs of dragonflies, mayflies, and of caddis-flies (Section 4.7.1), water-beetles and a variety of water-boatmen and other aquatic bugs. Also to be found are a number of species of gastropod snails and, particularly characteristic of the tropics, freshwater crabs and prawns. This littoral area amongst the water-lilies and patches of submerged *Hydrilla* and *Ceratophyllum* proved to be by far the most varied in Bung Borapet, and the marginal beds of water-hyacinth, *Eichhornia*, also provided a basis for a very diverse community in Ranu Lamongan, Java (Green *et al.*, 1976). The submerged roots of the floating meadows of the Amazon also provide a refuge for one of the richest communities. Almost every freshwater taxonomic group can be represented and the animals extracted from a 50×50 cm quadrat of the 'carpet' contained a sufficient variety of species to provide taxonomists with work for many years (Sioli, 1968). The vegetation which provides shelter for these littoral and marginal communities is rarely itself very diverse and tends to be made predominantly of a few species. A significant proportion of these, such as the reed *Phragmites*, which is common everywhere except in the Amazon, the submerged plants *Ceratophyllum*, *Utricularia*, and *Hydrilla* and the grass *Ipomoea*, occurs widely in the tropics, as now do many of the floating types.

Rocky areas can also contain a rich variety of animal types, and in Lake Tanganyika for example, where the precipitous nature of the basin has provided extensive underwater rocky areas, a considerable number of gastropod snail species have evolved to an extent reminiscent of the fishes (Section 7.1.1).

In rocky headwater streams, the diversity of organisms also seems to be high. The Gombak stream in Malaysia showed indications of a wide variety of algae as mentioned earlier but the number of animal species was exceptional. The total number recorded in this stream was 344 which did not include an additional but uncertain number of beetles from the family Elmidae and the caddis-flies (Trichoptera). The latter were extremely varied and could include as many as a further 79 species. As with the zooplankton and phytoplankton referred to previously, the types of animals are basically very similar to those found in equivalent habitats from temperate regions and differ mainly at the specific or generic level. The majority of the species in the Gombak stream are insects, with only 15–25% of the types present being non-insect, most of the insects belonging to the various flying species which pass their juvenile phase in the water. Amongst the insects the majority, including the dragonflies (Odonata), beetles (Coleoptera), caddis-flies (Trichoptera), and mayflies (Ephemeroptera), showed a considerably greater species richness than would be found in their equivalent temperate habitat (Bishop, 1973); for example, amongst the mayflies the average number of species in a temperate stream is between 15 and 20 with a maximum around 22 (Ulfstrand, 1977), whilst at Gombak, 51 species were recorded. The insect diversity can sometimes be astonishing, as reflected by the collecting of as many as 79 species of caddis-fly nymphs from this small Malaysian stream. More than 100 species of chironomid midge larvae can also be found in the smaller forest streams of the Amazon (Fittkau, 1964), whilst up to 1000 species of this group occur throughout the whole Amazon basin (Illies, 1964).

From the information available the non-insect groups do not all seem to possess this same richness but there are some important exceptions. For example, amongst the freshwater prawns, which scarcely exist in the temperate zone, those of Dominican forest streams, which include six species from the Atyidae and six from the Palaemonidae as well as some freshwater crabs, are more diverse than the fishes which they may replace as the dominant animals (Fryer, 1977a).

The apparent trend towards high numbers of aquatic insect species in tropical streams may be a general reflection of increased insect diversity in tropical terrestrial systems, as observed within the rain forest. Many of the insects which spend their juvenile stages in streams lead their adult life on the land and in the air and it is this adult aerial and feeding phase, as seen in the dragonfly (Section 4.7.1), which must have a considerable bearing upon reproductive success. The species richness seen in the streams may just be a passive result of there being so many insect types in the surrounding trees and vegetation, which require water for their young. However, mayflies spend only a few hours as adults in the air and do not feed during this time (Section 4.7.1), which suggests that there are at least some attributes of the streams which promote and sustain diversity within the aquatic phase itself.

Not all aquatic insects show increased diversity in tropical streams. Even in Gombak, where so many groups appeared to be rich in species the numbers of stone-fly (Plecoptera) types, 19 in all, was rather less than the maximum of 40 which can be found in European streams but was closer to the average of 15–25 (Ulfstrand, 1967). In African streams stone-fly nymphs are generally rare (Chutter, 1972).

Amazon headwater forest streams in Peru may have a similar number of species to temperate streams with a maximum of only 15 non-insect and 102 insect species found in any single stream (Patrick, 1964) which is substantially lower than the overall total of 344 known from Gombak. More specifically, the Peruvian streams contained only 33 Diptera species and 17 caddis-flies which is some three times less than the Malaysian water. The Peruvian streams were, however, at 600 m—an altitude, which at least in the case of woodlands and other terrestrial systems, can restrict the number of species contributing to a community.

Even at lower altitudes in the wet tropics, considerably more restricted species complements occur. In the mountain streams of St. Vincent in the West Indies, no more than 41 species were found with only 4 mayfly and 8 caddis-fly (Harrison and Rankin, 1975). Similarly, the Arima River of Trinidad yielded only 38 species (Hynes, 1971). Harrison and Rankin (1975) thought that some of the streams, at least, may have suffered from the effects of deforestation. There is also the possibility that, on these islands, colonization by new species is an uncommon event and the benthic invertebrate fauna has been unable to accumulate to its full extent. However, a forest stream at Pawmpawm in Ghana only yielded 31 species (Hynes, 1975) which, again, is in contrast to the Gombak stream in Malaysia and cannot be attributed to the 'island effect'.

The analysis of the fauna from the Gombak stream has given some indication as to how diversity varies within the system. Generally, insect diversity decreases and non-insects become more numerous downstream. At the upstream forest sites, characterized by overhanging trees, a boulder-strewn bed, and submerged root entanglements, the highest number of species, 185, was found and there was a general reduction in number away from the forest edge as the stream opened out into more lowland areas, becoming wider and deeper with the more regular flow characteristic of the potamon zone (Section 4.7.5). The deposition of sediment in this region was apparently associated with a reduction in the variety of organisms, whilst those which were present became numerically abundant owing to increased availability of detritus. This was particularly evident in the lowermost station of the river which was below a sewage inflow. A considerable reduction in the number of species occurred here but the abundance of the dominant types, mainly oligochaetes and chironomids, increased significantly (Bishop, 1973).

The reduced number of insect species downstream was much less evident in the West Indian streams (Hynes, 1971; Harrison and Rankin, 1975), whilst, down a major river system, the Bandama of the Ivory Coast, the number of benthic species increased (Gibson and Statzner, 1985). Moreover, the types characteristic of the upper reaches, such as the caseless caddis, *Hydropsyche* and *Philopotamnion*, and the black-fly, *Simulium damnosum*, persisted throughout with additional species appearing downstream. The absence of major discontinuities of distribution was attributed to the modest gradients, the lack of a well-defined source, and the large numbers of seasonal tributaries which need recolonization at the beginning of each rainy season.

7.1.5 Some are Less Diverse than Others

To return now to the original question, the existing evidence suggests that not all communities in tropical fresh waters show the high diversities usually attributed to tropical systems. The phytoplankton, zooplankton, and possibly some benthic communities show species contents which are not markedly different from those found in equivalent temperate environments and which may, indeed, be somewhat lower. In the case of plankton, it might be considered that the relatively modest diversity could be due to the uniformity of the open-water environment, but this is not necessarily the result of a planktonic existence in tropical waters as shown by the high numbers of species which are found in the central regions of tropical oceans (Raymont, 1980). On the other hand, those situations in which the diversity does tend to be high, such as amongst marginal vegetation and rocky areas in lakes or rivers, do possess a physical, three-dimensional complexity which complements and sustains perhaps a disproportionately varied biological diversity. In some of these cases the greatest proportion can be made up of insects, which are also very diverse in neighbouring terrestrial systems. How important the aerial adult phase of those which spend their developmental stages in the water is in contributing to the aquatic diversity is an area for speculation.

However great the number of algal or invertebrate species proves to be, the types which are found in tropical waters are essentially recognizable as being common also to temperate areas (Section 4.5.2) and the main differences occur at the generic or even specific level. There is often a significant proportion of cosmopolitan, or at least pantropical types amongst the phytoplankton, marginal, and submerged plants and also amongst the invertebrate animals. This is much less true of the fishes where the degree of difference extends to the level of family or even order and this makes tropical fish assemblages distinctly different from those of the temperate areas. Moreover, it is evident that, under most circumstances, they are considerably more varied with regard to the number of species present. The role of fish in adding to the richness and character of the tropical freshwater communities may depend in part upon the fishes being able to exploit to the full the three-dimensional possibilities of the water body, and also to the fact that they contribute to every trophic level occupied by heterotrophs whether based upon the consumption of live or dead material (Figs. 4.2, 4.3, 4.4; Section 4.8.1).

An examination of freshwater communities demonstrates that high diversity in the tropics cannot be taken for granted and that each case must be considered individually. This presents difficulties in some respects in that many tropical groups are poorly known taxonomically and will require considerable work to bring our state of knowledge up to that available for temperate situations. This is compounded by the current unfashionable status of taxonomy, even though the first steps in any survey should be to give curiosity a free rein and 'see what's there'. Diversity itself is also a complex property which in some ways represents the ultimate step in biological organization and its role in community function and regulation is only just beginning to be understood.

7.2 THE MEANING OF DIVERSITY

In the previous section the diversity of tropical freshwater systems was assessed in terms of how many species had been found in a particular case. If many species are present then it suggests that there are a large number of niches or ways of exploiting the system available.

That is not to say that there can be no overlap between niches. A species can be seen to possess the optimal adaptation to exploit one particular combination of resources to a maximum, but nevertheless its morphological, physiological, and behavioural characteristics may also allow it to exploit other combinations to a lesser extent. This might be indicated by fewer individuals inhabiting a particular place or by fewer individuals visiting a particular place or pursuing a particular activity for a shorter period of time. For example, in Fig. 7.1, species 1 is particularly suited to resource category C but can also utilize resources to a lesser extent from categories A, B, D, and E. Species 2, by contrast, does best in category F but can also exploit categories D, E, and G, although it obtains less from these. Categories C and F are, therefore, both exploited very efficiently by single species. However, the combined attentions of both species in D and E ensure that these, too, are exploited with a high degree of efficiency. The existence of species as stable gene complexes maintained by the absence of inter-breeding allows specializations to develop which improve efficiency of resource utilization in particular areas.

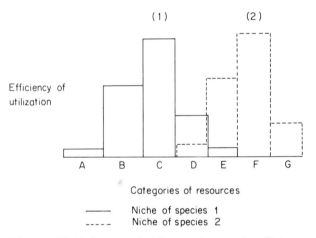

Fig. 7.1 The influence of niche overlap on the efficiency with which resources are utilized

The presence of a large number of species, therefore, implies a very efficient use of resources. The number of niches in a community or ecosystem can be seen as the number of ways the resources can be parcelled out amongst the organisms living there. The niche is the way a species operates to supply its needs within the community. It can be seen as the sum total of all the interactions between the organism and its environment (Hutchinson, 1957a,b, 1965).

Energy, materials, space, and time are the fundamental resources available to a species. Energy and materials are required to allow individuals to grow to the requisite size for reproduction. The ultimate index of size and success of the niche is the amount of energy processed by the population and the efficiency with which this is converted to reproductive products or, more accurately, into

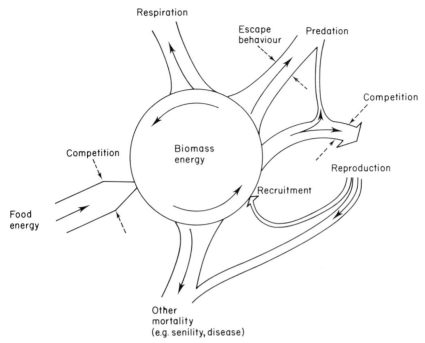

Fig. 7.2 Bioenergetic view of niche function. Dotted arrows represent potential points of constraint

recruits to sustain the population within the community (Fig. 7.2). The effectiveness with which this is accomplished is also a measure of evolutionary fitness.

The dimensions of space and time influence the efficiency of food gathering, growth, predation, and reproduction. Space can mean the distribution of the food supply, the density of prey to predatory organisms, the availability of a restricted number of reproductive or shelter sites, or even just a place to be whilst awaiting events.

Time increases the probability of a given event occurring, such as the location of a new food supply or a mate (Watt, 1972); also the use of time to stagger feeding between day and night, for example, or reproduction between the seasons allows resources to be more efficiently partitioned between the species of the ecosystem. Space and time are important elements in the behavioural interactions between species, such as predation and competition, which are modifying factors in the availability and conservation of energy and materials within the niche of the population. If the net effect of these influences is that recruitment into a population does not cover losses then this species will gradually be eliminated from the community and the resources in its niche may be appropriated by other species, whilst its predators must shift attention to other types. Many predatory fish can be fairly flexible in their diets (Lowe-McConnell, 1975).

Niches may vary not only in number but also in size. MacArthur (1960) used a simple model to demonstrate that, if resources are allocated between species strictly at random and there is no overlap between niches as a result of competitive interaction, then a few species receive a large proportion of those available whilst the majority have moderate or small allocations. This reflects a common pattern in many communities where a few species are common and the majority are rare.

The degree of discrepancy between the relative abundance of common and rarer species is termed the equitability. A truly diverse system would have a high equitability between species as well as a high species richness. A species list alone, as used in the previous section, is an incomplete indicator of diversity and one which varies significantly with sample size; the longer you look the more you find.

Diversity is a measure of the degree of organization and efficiency with which energy, materials, space, and time are used within the community and, as such, it is not only a number of species which is important but some aspect of their relative niche sizes within the community. This can be indicated by abundance or biomass or ideally by the proportion of the energy input they process. For this reason, a number of indices of diversity have been produced which take some of these other dimensions into account. Margalef (1968) produced one based on information theory since the community can be viewed as an organizational system for processing information. However, rather more flexible expressions which are less dependent upon sample size are those derived from Shannon and Weaver (1949) which can be summarized as

$$\text{Diversity} = -\Sigma \frac{n_i}{N} \log_2 \frac{n_i}{N}$$

where n_i is the number of species i in the sample and N is the total number of individuals in the sample ($\log_2 = 1.442 \cdot \log_e$).

The diversity index produced in this way can be expressed in information terms as bits/individual. Such an index takes into account the relative abundance of each species as n_i/N and also the sample size since N is the denominator.

It is curious that this approach has not been more widely applied to tropical systems, bearing in mind the usual association of the tropics with high-diversity. It has scarcely been used in tropical freshwaters but, in one instance where it has been employed, the phytoplankton of Lake Titicaca had a mean diversity index of 2.8 bits/individual with a range of 1.1–4.0 (from Widmer et al., 1975). Data in a comparable form from Gull Lake, Michigan, in North America showed a mean of 2.32 bits/individual with a range of 0.9–3.6 (Moss, 1973). These two lakes from very different latitudes do not, therefore, show a good deal of difference in their diversity considering the annual range in each case. This in itself would reinforce indications provided in the previous section that the diversity of tropical and temperate phytoplankton was of the same order, although Titicaca is a high-altitude lake in the Andes and may not be representative of lowland waters. There is obviously, however, considerable scope for using comparisons of this kind to assess the diversity of tropical waters rather than relying on species lists.

7.3 THE RELATIONSHIP OF DIVERSITY TO PRODUCTION AND TIME

Whilst latitude may have an effect on overall species richness and diversity, these are not unchanging properties of the community. With the ecosystem, ecological succession tends to occur, with one series of communities replacing another until the mature climax is achieved. A characteristic of communities early in the succession is a high rate of production from a relatively small biomass, that is, a high P/B ratio, whilst more mature stages often show a high biomass but with a small net production, that is, a low P/B ratio. The trend, therefore, is towards the support of a greater amount of biomass per unit production, in other words increased efficiency. As discussed above, high diversity promoted efficient use of resources, and increased diversity as well as a reduced P/B ratio is a frequent feature of the later stages of succession. Applying these generalized ideas to freshwater lakes, Margalef (1968) suggested that, during a successional series, lakes began in a eutrophic phase and moved to a more oligotrophic stage with time; that is, a decline in net production takes place as the system becomes more highly organized and more species contribute to the pattern of utilization. In the tropics, however, there is some problem in distinguishing oligotrophic conditions since the large, deep, clear lakes like Tanganyika and Lanao, which have many oligotrophic characteristics by temperate standards, maintain rates of production higher than many temperate eutrophic lakes by virtue of a very rapid nutrient recycling which permits the development of a deep euphotic zone (Section 5.2.4).

Margalef (1968), who worked frequently with marine phytoplankton, remarked that perturbations or fluctuations in the environment cause 'rejuvenation' of the system associated with a simplification of the species composition and eventually a resurgence in productivity. Such rejuvenation can be caused by physical turbulence or disruption, or a pulse of nutrients. Margalef also generally noted that species diversity was negatively related to turnover.

These ideas can be used as a framework to consider the interrelationships between diversity, production, and time in tropical waters. The less changeable or more predictable systems should show the highest species diversity, and since the resources are used most efficiently these should also possess a low P/B ratio. In Lake George, there are very muted fluctuations in productivity during the year and seasonal changes are small with little change from year to year at this equatorial location. The biomass of the phytoplankton in Lake George is very high with a total of $48.8 \, \text{g C m}^{-2}$ and the daily net production is $1.52 \, \text{g C m}^{-2}$ (Burgis and Dunn, 1978), giving a P/B ratio of 0.03 which is considerably lower than that which can be derived for Lake Lanao where, with an average daily net production of $1.7 \, \text{g C m}^{-2}$ (Lewis, 1974) and a biomass equivalent to some $2.37 \, \text{g C m}^{-2}$, the P/B ratio is 0.72.

Yet Lake George is eutrophic and its rate of production is high even by tropical standards, and the P/B ratio is low because of the exceptionally high phytoplankton biomass. For some reason the accumulated phytoplankton is unavailable to the grazers of the lake and is mainly exploited after death by bacteria in the water column, or to a lesser extent on the lake bed, or is lost through sedimentation. Rapid bacterial activity ensures a rapid recycling of materials through the production system (Section 5.2.4).

The species diversity of most types of organisms including phytoplankton, zooplankton, fishes, and benthic animals, is low and the low biomass of animals supported by the phytoplankton, $0.37\,g\,C\,m^{-2}$ out of a total $49.17\,g\,C\,m^{-2}$, indicates less efficient cropping by the grazing chain. In this eutrophic lake the rapid turnover does appear to be associated with low diversity, but one further feature is the diurnal fluctuating conditions. As outlined earlier (Section 3.5.1(e)), there is a daily stratification of the shallow water column in Lake George which breaks down at night rendering the diurnal changes of an equal or greater magnitude than the seasonal. The species involved must, therefore, have a wide tolerance to these daily fluctuations. Since these events show little seasonal change, there is no succession of species with different optima during the year and the number of niches is consequently limited. The continual perturbations may keep the system 'immature' and productive with a low diversity. The low P/B ratio, as a result of large accumulations of algae, may be uncharacteristic due to the apparently atypical grazing relationships in Lake George. However, another highly productive lake with a daily mixing cycle and very low species diversity, Lake Nakuru, also shows a very low P/B ratio of some 0.04 (from Vareschi, 1982) because of the enormous density of phytoplankton occurring there.

The role of physical stability of the environment in influencing diversity is further demonstrated by the benthic fauna of the Lake George sediments where the scarcity of species is a result of the regular disturbance of the upper 15 cm of the mud by wind action in this shallow lake, and also of the difficult physiological conditions created by the anaerobic reducing environment of the mud which relatively few animals can tolerate. Extreme physiological demands by an environment often drastically reduce the diversity, so that some of the Class III soda lakes with exceptionally high salt concentrations and a high pH often contain only one species of algae, such as the blue-greens *Spirulina* or *Anabaena*, although these are often extremely abundant. These shallow soda lakes are amongst the most productive aquatic systems and extreme examples of eutrophic lakes. As a rule any difficult environment such as polluted water or an estuary will show a reduced species diversity, but those types which are able to adapt and tolerate these conditions often become very abundant as they can benefit from the resources available, with reduced competition from other species.

With respect to phytoplankton, Margalef (1968) has suggested three responses of species diversity to different water conditions. Firstly, in turbulent, mixed, and often nutrient-rich waters, diversity is at its lowest and the phytoplankton may be dominated by diatoms, whilst secondly, in stratified nutrient-depleted waters, diversity tends to be high. Finally, where inflows occur, an intermediate degree of diversity may be found, particularly as a result of new species washed into the system. In temperate lakes, increases of diversity have been found between the period of the spring bloom of diatoms when the water is fully mixed after the winter and the phase of stratification with thermocline formation when production is low (e.g. Moss, 1973). Stratified conditions in the oceanic areas such as the Sargasso Sea are also associated with increased diversity (Hulburt *et al.*, 1960).

There has been no systematic study of phytoplankton diversity in tropical lakes but sufficient information is available for inferences to be drawn. For example, in Lake Victoria, major peaks in biomass occur following the principal mixing period in June and July and more temporary depressions of the thermocline in January and March. In each case one or two species of diatom totally dominated these blooms, *Melosira nyassensis* with *Stephanodiscus astrea* in the first case and *Nitzschia acicularis* and *S. astrea* in the temporary phases (Talling, 1965). A strong dominance by a small number of species indicates a low equitability of resource distribution typical of a low species diversity as discussed in the previous section. Conversely, the 3-month period of stratification which culminated in May is characterized by a very low biomass and a mixed species composition with no marked dominance which suggests an increased equitability.

A similar pattern was evident in Lake Lanao where the major period of complete mixing by the wind, beginning at the end of December, resulted in an enormous diatom peak dominated by *Melosira granulata* and *M. agassizii*, and subsequent periods of partial mixing by typhoons or strong winds produced other diatom peaks (Lewis, 1978b). By contrast, during a strongly stratified period in May, abundance was generally depressed and the major phytoplankton groups were much more equally represented.

In these cases, therefore, as with other aquatic systems examined, turbulence and mixing appear to be associated with high productivity and low diversity, whilst stratification brings progressively reduced production, as nutrients are depleted (Section 5.2.4(b)), but an increasing diveristy. Explanations for this remain speculative but it is probable that the turbulent conditions and the sudden pulse of nutrients can only be rapidly exploited by a few types, in this case most often diatoms. Under calm, stable, stratified conditions, more niches become available in this relatively homogeneous environment since the algal species have the opportunity of accumulating at their optimal depths in the water column in the calm conditions. It has also been suggested (Fogg, 1955) that extracellular secretions from one species might render the neighbourhood more or less favourable for other species and that this chemical conditioning of the water might create more niches. Bearing in mind that from 6 to 58% of the total production in Lake George was shown to be in the form of extracellular secretions (Ganf and Horne, 1975) and that blooms of *Lyngbia limnetica* and *Oscillatoria limnetica* reduced cladoceran numbers in Lake Valencia, Venezuela (Infante and Reihl, 1984), this must remain a possibility.

In the deeper reservoirs and man-made lakes stratification occurs during part of the year, which should lead to the development of a more diverse planktonic community. The persistently turbulent and changing conditions in rivers may account for the relatively low species diversity in river phytoplankton, although the spatially more complex bottom and vegetated areas appear to support greater numbers of algal species (Section 7.1.2). Reservoirs with a rapid water turnover will also present turbulent conditions which will produce a relatively low diversity in the phytoplankton, although the inflowing river may introduce additional species and therefore create the intermediate case outlined by Margalef (1968) mentioned above.

Following a period of mixing when the water column becomes isothermal, there is a transition to a more stable, stratified condition which is accompanied by a progressive nutrient depletion after the thermocline has been formed as the nutrients are used up during photosynthesis (Section 5.2.3(a)). During this sequence of events, not only does the number of species and equitability increase but the nature of the niche, as represented by the most common type of algae, also alters. The progressive change in conditions generally leads to a succession of algal types. Diatoms are usually first to dominate the phytoplankton following a mixing phase as indicated in the cases of Lake Victoria and Lake Lanao above. A careful analysis of the sequence of changes in Lake Lanao suggested that, as the diatom bloom subsided, the next to be favoured were green algal species, then blue-greens as stratification began, and finally, as nutrient deficiency became extreme during long periods of stratification, dinoflagellates, although algal numbers by this time were very low indeed (Lewis, 1978b). This sequence is very similar to that found in temperate lakes during spring and summer (Reynolds, 1982, 1984). The difference in niches occupied by species within these major lakes has been considered in terms of three simple factors, light, nutrient availability, and the rate of sinking (Fig. 7.2).

Immediately following mixing, much of the surface phytoplankton is carried down to greater depths where light is limiting and where photosynthesis is inhibited. Diatoms, however, because of their heavy silica frustules are more prone to sinking and may do so at an average of 2.5 m day^{-1} compared to the much lighter green algae, for example, where the rate may only be 0.8 m day^{-1} (Lehman et al., 1975). Turbulence, therefore, may recirculate a greater proportion of them back into the euphotic zone than under more stable conditions. In some such as *Melosira*, the increase in numbers observed in both Lake Lanao and in Lake Victoria is so rapid that resuspension of considerable numbers of cells or dormant stages from the bed of the lake seems probable. At this point the mixed layer is quite deep and the water is still being circulated so that individual cells may only be close to the surface for a short period of time, consequently light availability is generally poor. Diatoms, however, appear to be adapted to this provided nutrients are high and that sinking rate is reduced as a result of water turbulence (Fig. 7.3).

Once mixing starts to cease, the probability of sinking increases. At the same time, the light climate improves whilst nutrients are becoming depleted and the lighter algae such as the greens tend to become common. With the onset of stratification, nutrients become scarce and blue-greens often show a growth pulse, in some cases because they can fix nitrogen, and therefore a lack of nitrogen in the water is less of a problem. However, since the phenomenon is not restricted to nitrogen fixers alone, it is also possible that a further advantageous feature is their buoyancy mechanism (Section 4.5.2(d)) which enables them to adopt a stratified distribution, as was observed in Lake Redondo, Brazil (Reynolds et al., 1983).

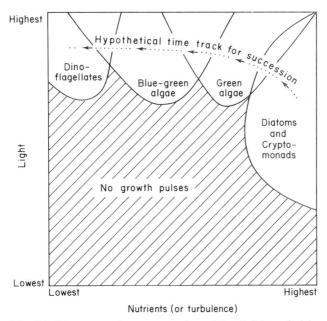

Fig. 7.3 Diagrammatic representation of the division of niche space between the major groups of phytoplankton during succession in Lake Lanao, Philippines. (*Reproduced by permission of Blackwell Scientific Publications Ltd. from Lewis, 1978b*)

Finally, under conditions of severe nutrient depletion with prolonged stratification, when numbers are generally low, the dinoflagellates such as *Ceratium* may be most frequent. At this time, when there is little water movement to offset the rate of sinking, it may be more than coincidental that a flagellate, with some degree of mobility, is most favoured. In addition, the great nutritional flexibility of this group will be of value under the conditions of nutrient impoverishment in the epilimnion (Lewis, 1978b).

This is the most common sequence of algal succession in Lake Lanao and it generally bears a close resemblance to changes through time in Lake Victoria (Talling, 1965) where diatoms, greens, and blue-greens also followed each other as they do in the better documented temperate lakes (Reynolds, 1982, 1984). It demonstrates that, as resource availability changes under the influence of the physical and chemical conditions and also as a result of the activity of the organisms themselves, so the nature of the niches alters as conditions shift from the optima of one group of species to another (Fig. 7.3).

Whilst the sequence of changes in the succession of phytoplankton types is apparently similar in tropical compared to temperate lakes, one difference observable, at least in Lake Lanao, was the frequency of these episodes. In Lake Lanao, fourteen sequences were observed during a year compared to five for some temperate lakes (Lewis, 1978b). This emphasizes the fact that tropical lakes are far from being stable, unchanging environments and often show several mixing events during the year culminating in lakes such as Lake George where the changes cycle takes place on a daily basis, a frequency which is too rapid to allow successional events to occur. This contrasts with the central regions of tropical oceans where, despite similar conditions of light and temperature, the thermocline is virtually permanent. Production is one-tenth that of comparable temperate seas but, owing to the physical stability of the water column, a high diversity of both phytoplankton and zooplankton can be attained. The phytoplankton diversity of tropical aquatic systems can be greater than that from temperate areas but in lakes, and probably also in rivers, this is rarely achieved owing to frequent mixing episodes.

Zooplankton are less liable to short-term successional changes in species abundance than phytoplankton since their longer generation times slow down the speed at which events of this type can occur. Nevertheless, as we have seen (Section 7.1.3), the species richness is generally no greater than in temperate waters and may even be a little less. As the zooplankton rely a good deal on the phytoplankton as a source of food, the moderate diversity of algae in tropical waters may restrict the number of niches available for zooplankton species and thus the one reflects the other. In considering this question of the rather modest zooplankton diversity in the tropics, Fernando (1980a,b) suggests that this is particularly due to the lack of cladoceran species and of the genus *Daphnia* in particular, which are so prevalent in temperate waters. This may be a result of reduced fertility or feeding efficiency of this genus at tropical temperatures, although it would be surprising that, if niches were left vacant by the absence of certain genera, others would not produce the evolutionary adaptations to occupy them.

Annual pertubations in rivers tend to be more extreme than those seen in lakes. In floodplain rivers, particularly, this leads to a seasonal pulse of nutrients as the terrestrial areas are flooded. In Margalef's terminology this should cause an annual 'rejuvenation' of the system which maintains 'immature' characteristics. Consideration of riverine fish communities (Lowe-McConnell, 1975, 1977, 1979) suggests that they have a higher P/B ratio, less pronounced specialization in feeding, and a greater reproductive potential to allow for increased opportunities during good flood years and high mortalities in dry years, compared to more physically stable lake communities.

Life cycles tend to be shorter in riverine populations and the number of age-groups in a population smaller, all indicating a more rapid turnover. There is a constant seasonal oscillation in resources in rivers but predictability in fluctuations can often lead to a community being very productive (Watt, 1972). Much of the ultimate source of organic production in rivers, however, originates outside the system in the form of allochthonous material washed in or submerged during floods within the catchment area.

Whether tropical rivers are less diverse than lakes is not clear. Certainly there is little difference in species richness (Section 7.1.1), although the more seasonal rivers such as the Niger and the Zambezi have fewer fish species than the equatorial Amazon and the Zaïre (Lowe-McConnell, 1977). Predictability of the environment, whether flowing or still water, allows speciation to take place. In any branch of the river system there may be less equitability between species and more obviously dominant types than in lakes (Lowe-McConnell, 1977) or there may be fewer interactions between species.

One further point which Margalef (1968) has made is that, within any system, the older animals of a population tend to move into the most diverse and least productive areas whilst the younger are to be found in the simpler, more productive regions. He illustrated this with reference to sardines in the Ebro estuary, the young of which were in the mixed waters close to the estuary itself, with a phytoplankton corresponding to his intermediate classification, whilst the older sardines moved out to the more stable open-water areas with the higher phytoplankton diversity and lower productivity. A striking parallel to this are the migrations of the adult Amazon fishes (Section 6.4.3) which move down the infertile clear-water tributaries to the richer nursery grounds in the flooded areas of the white-water channels for spawning and then move back to the clear-water rivers where they spend most of their adult existence. Moreover, the older fishes tend to move progressively to more upstream tributaries with each successive year (Goulding, 1981).

7.4 COEXISTENCE IN A CROWDED ENVIRONMENT

Whilst the small animals and plants which occupy the upper layers of the water, the loose sediments, or any physiologically extreme environment do not show the richness of species normally associated with the tropics, those organisms which occupy environments of greater physical complexity, such as invertebrates amongst vegetation, algae on rocks, and above all the fishes, whose mobility in three dimensions helps them create their own complexity of environment, do reveal considerable variety. It is those organisms which experience a long-lasting variety or patchiness in their environment that are provided with the opportunities to express the great potential for a high diversity which seems to exist in tropical systems. This quality, Hutchinson (1959) termed the 'environmental mosaic' and implied that the more pieces there were in the mosaic, even if some of them were very small, the more niches would be available. There is certainly a physical dimension to this mosaic idea in that the more complex the actual structure of the environment, the more places, and in particular the more sorts of places, there are for organisms to live.

The nature of this structure, however, may mean that some types of animals or plants are more able to profit than others. For example, the renowned diversity of the tropical rain forest, which provides a three-dimensional environment for other organisms to a height of up to 150 m, is more strongly reflected in the bird species, which are particularly numerous, than in the numbers of mammals (MacArthur, 1957) since the birds have just the right basic adaptations to make the most of the opportunities provided by this structurally complex aerial system. In the same way, the high diversity of the rocky headwaters of a tropical stream does tend to be reflected in the great variety of insect nymphs and larvae which can crawl over, under, or between rocks rather than, for example, oligochaete worms which live in sediment of which there is little in the eroding phase of the stream. Further down, however, as was shown by the Gombak stream, where sedimentation begins to take place, insect diversity decreases whilst that of oligochaetes rises.

For mobile animals such as fishes which can control their own distribution in the water, a rich mosaic of environmental possibilities becomes available. Moreover, the fishes have proved, like the birds in the tropical forest, that they have the evolutionary flexibility to produce species to fill the spectrum of niches presented. They can be very big or very small, inhabit open waters, or stay close to the bottom and they are present at every consumer trophic level in both the grazing and decomposer chains (Section 4.8.1). Naturally the purely structural diversity of habits open to fishes is the same in temperate as it is in tropical regions yet the number of fish species, even in small waters, can be so much greater (Section 7.1.1). There does seem to be some dimension of life in the tropics that promotes diversity which, in lakes and rivers, is evident amongst the fishes if not in the plankton. In this respect they contravene the generalization of Margalef (1968) that the level of diversity is reflected right through an ecosystem.

This is in contrast to the tropical forest where the growth form of the primary producers, the trees, gives a three-dimensional enhancement to the niches available. The number of tree species evolved in response to the patchy distribution of resources in the soil and the microclimate of the surrounding air further promotes the variety of niches available to dependent organisms. Only in certain circumstances, such as macrophyte-dominated floating meadows and swamp communities, are the primary producers large enough to have this sort of influence on diversity, although the chemical environment they create around themselves, such as low oxygen concentration and acidic pH, often tends to inhibit diversity. The depth of the water column, however, does allow the possibility of niche stratification as with the trees in the forest, but the biological diversity amongst the macrophytes is much less than amongst trees; the complexity they engender is, therefore, mainly physical.

The profusion of fish species in tropical lakes and rivers compared to their temperate equivalents does, then, pose the question of whether the greater number of niches within a similar physical framework leads to an increased degree of niche overlap. Significantly more niches could produce little overlap but require a high degree of specialization or, alternatively, the niches could remain quite broad although with a considerable degree of overlap of the type shown in Fig. 7.1. The niche is a multidimensional entity (Fig. 7.2) within which there is considerable scope for differences to arise. The elements which have received the most attention as regards the distinctiveness of the niche are food sources, living space, and opportunity for reproduction.

There have been many studies on the diet of fishes (Sections 4.8.2 and 4.8.3) and a high proportion have concluded that there is a considerable overlap in the types of food consumed by fishes within tropical communities (Lowe-McConnell, 1975). For example, 54% of the sixteen species found in streams in the Zaïre basin took solely terrestrial insects (Gosse, 1963) and similar observations were made from clear-water streams of the Amazon (Knoppel, 1972). Even in the large lakes, where a greater environmental stability has led to the evolution of a high degree of specialization, many of the most specialized habits, such as scale eating or egg eating amongst the cichlids of Lake Victoria, have evolved several times (see Table 4.2, Section 4.8.3). This could suggest that food supply is not a limiting factor in tropical waters and that the difference in food sources is not a major feature of maintaining the distinctiveness of niches, particularly within rivers.

A considerable degree of ecological segregation was found amongst the fishes living in a series of forest streams in Sri Lanka (Moyle and Senanayake, 1984). In addition to diet and morphological specialization, five characteristics of the microhabitat of each fish were estimated: depth of water column, distance of fish from bottom, water velocity at the station of the fish, mean water column velocity, and substrate composition. The most significant differences amongst the species with respect to these were the relative depths of the fish in the water column which segregated the greatest number of species rather in the same way described for some Panamanian fishes (Zaret and Rand, 1975; Fig. 7.4) and which has also been found to be important in temperate stream communities (Schlosser, 1982). Other factors of the microhabitats appeared less important, such as current speed in the vicinity of the fish, since most of the species, even in generally rapid water, tended to lie in calmer areas in the lee of banks, roots, or rocks (Moyle and Senanayake, 1984). Moreover, only two of the Sri Lankan species could be separated on the basis of the type of bottom over which they were found, the rest being rather more wide-ranging or discriminating on too fine a scale to be determined.

It was possible to divide the twenty species into four feeding groups depending upon the types of food taken, including terrestrial insects, aquatic insects, filamentous algae, and diatoms with detritus. Within these groups significant overlap in diets occurred but between the groups, the overlap was small. Taking all characteristics of the niche of each species into account showed that the large majority of species, when compared with all the other species, over 85% in fact, showed a low degree of overlap with respect to at least one or often more of the

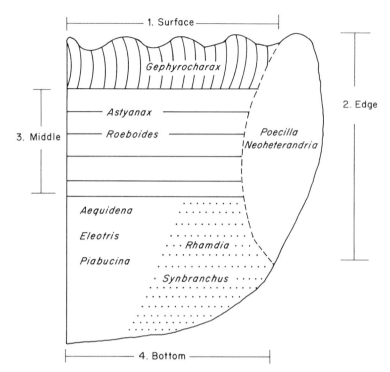

1. Surface

2. Edge

3. Middle

Gephyrocharax

Astyanax

Roeboides

Poecilla
Neoheterandria

Aequidena

Eleotris

Rhamdia

Piabucina

Synbranchus

4. Bottom

Fig. 7.4 Distribution of fish species in a dry season of a Panamanian stream showing five groupings; stippled areas indicate nocturnal species. (*Reproduced by permission of the Ecological Society of America from Zaret and Rand, 1971*)

habitat or dietary characteristics, so that each of these species had a niche which was in some significant way different from that of other species even though some aspects could be shared to a considerable extent.

Three species, however, could not be easily segregated from the others and showed considerable overlap in most aspects of their niche. These, however, were frequently to be found as the only species above waterfalls which suggests that they are good colonizers and that their generalized life-style suits them particularly to more marginal and difficult habitats. When mixed with others which are strong competitors, their niche size contracts and, as more species are added downstream, these generalists may be eliminated or replaced. One of these habitat and food generalists in the Sri Lankan forest streams is a percid, *Belontia signata*, which appears to be a rather aggressive feeder and therefore, when in competition with other species, appears to create its own feeding space by driving other fish away from their own areas.

The information collected from the Sri Lankan community was sufficient to produce estimates of niche breadth for each species in relation to the status of the six major resources or dimensions measured. These generally proved to be rather narrow; for example, of the 11 resource states estimated, an average of only 3.18 were occupied by each species, although the range was from as little as 2.04 for *Ophiocephalus gaucha* to 4.8 for *Barbus sarana*. Generally the niche dimensions indicated that most species used a relatively narrow spectrum of the resources available, although for each dimension there appeared to be both specialists and generalists.

The niches within these forest streams appear, therefore, to be quite narrow and well defined although often with some degree of overlap, at least in some dimensions. It is also apparent that the place in which the fish feeds is as important as the type of food it takes in the sharing out of the food resources between the species.

Observations from other streams and rivers in the tropics indicate that the degree of overlap between niches may be more pronounced than that measured in the forest streams just considered, although few have been analysed to the same extent. Moreover, niche number and size may vary seasonally. Fishes left in the dry-season pools of the Rupununi River in Guyana showed considerable overlap in the food taken, largely due to all sources being in very short supply. Consequently the fish had to take whatever they could get (Lowe-McConnell, 1964).

A similar but less dramatic situation has also been found amongst the aufwuchs grazing loricariid catfishes of a Panamanian forest stream (Power, 1983).

These fishes graze the algae growing on a variety of different surfaces including mud, sand, cobble, leaf mats, submerged grasses, and bedrock, and during the rainy season the four loricariid species which occur together show a marked tendency to feed off different surfaces. In the dry season, however, as the river level falls, competition for grazing intensifies and a much greater degree of overlap—up to 86% in some cases—can occur in feeding sites and growth rates can be curtailed. One of the pressures which further restricts the feeding sites available in this case is the expanded areas of water with a depth of less than 10 cm during the dry season, which the fish avoid owing to terrestrial predators such as kingfishers and herons.

Those rivers which flood their banks during the rains must show a dramatic change in niche structure as the terrestrial vegetation is submerged. Specialization may be more appropriate at times such as these when food sources are varied and there is considerable latitudinal and vertical spatial diversity in the newly flooded habitats. The tambaqui, *Colossoma macropomum*, of the Amazon, for example, is particularly adapted to eating fruits and seeds, upon which it grows fat during the floods. These are in short supply, however, during the dry season when the fish must survive on a more generalized diet (Goulding, 1981). Once more, therefore, specialized, highly segregated feeding in the rains appears to give way to a more generalized approach in the dry season when food is relatively scarce.

In contrast, the fishes of a community from a Panamanian forest stream showed a reverse trend, with segregation amongst the various species being greatest at low water and overlap being most evident during the rains (Zaret and

Rand, 1971). During the dry season, it was found that, as well as separating the fishes into major feeding categories, it was also possible to divide them into five habitat groups (Fig. 7.4). Those pairs of species which encroached significantly on each other's food sources tended to be located in different parts of the stream. In addition, two of the species were nocturnal which naturally prevents direct competition with those which feed during the day. The food resources were, therefore, partitioned amongst the fishes in space and time, as well as by the nature of the food; particular types of food were sufficiently abundant to support more than one species of fish.

In the rainy season, the degree of overlap tended to be greater and eight pairs of species showed a significant overlap in diets. Some of these also existed in the same habitat groups and therefore foraged in similar areas.

The implication in this case is that, during the dry season, the reduction in food supply increases the need for competition and leads to less overlap through more specialized feeding (Zaret and Rand, 1971). This might be seen to support the competitive exclusion principle, or Gaus's axiom, that no two species can occupy exactly the same niche as one will always be eliminated through competition. However, this principle would appear to be contradicted when a reduction in the food available produces a greater overlap (Lowe-McConnell, 1975) as it does in the three examples given above (Lowe-McConnell, 1964; Goulding 1981; Power, 1983).

An organism would be fortunate to experience the conditions to which it is optimally adapted, throughout an entire year. Provided, however, those conditions arise for a reasonable proportion of the year, whether in the wet or dry season in this case, the organism may still share the resources within its niche with other species for the rest of the time, yet possess a niche which is unique. Even if conditions for food gathering were identical, those governing reproduction or predation may be different. Moreover, it must also be true that the more diverse the environment, both biotic and abiotic, the less chance there is of niches coinciding.

It is, however, an interesting question as to why some communities show the greatest overlap when food is scarce whilst others show it at times of abundance. In the forest stream assemblage examined by Zaret and Rand (1971), although there might be some reduction in the food supply in the dry season, the amount still remains considerable owing to the rain of debris from the trees (Angermeier and Karr, 1983). This reduction may sharpen competition between the species leading to a more elaborate ecological segregation, but survival of the species themselves is not jeopardized to a great extent.

In rivers of the savanna regions, such as the Rupununi, where flow reduction during the dry season is extreme, the food resources themselves become very scarce and also lack the variety to support a highly structured fish assemblage. All fish must therefore be opportunists during this period.

The streams of Panama provide examples of opposite responses. The loricariid catfishes are bottom-dwellers and depend upon autochthonous production on benthic surfaces (Power, 1983). The surface area for grazing becomes reduced during the dry season and appears to force the species to share feeding sites to a greater extent. In the more complete assemblage examined by Zaret and Rand (1971), although the availability of the largely allochthonous food was reduced, the fish were able to use the full three-dimensional opportunities of the water column (Fig. 7.4) to assist in partitioning out the more varied types of food between the species.

In those streams and rivers where the increased discharge of the rainy season can be equally as marked as the lack of flow in the dry season, the tremendous amounts of water which come down would make the elaborate exploitation of different parts of the water column—to the extent found in Panama and Sri Lanka where seasonal differences are not so great—extremely difficult. Perhaps, therefore, as with the phytoplankton, the highly structured utilization of the water column is encouraged by more stable conditions and depressed by turbulence.

The greater degree of stability which characterizes lakes compared to rivers allows a structured exploitation of the water volume and bottom areas uninterrupted by regular major hydrological upheavals. As a result of this, in some of the older lakes there has been a great proliferation of often closely related species which appear to differ only slightly in their microhabitat and food preferences (Fryer and Iles, 1972). For example, the haplochromid fishes of Lake Malawi have radiated into a tremendous variety of niches often showing only small degrees of morphological modification and complex food webs can be built on relatively few food sources (Lowe-McConnell, 1975).

Certain ways of life, however, appear to support particularly large numbers of species; for example, the rocky areas around the shores are inhabited by over 200 species of closely related haplochromines which have become known collectively as the 'mbuna'. All feed on the aufwuchs, and although some may occur in different geographic regions of the lake, a number usually coexist at any one point. Many are territorial and some of the aufwuchs feeders, such as *Pseudotropheus* and *Petrotilapia*, aggressively defend 'algal gardens' of their own use (Ribbink *et al.*, 1983). There is, therefore, a considerable behavioural component to the partitioning of energy and materials.

The rocks themselves provide that physical structure which promotes habitat diversity and is conducive to the ecological separation of niches. Being relatively small animals of a few centimetres in length, the mbuna rock cichlids also experience fully the effects of the environmental mosaic (see Hutchinson, 1959) which they seem to have responded to. As well as feeding sites, the rocks provide cover which can be a key factor in permitting diversity to develop and in allowing it to be maintained (Lowe-McConnell, 1975). However, this proliferation of niches has not been confined to the spatially diverse rocky areas in Lake Malawi because there is also a group of some twenty similar species, known as the 'utaka', which have occupied the more homogeneous upper waters. An equivalent group of zooplankton-feeding haplochromines has been discovered recently in Lake Victoria, where a large number of cichlid species have also evolved (Section 7.5). These, however, appear to live in deeper water and migrate upwards to feed.

Younger lakes do not have the benefit of a species complement tailor-made by evolution to their own particular range of microhabitats and food sources. They receive their species through colonization from nearby water bodies or inflowing rivers. Nevertheless they do come to support a relatively large number of species, richer than their temperate equivalents. In recent or new lakes major niches may remain unexploited. A particularly good example was the zooplankton-feeding niche in the man-made Lake Kariba which was filled by the introduction of the freshwater sardine *Limnothrissa miodon* from Lake Tanganyika with the result that the fishery catch went up from 2000 to 10 000 tonnes (see Section 8.2.3). Similarly the crop from both artificial and natural lakes in South-east Asia increased when the algal- and plankton-feeding *Oreochromis mossambicus* was introduced, probably because no indigenous South-east Asian fish is able to exploit this major food source (Fernando and Holcik, 1982; De Silva, 1985).

Although the relative stability of the environment in lakes may allow specialization in feeding mechanisms, there can still be considerable overlap and flexibility even though many of the species show quite marked morphological specializations. For example, two of the Lake Malawi mbuna cichlids, *Pseudotropheus zebra* and *Petrotilapia tridentiger*, have morphological adaptations for scraping algae off the rocks. When observed under natural conditions, however, both could also be seen plucking at zooplankton and, in fact, *Pseudotropheus zebra* spent almost equal amounts of time on these two activities as well as nibbling detritus (McKaye and Marsh, 1983). Therefore, although specialized, these two species do not confine themselves to scraping algae and *P. zebra* was observed switching food sources on average twenty two times per hour whilst *Petrotilapia tridentiger* was less prone to this but still switched on average six times per hour. The importance of the specialization may arise when the high-energy zooplankton is least available. Use of the alternative source becomes an advantage when food availability is depressed in a rather similar way to that observed in the Panamanian stream when the fish species became more clearly segregated in the dry season (Zaret and Rand, 1972). The specialization may be of particular value to the male cichlids when they are holding territory and cannot move in search of zooplankton but must feed on what is most readily available within the territory, notably the algae on the rocks. Certainly, territorial male *P. tridentiger* fed on nothing else at this time.

Food switching to give flexibility to a niche can be seen amongst some of the most specialized fishes of Lake Malawi, The cichlid *Melanochromis crabro* acts as a cleaner fish, removing the ectoparasites from the catfish *Bagrus*, but when the catfish is spawning and guards its eggs the cleaner fish benefit from the proximity of this rich energy supply by switching to consumption of the eggs (Ribbink and Lewis, 1982).

Many of the cichlid fishes have proved adaptable in their feeding arrangements even when morphologically specialized, and their food-switching tactics are obviously beneficial. Whether this is generally the case amongst aquatic animals in situations of high diversity or whether it is just a fortuitous combination of morphological structures which allows considerable feeding flexibility to the cichlids (Liem, 1980) remains to be seen.

The rate of reproduction maintains the continuity and status of a species within its niche in relation to others in the community. Where many species coexist, therefore, competition for the opportunity to reproduce as an expression of the energy and materials being appropriated and husbanded within a particular niche or way of life (Fig. 7.2), adds another dimension to the niche itself. The period of reproduction is usually geared to the maximum period of food availability, but not all species take the same food in the same place in the same way. Since the fluctuations in different types of food, or the environmental opportunities for the young may vary seasonally, then some segregation in the periodicity of reproduction is likely to occur (Lowe-McConnell, 1979).

In rivers, some fishes spawn during the dry season, most spawn in the early rains, and a few at the end of the rains. In the early floods when many species require spawning sites, there is often a great increase in the space available in the newly flooded areas so that competition for sites may not be too severe. In other situations, however, space for reproduction can become a considerable problem. In Lake Jiloa, Nicaragua, the cichlid fishes of the lake all preferred the limited rocky areas for breeding, which typically involved the holding of a territory and the construction of a nest (McKaye, 1977). The restricted areas of rocks meant that reproductive sites were a potentially limiting factor. If one of the species were to lose access to the breeding sites through competition with the others then its numbers and biomass would decline and the food resources it previously exploited would be taken over by one or more of the other species. In this instance, the rocks are shared out in time. During the rainy season, at a time when its own preferred food of terrestrial insects is particularly abundant, *Cichlasoma citrinellum* aggressively excludes all but the two smallest species from breeding in the rocky area. The remainder of the nine species use the rocks during the dry season although they are segregated within that period by both depth and timing. The extent of competition under these circumstances is demonstrated by the fact that, of 209 pairs of cichlids observed in Lake Jiloa, only 29 held territory long enough to reproduce successfully (McKaye, 1977).

There are several instances in which species of similar reproductive requirements have staggered their reproductive periods in a way which minimized competition. The small group of six tilapia species in Lake Malawi all make sand scrape nests but spawn at different times. Amongst the three most similar species, *Oreochromis saka* spawns in the hot weather before the rains at the same time as *O. lidole* although in rather more shallow water. *Oreochromis squamipinnis* spawns later in the year, during the rains, in the deeper water (Lowe-McConnell, 1975). The segregation of spawning in this way also helps to maintain the reproductive isolation between closely related species.

This separation of spawning in time and space in the Lake Malawi tilapias occurs even though seasonality in the lake is quite muted. A similar situation is to be found in the forest streams of Panama where differences in stream conditions between the wet and dry season are moderate compared to many tropical rivers. Amongst the characoid fishes of the Rio Frijolito in Panama, the breeding periods ranged from extremely brief for *Bryconamericus* and *Piabucina*, where the whole population spawned only during 1–2 days in the year, to moderate duration of 2 months in *Brycon* and 4 months in *Hyphessobrycon*, through to more or less continuous breeding in *Reoboides* and *Gephyrocharax* (Kramer, 1978). Although some spawned primarily in the rains as in more seasonal rivers, *Hyphessobrycon*, *Brycon*, and *Reoboides* showed the greatest reproductive activity in the dry season. Therefore, although reproduction would appear to be possible for all species throughout the year, in fact segregation of spawning takes place. Moreover, the three species most similar in diet, *Piabucina*, *Bryconamericus* and *Brycon*, tend to spawn at different times, two during the floods and one in the dry season, in a way that will give a more equitable partitioning of the food resources around the crucial time. No doubt this segregation will reduce competition for food or reproductive sites amongst the adults as well as competition for food between the juveniles.

Spatial factors may be equally as important as those of time, in allowing a considerable number of species to coexist and reproduce within a given environment, as has been demonstrated for the *Barbus* species of Sri Lanka (Kortmulder, 1982a,b). Two important features in this respect are the relative abundance and the absolute size of such sites. The abundance of sites may depend upon such things as the nature of the bottom, availability of rocks and quiet backwaters, or the right conditions of current flow. The size of the site will depend upon whether territory is held or not and the optimum size required. In non-territorial types which aggregate, it will depend on how many individuals must associate and how much space the spawning group needs to accomplish the process. At the other end of the spectrum is the opportunist who is a loner and holds no permanent territory but attempts to mate with a female whenever the occasion presents itself. It is possible to share out the space for reproductive activity at any particular point in time, providing the animals adopt different strategies.

In reproduction as with feeding, therefore, the sharing out of resources in space and time as well as the nature of the resource is a most important feature of maintaining a large number of species within a community in the tropics. The main limitation each species has to contend with is not physical factors as in the temperate zone, but all the other species it mingles with. The most important thing is, then, not just to obtain a certain type of resource but, as Lowe-McConnell (1969) summarized it, to find 'a place to live'. Although there can be overlap between niches, there is usually one aspect of its life-style which, at certain times, a species can exploit more successfully than any other species. Whether this area of overlap is any greater or smaller than the equivalent temperate species is still difficult to say, but it is evident that there is a good deal of flexibility between niches. Even the most specialized animals such as the cleaner fish, *Melanochromis* from Lake Malawi, can produce a compensatory response to changes in their environment.

There are situations in the tropics where the packing of species into an environment appears to be so dense that considerable overlap would seem to be inevitable. It has been argued, with reference to the very varied pomacentrid fishes found in coral reefs, that even if species do coincide closely in their requirements, space may be partitioned on a lottery, first come first served basis (Sale, 1977, 1978). The reef fishes themselves are territorial and depend upon this territory for many of their needs; the eggs and larval stages, however, are pelagic and the young of all species mix in the upper waters. As a territory becomes vacant, a new young fish can then occupy it, although which species is a matter of chance. Mathematical modelling has shown that this lottery system can persist for long periods of time given the assumptions of territoriality, a random mixture of pelagic larvae, and a curvilinear relationship between recruitment and parental stock size (Sale, 1982).

An equivalent system may occur amongst the freshwater mbuna rock-fishes of Lake Malawi, where as many as 300 species may eventually be recognized (Ribbink *et al.*, 1983). Most are territorial and appear to compete for sites as individuals rather than as species (Fryer and Iles, 1972; Lowe-McConnell, 1975) in a fashion reminiscent of Sale's reef-fish. The lottery hypothesis suggests a mechanism whereby, under certain circumstances, species with similar requirements need not necessarily compete to the eventual exclusion of all but one. The implication is that there can be considerable overlap of niches in this system.

The true measure of the size of a niche which a population occupies is the amount of energy it processes per unit time and the efficiency with which this is used to promote growth and produce recruits, since these factors reflect its share of the resources within the community. To obtain an estimate of the amount of energy processed per species is, however, a difficult task. A more accessible but much more indirect measure is that of abundance or biomass of the species supported within the community, although the relationship between biomass and energy turnover is not constant. However, the relationship between biomass and the number of species does allow some anticipation of how more diverse communities might be supported in tropical compared to temperate environments.

In temperate North America, Carlander (1955) showed that there was a significant increase in fish biomass in Midwestern reservoirs with higher numbers of species. Resources seemed to be used more efficiently when a wider variety of species was present which consequently allowed a greater standing crop to be maintained. From the general relationship obtained by Carlander, it is apparent, therefore, that the average biomass per species declines as the number of species increases (Table 7.2). A similar effect also seems to occur in North American streams (Table 7.2).

There are not very many estimates of fish biomass and species numbers from tropical lakes which are comparable to the measurements of Carlander.

Table 7.2 The relationship between biomass and number of fish species in some temperate and tropical waters

	Biomass (g m^{-2})	Species number	Biomass per species (g m^{-2} sp^{-1})	Author
Midwest reservoirs, USA	3.6 to 7.2	1 10	3.6 to 0.7	Carlander (1955)
Lake George, Uganda	29.0	26	1.1	Gwahaba (1975)
Kentucky Stream, USA	5.5 6.35 7.15	1 8 15	5.5 0.79 0.48	Lotrich (1973)
Illinois stream, USA	10.17	26	0.39	Schlosser (1982)
Sierra Leone stream	10.3	1	10.3	Payne (unpublished data)
Nigerian stream	15.1 40.2	10 11	1.5 3.7	Sydenham (1975)
Thailand streams	8.1 11.8 18.6	11 17 18	0.73 0.69 1.03	Giesler et al. (1979)
Malaysian stream	18.0	16	1.12	Bishop (1973)

Estimates from Lake George show that, although the number of species found here greatly exceeds that from the North American lakes, the biomass per species is still rather more than a lake with less than half the number from the temperate locations.

Observations from tropical streams also show that the mean biomass per species is higher than from the North American series (Table 7.2). For example, the Nigerian stream with ten species supported a biomass per species almost twice that of the Kentucky stream with only eight. The implication here is that the energy available to the average niche appears to support a rather larger amount of biomass in the tropics. To maintain a higher average biomass for each species means that they either use energy more efficiently, that there is a greater total amount of energy available, or that both apply.

There is evidence that populations of both tropical crustaceans and fishes are made up of predominantly younger animals than those of temperate systems (Sections 5.4, 5.5) which would, therefore, process energy much more rapidly and much more efficiently. In addition, tropical lakes can be shown to have on average a primary productivity two to three times greater than that of temperate lakes (Section 5.2.3), which means that there is more energy passing through these ecosystems. This could be shared to make larger niches, more niches, or a little of both of these options which appears to be the most probable case on the fragmentary evidence available. One further aspect of this higher productivity is that it is more predictable than in temperate regions. There is a more continuous supply not only of algae but also of other food sources, including young fish for predators, throughout the year which helps to create more niches (Lowe-McConnell, 1975, 1977; Slobodkin and Sanders, 1969).

Paradoxically, Margalef suggested that highly productive, eutrophic conditions are associated with a lower diversity whilst in oligotrophic mature lakes the diversity is high. With regard to fish in tropical and temperate environments, however, the greater productivity within the ecosystem appears to be associated with the greater diversity. Possibly, therefore, the relationship between diversity and production suggested by Margalef is only true of successional series within latitudinal zones.

7.5 SOURCES OF DIVERSITY

The numbers of species to be found in a system is partly a product of the size of that system. Direct correlations have been found between fish species and area in tropical lakes (Barbour and Brown, 1974), whilst for river basins, a predictive equation estimating species richness from area has been arrived at (Welcomme, 1979; Section 4.8.1). The larger the area covered, the greater the potential for habitat diversity which, as we have seen in the previous section, is one of the most potent factors in the maintenance of a large number of species. Species can colonize a lake or river from other water bodies or, in the case of insects, from the air. An obvious source for lakes will be inflowing rivers whilst, amongst rivers, the constant capturing of one river system by another through geological time as the elevation of the earth's crust alters is a frequent process whereby species are introduced from one river to another. Ultimately, however, a community rich in species must result from a high rate of production of new types or a low rate of extinction or both.

The initiation of a new species is commonly associated with the geographical isolation of a population of an existing species in a new environment with the result that, after thousands or millions of years, natural selection will produce adaptations to the new possibilities and the isolated population will have become different from the original stock. The geographical isolation, caused by perhaps a land barrier or a waterfall, gives a restricted gene flow between the population in the new environment and the parent stock which allows the specializations to stabilize genetically without being diluted by reproducing with the original parental population. The physically isolated population is therefore also reproductively isolated and can strictly only be considered a new species if, when the opportunity arises in the future, the new species and the parental types do not recognize each other as potential mates or do not produce viable offspring.

Speciation takes time and is only a significant factor in topping up the complement of species in the older water bodies. Most lakes are only a few thousand years old and have received their species from elsewhere, although in larger, older lakes such as Lake Tanganyika, which is probably around 2 million years old (Bannister and Clarke, 1980), a tremendous number of new, endemic fish and mollusc species have arisen. Speciation in large stable lakes is not confined to the tropics; Lake Baikal from Siberia contains over 1000 species of animals of which 240 alone are crustaceans belonging to the family Gammaridae. In this case, however, it is the age and stability of the lake which have led to this situation in that the great majority are confined to the abyssal zone of this very deep lake where temperature and oxygen vary only slightly. In fact, temperature on its own probably has little effect upon speciation; it is the predictability of the environment and the absence of physiological extremes which permit this.

The rather modest diversity of both phytoplankton and zooplankton in the tropics suggests a low degree of isolation. A significant proportion of cosmopolitan types are shared with the temperate regions. Indeed, some of the zooplankton species of the high-altitude lakes of the Andes can be traced down through the southern temperate zone to the Antarctic (Löffler, 1964). Both phytoplankton and zooplankton seem capable of wide and efficient dispersal, whether by the wind or by waterfowl, and so isolation must often be incomplete, whilst the largely asexual reproduction amongst the algae of the phytoplankton must reduce the variability upon which natural selection can act to produce new species.

Even given a degree of geographical isolation, speciation does not automatically proceed nor is the rate the same from place to place. A certain degree of isolation is always possible in a river basin because the tributaries are, to some extent, geographically isolated and this is particularly true when waterfalls interrupt the upstream colonization. For example, Daget (1962) in the Fouta Djallon massif in Guinea described a number of unique species of the cyprinid fish *Barbus* from headwater streams above major waterfalls, where their evolution was apparently promoted by the isolation caused by these barriers. When conditions fluctuate predictably with seasons, as they do in many rivers, this can further augment the range of niches available (Lowe-McConnell, 1975).

Within lakes the factors causing isolation are rather more difficult to envisage. The origin of species within a single location, with no apparent barriers between them, is termed sympatric speciation and there has been a good deal of debate as to whether this can actually occur. In Lake Malawi there are more than 245 species, mainly cichlids, 93% of which originated within the lake. A contributory factor is thought to be the isolation of populations of rather sedentary types on rock outcrops separated by sandy areas over which the fishes are reluctant to move (Fryer, 1959, 1977c). Sedentary species often seem more prone to speciation, as also exemplified by the loricariid catfishes of South American rivers (Lowe-McConnell, 1975). In these cases, separation by microhabitat may be equally as effective as isolation on a geographical scale. In addition, Lake Malawi appears to have enlarged gradually during its early history and to have captured inflowing rivers progressively from north to south

(Bannister and Clarke, 1980) which would have allowed sequential invasions by different species (Fryer, 1977c).

A similar range of diversity in fish species can also be seen in Lake Tanganyika, where there are 214, 80% of which are endemic. A certain degree of isolation may have been imposed within the lake by considerable fluctuations of water level leading to isolation of the south basin from the north basin on a number of occasions. Fluctuations in lake level may have more subtle effects, however. Within Lake Malawi, there are numerous rocky islands and underwater plateaux and many of the 200+ mbuna rock-fishes are extremely localized, often occurring only in the vicinity of one of these islands (Ribbink et al., 1983). The depth to which these rather sedentary, bottom-dwelling fishes can descend is limited owing to the restricted range of compensation which can be made by their swim bladders. Consequently, when the lake level is high, scope for movement between the islands is more limited and greater isolation is imposed because the fish are restricted to shallower parts (Ribbink et al., 1983). Moreover, species differ in their potential maximum penetration depth. For example, three sympatric species of Petrotilapia showed maximum depths of 39.0, 35.5, and 30.4 m although they were normally to be found in shallower regions (Marsh and Ribbink, 1981). It is possible, therefore, for a given change in lake level to cause a differential effect upon the isolation of populations.

Lake Victoria presents a much more uniform environment, with many fewer rocky areas, than Lake Malawi and Lake Tanganyika. It is possible that isolation of populations in a series of smaller lakes formed earlier in the Lake Victoria basin may have led to allopatric speciation of populations from one or a few original species (Greenwood, 1974). Some evidence is provided for this by the presence in Lake Nabugabo, only relatively recently cut off from Lake Victoria, of five endemic species of haplochromines evidently derived from closely related types in the main lake.

Genetically stable and discontinuous differences, that is polymorphic variation, in populations of nominally the same species can occasionally be found in the same environment. For example, the mbuna cichlid, Pseudotropheus zebra, occurs in several distinct colour forms in Lake Malawi (Fryer, 1959). In addition, an isolated population of cichlid fishes living in the Cuatro Ciengas basin in Mexico, apparently belonging to the same species, shows three distinct forms of pharyngeal teeth associated with very different feeding habits: detritus eating, piscivory, and mollusc eating (Sage and Selander, 1975; Kornfield et al., 1982; Liem and Kaufmann, 1984). If this is true polymorphism, it may represent incipient segregation to produce other new species.

It is not clear what stimulates the production of new species in the absence of a heterogeneous habitat. The small isolated crater lake of Barombi Mbo possesses 17 species of fish, 12 of which are endemic (Trewavas et al., 1972), whilst in a similar crater lake of similar age, Lake Bosumtwi in Ghana, only one of the 11 species is endemic (Whyte, 1975; Beadle, 1981). Possibly the degree of isolation was not quite the same and the subsequent waves of colonization filled most of the niches in Bosumtwi whilst those of Barombi Mbo had to be filled by the few species which managed to reach it. Within this crater lake, however, it remains difficult to imagine how the populations become isolated from each other since there are no apparent barriers.

The ability or rate at which new species are formed can also be a product of the nature of the colonizing species. The haplochromine cichlid fishes of the East African Great Lakes originated from a very limited number of *Haplochromis* species occurring in the river systems that originally filled the basins (Greenwood, 1979) but they have produced extensive species flocks. A species flock strictly means an aggregate of several species which are endemic to a circumscribed area and which are each other's closest living relative; that is, they are of monophyletic origin (Greenwood, 1984a). Each of the large East African lakes, therefore, contains not just one but several haplochromine species flocks whilst other genera have scarcely produced any new types.

Different genetic groupings differ in their ability to provide sufficient variation upon which natural selection can act. The genus *Barbus* does speciate quite readily but probably at a much slower rate than the cichlids (Bannister and Clarke, 1980). Lake Lanao has a small group of eighteen endemic cyprinids which have been considered a species flock derived from a single species, *Barbus binotatus* (Myers, 1960). It now appears that there could have been more than one species initially and that the cyprinids therefore do not constitute a single flock (Kosswig and Villwock, 1965; Greenwood, 1984b).

In the case of cichlids the rate of speciation can be very rapid. For example, five of the nine species of cichlids in Lake Nabugabo close to Lake Victoria in Uganda are endemic yet this small lake was only isolated from the main lake 3700 years ago. Yet even amongst the cichlids there are differences in the possibilities for speciation since both of the Lake Victoria tilapias, *Oreochromis esculentus* and *O. variabilis*, are present in an unmodified form in Nabugabo whilst all the endemic species are haplochromids (Greenwood, 1965).

There can be no doubt, however, that where there is a complex habitat, and given an appropriate organism with a degree of genetic plasticity, speciation in the tropics can be explosive. This is exemplified most dramatically by the cichlid fishes from the East African lakes but the molluscs, atyid shrimps, and harpacticoid copepods provide other illustrations (Fryer, 1969). Whether the rate of speciation is equally high in rivers is difficult to tell since it is often not easy to date the various stable phases of the development of the system. The Amazon, in its present form, is probably relatively recent and the tremendous diversification of its fish fauna, particularly the incredibly rich characoids, may only be the product of a few million years of evolution from an original stock of 200–300 founder species (Roberts, 1972). The relatively sudden formation of new lakes or rivers will provide many empty niches for the small group of initial colonists and some will possess fortuitous combinations of features capable of modification through natural selection to fit a wide range of habitats. The haplochromine fishes of Lake Malawi, for example, occupy niches as diverse as algal scrapers, predators, scale eaters, and egg suckers but, whilst they vary considerably in tooth structure and jaw shape, a careful experimental analysis of these fishes has shown that all these feeding actions can be accomplished within a framework of no more than eight modes of action of the jaw and pharyngeal mechanism (Liem, 1980). In fact, it has been suggested that this versatility and the enormous diversification it has led to may depend upon one morphological modification, the uncoupling of the connection between the maxilla and premaxilla to produce a mechanism which is particularly adaptable (Liem, 1980).

In a recently formed and not wholly occupied environment new species may themselves give rise to others and multiple splitting of species appears to be a possibility. The mbuna rock-fish *Tropheus moorii* occurs in several isolated populations around Lake Malawi and has strikingly different male reproductive coloration which appears to provide a particularly good example of such multiple splitting in action and may show that, at least in some cases, reproductive differences precede trophic changes (Fryer, 1977b). In the sense that the production of a new species increases the likelihood that other species will also subsequently appear, diversity begets diversity in the tropics (Lowe-McConnell, 1969).

One other factor which may promote diversity is predation. In other environments it has been shown, for example, that the exclusion of starfish predators from sedentary animal communities on marine rocky shores reduces the number of species. The grazing of grasslands can also increase the diversity of herbaceous plants. There has been a good deal of controversy as to the role of predatory fishes in initiating and maintaining diversity in tropical lakes. The presence of the predatory Nile perch, *Lates*, and tiger-fish, *Hydrocynus*, in some lakes such as Lakes Albert and Turkana has been said to suppress species diversity (Jackson, 1961), whilst their presence in Lake Tanganyika may well have promoted it (Fryer, 1965). The continual interplay between the predator and its prey will apply considerable selective pressure for the prey to out-evolve its predator, and also between prey species to minimize the share each receives of the predator's attention. Predators can be a factor promoting isolation by preventing relatively sedentary and slow-moving animals from crossing open spaces such as stretches of sand between rock outcrops. In open waters, however, where uniformity is the main defence, predation will tend to produce more uniform genotypes and to control speciation (Lowe-McConnell, 1975). There are instances, in fact, where the lack of predators has allowed speciation to take place. For example, the *Barbus* species which have evolved cut off above waterfalls in the Fouta Djallon have only developed in the absence of predatory *Hemichromis* to which they are very susceptible (Daget, 1962).

7.6 STABILITY AND CHANGE

The number of species in a community or an ecosystem, the frequency and degree of interactions between the species, and the evenness of abundance amongst the species all contribute to the complexity of the system. Complex ecosystems were generally considered to be more stable but recent work, particularly on ecological models, suggests that the reverse may be true (May, 1975; Pimm, 1984). There are elements of tropical inland aquatic ecosystems which may be no more complex in these terms than their temperate equivalents, such as the phytoplankton, zooplankton, and benthic infauna. However, when a small bay in Lake Victoria can contain fifty-five closely related fish species and more than eighty species can be taken from the Amazon in one haul with a net, this suggests a very high degree of complexity in at least some environments, particularly bearing in mind the intricate spatial and temporal divisions between species discussed in the previous section.

Two aspects of stability which it is useful to consider are the resilience and persistence of the system (Pimm, 1984). Resilience is the rate at which change is absorbed and the original equilibrium re-established. Persistence is the time any one variable lasts before it is altered to a new value. In the short term, the primary producers of tropical lakes, the planktonic algae, are much less persistent than the trees of the tropical forest mainly because they have much shorter life cycles and therefore can respond to relatively small environmental changes over brief periods of time by changes in species composition and species abundance. They may even be less persistent than temperate phytoplankton in that the changes can occur more frequently during the year and sometimes less predictably.

In the longer term, however, the perturbations in the phytoplankton are seen to be cyclical in a way that does not necessarily lead to lasting change. Probably the most useful information on the persistence of a whole ecosystem is that gained from two expeditions to the crater lake of Ranu Lamongan in Indonesia which was first visited by the German Sunda expedition in 1928 (Ruttner, 1931) and then again, more than 45 years later, in 1974 by Green and his co-workers (Green *et al.*, 1976). Comprehensive species lists were compiled for most of the communities and data on the abundance and feeding interrelationships of some were collected.

Remarkably few differences appeared. The relative abundance of some of the phytoplankton species was slightly different in that two blue-greens were common in 1974 which were not found in 1928, and one which was common then was rare in 1974. However, these are changes well within the realm of normal cyclical oscillations of the phytoplankton and the great majority of zooplankton, marginal and benthic invertebrates were still to be found. This ecosystem, at least, has been shown to persist for a reasonable length of time, at least on a human scale, although this is not uncommon in many natural communities (Pianka in Pimm, 1984).

Perhaps an example of relatively short persistence time is demonstrated by some of the equatorial soda lakes where dense algal suspensions consisting often of a single blue-green algal species can suddenly change after several years to one of mixed green algae and diatoms although the original situation can return after a year or two (Melack, 1979).

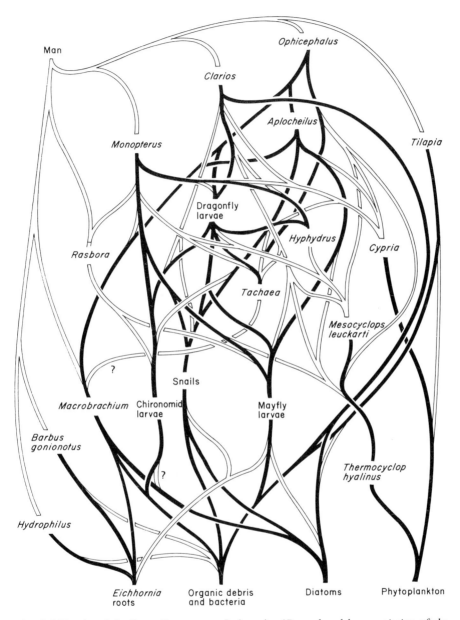

Fig. 7.5 Food web in Ranu Lamongan, Indonesia. (*Reproduced by permission of the Zoological Society of London from Green* et al., *1976*

   The truly remarkable features of the persistence and the resilience of the communities in Lake Ranu Lamongan is that considerable changes have occurred in and around the lake. Much of the forest within the crater catchment area has been removed and the human density and agricultural activity has increased considerably in the vicinity, both of which must affect the nutrient budget. In addition, four species of fish have been introduced into the lake. Two of these, *Oreochromis mossambicus* and *Barbus gonionotus*, have evidently established a considerable biomass since these are now the most important food fishes. Therefore, even though these two species occupy very prominent niches within the lake system—the former feeds on detritus and phytoplankton and the latter on the roots of *Eichhornia*—the initial structure of the communities has persisted (Fig. 7.5). However, the incorporation of these two species does mean that the system has changed, probably permanently, and that the interconnections between the existing species were insufficient to resist this change entirely although its impact on the rest of the community appears to have been virtually undetectable at the level at which the examination was made.

It could be said that a small lake such as Ranu Lamongan may have been too isolated to receive a full complement of species to exploit all its niches. The same could scarcely be said of Lake Victoria where not only colonization but almost unrivalled speciation has taken place to the extent that there are more than 200 species of cichlid fish present. Nevertheless, four more have been successfully introduced, in the form of four tilapia species, and survived to become significant features of the communities (Welcomme, 1967). Moreover, the introduced *Tilapia zillii* had a deleterious effect upon the indigenous tilapia, *Oreochromis variabilis*, through competition for nursery sites. It would seem, therefore, that although all niches appear to be filled, diverse and complex systems like Lake Victoria are not entirely stable in that their persistence can be interrupted by the addition of new species. They may also show a limited resilience to these changes which appear permanent and, in some cases, have caused alteration to the abundance of some of the original species. It must be said, however, that the introduction of the new species did coincide with an enormous increase in the lake level, but they have none the less maintained or expanded their position as the level has fallen. These minor adjustments following the introduction of herbivorous species are in strong contrast to those provided by the introduction of a large predator, the Nile perch.

Whilst addition of herbivorous species or species occupying lower trophic levels to these tropical communities can cause some adjustment amongst the existing members, the addition of a predator appears to have had a rather greater effect in several cases. Amongst the species of fishes introduced into Lake Lanao, the goby, *Glossogobius*, was a predator. As a result of its activities the abundance of some species of endemic cyprinids has declined, to critical levels in some cases (Fernando and Holcik, 1982). The Nile perch, *Lates niloticus*, was originally absent from Lake Victoria but was introduced from Lake Albert around 1959. Its major prey are the haplochromids and there are indications from fishing catches that in parts of the lake the abundance of the haplochromids, which as a group are the commonest fish in the lake, has declined considerably in recent years, whilst that of *Lates* has increased (Okemwa,

(i)

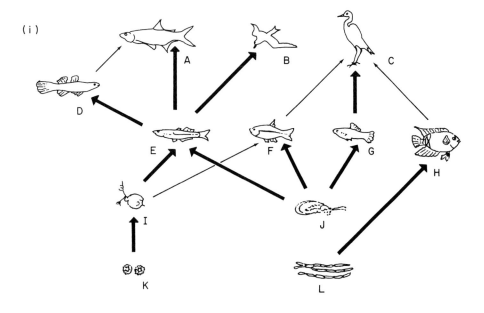

(ii)

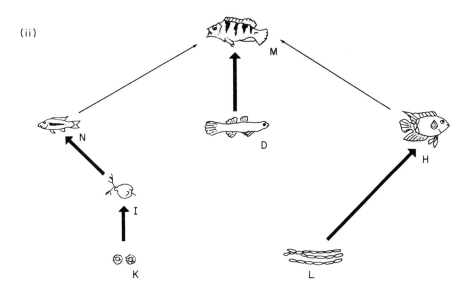

Fig. 7.6 Generalized food web of common Gatun Lake populations, contrasting structure before (top) and after (lower) introduction of the predator *Cichla ocellata*. (*Reproduced by permission of the American Association for the Advancement of Science from
Zaret and Paine, 1973. Copyright 1973 by the AAAS*)

1984). Currently *Lates* can account for up to 30% of the total annual catch. In addition, before the introduction of *Lates*, the main predator of the haplochromids was the catfish *Bagrus docmac* which consumed up to 75% of the mean haplochromid standing crop of some 600 000 tonnes annually (Chilvers and Gee, 1974). It is anticipated that the more efficient, specialist predator *Lates* will be able to consume an even greater proportion and that consequently the *Lates* numbers may stabilize in association with a lower haplochromid stock and also a reduced *Bagrus* abundance as its share of the haplochromid resources is restricted.

A similar situation can also be seen in the nearby Lake Kyoga where *Lates* was introduced in 1955. The expansion of the predator was initially at the expense of the haplochromids but, more recently, the tilapia, *Oreochromis niloticus*, has formed the largest component of the diet of *Lates* in this lake. One further notable feature of this predator was the early rapid growth rate and the particularly large size attained, both of which declined as the population reached equilibrium over the first 10 years in its new environment. Conditions were evidently so favourable in the early phase of colonization that the rapid growth produced individuals up to 250 kg, but as *Lates* became more numerous and the pressures on the prey species increased, these early trends were reversed.

Perhaps the best-documented example of the effects of a predator on a tropical aquatic community was the accidental addition of the cichlid piscivore, *Cichla ocellaris*, to Gatun Lake, Panama, around 1967 (Zaret and Paine, 1973). In less than 5 years the presence of this predator, which can grow up to 2 kg in weight, led to the elimination of six of the eight common fishes and decimated a seventh. One of the commonest, the atherinid silver side, *Melaniris chagrensis*, showed a 90% reduction in numbers over a 2–3-year period between 1969 and 1971–72. Only one species, *Cichlasoma maculicauda*, showed a small increase, probably due to the elimination of species that had previously fed upon its fry. The removal of so many elements of the second and third trophic levels considerably simplified the system as a whole (Fig. 7.6). The shoals of *Tarpon* and flocks of terns which habitually sought out the *Melaniris* congregations disappeared and increased numbers of mosquitoes appeared, no doubt owing to the great reduction in insect-eating fishes. Even the disease pattern of malaria in the surrounding districts altered in a way which may have been caused by the ecological disturbance. Some changes could also be observed in both the zooplankton and phytoplankton so that virtually the whole system was affected by the addition of this one species. This provides a very sharp contrast with the introduction of the plant- and detritus-feeding cichlid, *Oreochromis mossambicus*, to the Ranu Lamongan lake in Indonesia where very few changes could be found and demonstrates the importance of predators in regulating tropical aquatic ecosystems.

The effect of a single predatory species on these complex tropical communities reveals something of their relative stability. The communities are not sufficiently stable to absorb a predator and they show little resilience and a limited persistence with such an addition, certainly much less than when a grazer or detritivore is added, judging from the restricted amount of information available. Certainly there are distinct changes in species abundance and perhaps also in species richness. Of course, successful introductions will always receive more attention than unsuccessful ones, so that it is quite possible that many tropical aquatic communities are very stable and resist invasion by other species although this is not reported in the literature. This, however, is not the impression given by existing records.

To what extent the degree of destabilization caused by the introduction of a predator, for example, is characteristic only of more diverse, more complex tropical systems is difficult to say. The impact of *Lates* in Lake Victoria, for example, is perhaps of a comparable extent to that caused by the invasion of the North American Great Lakes by the sea lamprey, *Petromyzon marinus*, following the construction of the Welland Canal between the sea and these lakes. One of the results of this was a drastic fall in the abundance of whitefish (*Coregonus*) species which had previously dominated the lakes, with a compensatory replacement by other invading species. Perhaps, therefore, there may be little difference between the tropics and temperate situations but this requires more attention. The outcome of such a consideration will probably hinge upon whether the constituent populations within the communities are resource limited or whether it is the degree of predation which dictates their abundance and composition. The far-reaching consequences of predator introduction suggest that these do have a quite fundamental regulating influence within the communities. The dissimilar effects which predators have on benthic and pelagic systems, mentioned earlier, may mean that they play different roles in different communities.

The reduction or deletion of a species from a system also says something about its stability. An example of this is again provided by the fishes of Lake Victoria. Until relatively recently, the major fishery on this lake was for the tilapia *Oreochromis esculentus*, but the fishing intensity has been such that this stock has declined until it is now only some 1–2% of the catch. There has, however, been a rise in the proportion of *O. niloticus* caught. This is one of the newly introduced tilapia into Lake Victoria and it has similar requirements to *O. esculentus* which have apparently allowed it to expand as *O. esculentus* has declined. The combination of fishing pressure and competition has, therefore, led to a partial replacement of *O. esculentus* by *O. niloticus*. This is not an uncommon response within animal communities where several species are capable of exploiting an abundant resource, although it would be less likely to happen in a more recent system, such as a reservoir, where a full complement of colonizing species may not be present. It does seem likely, however, that even if the fishing pressure on *O. esculentus* were removed the community would not revert to its original equilibrium with *O. esculentus* predominant amongst the grazers since its place has now been taken, at least in part, by *O. niloticus*. Other changes within the community, however, appear to be minimal.

Deletions of predatory species are not very common, particularly as far as fish are concerned, since it is usually the substitution of the natural predator for a new one, man, which takes place. However, some indication of the possible implications can be obtained from the instance of the elimination of the Amazon crocodile, the caiman (*Melanosuchus niger*), which was hunted to virtual extinction over a period of 15 years from 1940, involving many millions of animals (Fittkau, 1970). During this period, fishermen reported a considerable decline in their catches which is perhaps surprising since competition from the caiman was being removed. A probable explanation for this, however, has been suggested by Fittkau (1970). In the Amazon basin many of the fish spend most of the year in the rich varzea lakes where they grow and become mature. To reproduce, however, they need to migrate into the main Amazon and accumulate at the mouths of various tributaries in enormous shoals prior to spawning. At this time, they were particularly vulnerable to the caiman and they could be taken in large numbers. The waters of the Amazon River are nutrient-poor compared to those of the varzea lakes and the river lacks the extensive floating meadows (Section 4.9.2) and marginal vegetation which contribute so much organic material to the lakes. The consumption of a significant percentage of the fish by the caiman in the river leads to a recycling of the nutrients incorporated into the fish whilst they lived in the varzea lakes, via the waste products of the crocodiles, which must have had a positive effect on the fertility of the system, particularly in the bays around the mouths of the tributaries which are so important to the young fish after spawning. Of course, the fishes may be harvested by man rather than the crocodiles but, since man removes the fish entirely from the system, no nutrient recycling is possible and the fertility is removed with the fish. The potential interruption in the transfer of nutrients from the varzea lakes to the nursery grounds in the river due to the elimination of the crocodile emphasizes the importance predators can have within the system.

There are equivalent cases to this, such as the elimination of the African crocodiles from most lakes and rivers on the continent, but detailed studies of the impact on community structure in terms of changes in species richness or species abundance are not recorded.

Tropical aquatic communities, therefore, are by no means entirely resilient to change. They are not uniformly more diverse or more complex than temperate systems although some elements, such as the fish, certainly are. Complex fish communities in tropical lakes appear to show considerable elasticity in accommodating new species from lower trophic levels into what is already a crowded environment. The addition of a predator, however, can be almost catastrophic. The more stable environmental conditions in tropical lakes may permit the evolution of very diverse fish communities incorporating considerable specialization in feeding and reproduction; this specialization may, however, reduce the adaptability and render the community more fragile (Lowe-McConnell, 1975, 1977). Riverine fish assemblages, by contrast, consist of populations adapted to greater seasonal and year-to-year fluctuations in resources and conditions and therefore may be more resilient or less fragile (Lowe-McConnell, 1975).

The extent to which the generally greater complexity of tropical aquatic communities makes them more or less resistant to change, or influences the rate at which changes are accommodated, is not possible to say at present. It is, however, a question of considerable practical importance since major changes are currently being imposed upon lakes and rivers by man acting as a predator and as an agent of environmental change through increasing water requirement for enlarging populations and through pollution. Whether the communities will prove particularly vulnerable to the predatory attentions of man or the simplifying biological effects normally associated with the creation of more extreme chemical and physical environments is a point which will no doubt become clear over the next few decades.

GENERAL READING

Fernando, C. H., 1980. The species and size composition of tropical freshwater zooplankton with a special reference to the oriental region (South East Asia). *Int. Rev. Ges Hydrobiol.*, **65**: 411–426.

Fryer, G., and Iles, T. D., 1972. *The Cichlid Fishes of the Great Lakes of Africa: their Biology and Evolution.* Edinburgh: Oliver and Boyd.

Green, J., Corbet, S. A., Watts, E., and Lan, O. B., 1976. Ecological studies on Indonesian lakes, overturn and restratification of Ranu Lamongan. *J. Zool. Lond.*, **180**: 315–354.

Hutchinson, G. E., 1959. Homage to Santa Rosalia or why are there so many kinds of animals. *The American Naturalist*, **93**: 145–159.

Lewis, W. H., 1978. A compositional, phytogeographical and elementary structural analysis of the phytoplankton in a tropical lake: Lake Lanao, Philippines. *J. Ecol.*, **66**: 213–226.

Lowe-McConnell, R. H., 1969. Speciation in tropical freshwater fishes. *Biol. J. Linn. Soc.*, **1**: 57–75.

Margalef, R., 1968. *Perspectives in Ecological Theory.* University of Chicago Press.

Pimm, S. L., 1984. The complexity and stability of ecosystems. *Nature*, **307**: 321–326.

Watt, K. E., 1973. *Principles of Environmental Science*, Chs 1 and 2. New York: McGraw-Hill.

CHAPTER 8

THE USE AND CONTROL OF AQUATIC BIOLOGICAL RESOURCES

8.1 MAN AND WATER

For man, the most important resource to be gained from lakes and rivers is the water itself, for drinking, irrigation of crops, livestock, or for waste disposal. From the point of view of biological production, however, which is the main emphasis of this book, the most readily utilizable commodity is fish. Natural levels of production can be exploited through capture fisheries in lakes and rivers, whilst there also exists the possibility of manipulating the environment to approach optimal conditions for the growth of the animals in some form of culture or fish farming. By no means all man's activities improve the environment, however, and with increasing intensity of agriculture, industrialization, and urban development many by-products can be detrimental to the water, both for the naturally occurring organisms as well as for the reuse of the water itself. In this case the capacity of the aquatic communities to carry out self-purification of the water is of considerable importance and the ability to accomplish this often depends upon the decomposer system.

Increasing human population sizes and the pressures for speeding up the rate of economic development in tropical countries have led to an even greater consideration of the extent of aquatic resources. The diverse uses to which water and the resources it supports can be put means that any change in one aspect of the exploitation of those resources can have consequences for others. A good example of this is the construction of a dam across a river. This has the immediate effect that it increases the availability of water which may be used for drinking, irrigation, or to generate electricity. At the same time the fishery in the river will be disrupted both at the immediate dam site and also downstream. The physical presence of the dam will interrupt fish migration and may prevent the fish from reaching their spawning areas, whilst the change in flow pattern down the river may reduce the environmental information available to the fish. Moreover, the heavy suspended load, much of which will be organic material, will sediment out in the lake greatly reducing the load in the outflow. Since the organic material in rivers is an intrinsic part of the production cycle (Section 5.2.3(b)), the productivity of downstream fisheries can be affected. Downstream effects can be detected some way below the dam, even as far as the estuary where a lucrative prawn fishery could be located, providing an exportable 'cash crop' bringing in valuable foreign exchange. It is important to know, therefore, if the losses to the downstream fisheries will be offset by gains from the reservoir fishery which will develop following impoundment of the river. If the reservoir fishery is likely to be substantial, how many fishermen will it support, where will they come from, and where will they live? The provision of a new major water source, particularly if it supports industries associated with power production, will attract people, and if new towns and villages are required where will the waste go? Into the reservoir or into the outflow? How severe will be the effects from the domestic and industrial waste? All of these questions and many more must be taken into account when increased utilization of aquatic resources is being planned.

Fig. 8.1 The Kariba Dam across the Zambezi River, eastern Africa

The construction of reservoirs and man-made lakes (Fig. 8.1) is proceeding at an increasingly rapid rate. Nearly all of the standing water in South-east Asia is in this form (Fernando and Furtado, 1975), whilst India probably has the largest number with some 1554 listed, covering 1 million ha. The complex problems which these projects entail require careful planning to obtain the best results and to minimize the detrimental effects. The situation becomes more complex when a water body spans more than one national territory. The Nile, for example, runs for most of its length through Sudan and Egypt, but although Sudan is upstream and could therefore have first option on the river water, the largest impoundment is actually in Egypt, although the resulting lake, Nasser/Nubia, tails back into Sudanese territory. Through mutual agreement, however, it has been acknowledged that Egypt's requirements are the greater; consequently Sudan only takes 17% of the Nile water.

The biological, economic, and political implications behind the increased efficient use of water and aquatic production require an integrated, multidisciplinary approach to planning. Answers and estimates of variables are generally required quickly but, as we have seen in earlier chapters, with respect to production in tropical waters, our understanding is only in its infancy. Research and development must of necessity, therefore, go very much hand in hand.

From a biological viewpoint there is one other aspect of man and his association with water that is of particular significance to the tropics. Water plays a part in the transmission of many human diseases either as a direct agent itself, particularly of enteric bacterial infections such as cholera and typhoid, but probably more often through animal vectors such as mosquitoes and black-flies which spend at least part of their life cycles in water. Any project which increases the degree of contact between man and water must also take this into account.

8.2 FISHERIES

8.2.1 The Basis of Production and the Prediction of Yields

Fish appear in most trophic levels of tropical aquatic communities and are an important component in secondary production. Since they are the part of the community most commonly harvested, yields from fisheries are most frequently the only index, however imperfect, of the productivity of the community as a whole. For African lakes and lakes in Madras, southern India, Melack (1976) was able to demonstrate positive correlations between fish yields and gross photosynthetic production. The nature of the relationship was similar in both cases with primary production being proportional to the log of fish yield; that is, fish yield increased exponentially with gross production. The implication of this is that as primary production increases the fish are able to take a greater proportion and convert it into their own growth, although an additional factor, since we are dealing with fish harvest and not fish production, is that fishing might be more effective at higher fish densities and thereby remove a greater proportion of the fish present. The percentage of net primary production taken by fish in African lakes varied between 0.1 and 1.5%.

For the varzea lakes and the rivers of the Amazon system, Schmidt (1973a,b) suggested that around 2% of net production might be converted to fish production although application to other situations suggested potential fish yields which appeared rather high (Schmidt, 1982). It is, however, probably unrealistic to use a single percentage conversion of primary to secondary production in this way since the relationships just outlined appear to be non-linear.

Relationships between the primary producers and fish yields (Melack, 1976; Oglesby, 1977) are ecologically enlightening and also offer the possibility of predicting yields from unknown waters, providing the data on primary production are available. However, the accuracy of such predictions is probably limited by the availability of appropriate data and by our restricted understanding of tropical systems. Application of the relationships derived by Melack (1976) and Oglesby (1977) to primary production estimates from Lake Tanganyika suggested fish yields of up to ten times lower than the current fish catch sustained by the lake (Hecky and Fee, 1981). The freshwater sardines which make up the bulk of the fishery appear to be very efficient producers, since fish yields amount to 0.45% of carbon fixed in gross primary production which is very high compared to the range of 0.001–0.1% previously found (Hecky and Fee, 1981). It remains to be seen if this proves to be exceptional for tropical systems.

The characteristics which ultimately control productivity in communities are the abiotic factors of the environment such as light and nutrient availability. There has been a feeling amongst fisheries scientists for some time that lake depth may be an important influence on fish production and for African lakes some relationship has been shown, with higher yields being associated with shallower lakes (Fryer and Iles, 1972). In North America a good positive correlation was found between yields and the ratio of TDS to mean lake depth (Ryder, 1965). This ratio, which became known as the morphoedaphic index (MEI), can be summarized as

$$\text{MEI} = \frac{\text{TDS}}{\text{Mean depth}}$$

Other measures related to total ionic concentration, such as conductivity or alkalinity, can be used instead of TDS. The MEI has also been shown to give a positive correlation (Fig. 8.2) with fish yields from African lakes (Henderson and Welcomme, 1974; Marshall, 1984). This means that as the lake becomes shallower and/or the ionic concentration becomes greater, the higher will be the MEI and the greater will be the catch. Since a significant regression line can be drawn to fit this relationship (Fig. 8.2) this means that an estimate of potential fish yield can be predicted from relatively simple physical and chemical determinations.

The correlation can be viewed as an empirically useful but obscure relationship or it can be regarded as a real 'black box' model of the ecosystem; the elements of the MEI are a function of the limiting inputs and the fish production represents the eventual output (Ryder et al., 1974). For example, the alkalinity, which is largely a measure of the bicarbonate ions in the water, is closely correlated with conductivity or TDS (Talling and Talling, 1965). All of these will

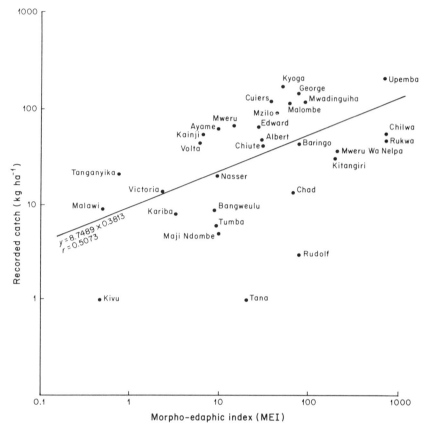

Fig. 8.2 The relationship of the morpho-edaphic index (MEI) to fish yields from African lakes. (*Reproduced by permission of the Food and Agriculture Organization of the United Nations from Henderson and Welcomme, 1974*)

therefore give an indication of carbon availability for photosynthesis in the water. The mean depth, however, is probably related to a complex of factors. In shallow lakes, a greater proportion of the water is within the euphotic zone than in a deeper lake of the same size. In addition, long-term stratification becomes less of a problem and the proximity of the productive zone to the lake bottom will speed up nutrient recycling. In these terms it could be seen why increasing TDS and decreasing depth, when combined into the MEI, are associated with higher production.

Irrespective of the theoretical background, however, the use of MEI does allow a rapid first estimate of potential fish catches from an unknown lake or reservoir, although the accuracy of the estimate is only as good as the original data from which the relationship was calculated. Amongst the original data for which the predictive line for African lakes was determined were some which had rather low numbers of fishermen and were likely, therefore, to be yielding less than their maximum. Omission of these would, therefore, give a better representation of maximum potential yield. Better correlations will also be obtained the more homogeneous the series of lakes are for morphology and chemical composition. Predictions from such relationships will never be entirely accurate but they will indicate an order of magnitude—whether, for example, 10, 100, 1000, or 10 000 tonnes are to be expected from a given body of water. This is essential for the first stages of planning in order to have some suggestion as to the level of development the fishery will sustain in terms of mechanization, new roads, harbours, or processing plants (Henderson *et al.*, 1973; Ryder and Henderson, 1975). A comparison of three methods of estimating yield using the MEI, primary production, and standing crop as indicators in Lake Bangweulu, Zambia, showed that they all lay fairly close together over a range of 10–20 kg ha^{-1} year^{-1} which coincided with the directly estimated yield of 19 kg ha^{-1} year^{-1} (Toews and Griffiths, 1979). Only with the selective relationship for MEI against maximum yield was some deviation observed when an estimated possible catch of 35 kg ha^{-1} was indicated.

Fish catches in rivers do not seem to show such an apparently simple relationship to physical and chemical variables. Chemical factors are much less variable compared to those of lakes, and depths are usually relatively shallow and also seasonally variable. Much better correlation is obtained between fish catches for African rivers and the area of the drainage basin; distance from source is also significant (Welcomme, 1976). Quite simply, the larger the catchment area the larger the yield, although those from rivers with extensive floodplains do have yields of the order of ten times greater than those more confined to the main channel. This disparity is due to the enormous increase in resources available to the fishes from the large areas of land which are submerged during the floods. Another feature, however, is that major floodplain fisheries can become centralized and are more likely to be monitored than in more diffuse river systems, and consequently this recording effect may contribute to the apparent difference in catches between the two types of rivers.

The fish catches per kilometre of river appear to increase with distance from the source and can vary between 1 tonne km^{-1} at 10 km from the source to 20 tonnes km^{-1} at 1000 km (Welcomme, 1976). This may reflect the accumulation of organic material and other food sources down a river system. Since river production is much less dependent on primary production it is not surprising that chemical composition and inorganic nutrient levels are less useful indicators of fish catches than in lakes where the influence is more direct. It would be interesting to know, in fact, if secondary production including fish yields were related to the organic carbon content of river waters as a measure of allochthonous and autochthonous detritus availability.

The relationship of fish yields to some readily estimated chemical or physical factor can give a valuable initial prediction of the order of catches to be expected from a given water body in the preliminary planning stage of any development. However, to estimate catches from existing fisheries more directly can be a difficult task. Where recognized landing grounds are to be found, monitoring catches can be carried out effectively but often fish are landed in a large number of widely dispersed areas. To overcome this difficulty, at least in part, a 'frame survey' can be carried out where the shoreline is divided into sections which are sampled at random. These samples can then be treated statistically to produce an estimate of catch for the whole water (Bazigos, 1974).

River systems are rather harder to deal with than lakes owing to the fact that the network of tributaries can be spread over such a wide area. In these circumstances it becomes very difficult to assess how much fish is being taken because many people will fish casually in the tributaries, purely on a subsistence basis, and often there are few recognized landing areas. At the same time the extensive nature of the river drainage basins means that the fish supply influences more people than does a more localized lake fishery. The river basins of Africa cover 75% of the sub-Saharan region of the continent. At present only first estimates of fish catches from rivers are possible so that, for example, a theoretical estimate of production from all African rivers suggests an annual yield of some 1.2 million tonnes of fish as opposed to a recorded estimate of some 400 000 tonnes (Welcomme, 1976). In the Amazon basin recorded production is currently 850 000 tonnes although the estimated potential may be 212 000 tonnes if currently underutilized species are included (Bayley, 1981). Monitoring of river fisheries is, therefore, a real problem.

8.2.2 Fishery Management

Estimation of catches and monitoring fisheries is only a starting-point. Fisheries are dynamic production systems which need to be managed to obtain the maximum return. The population characteristics which determine the level of production can be put most simply as

$$S_2 - S_1 = R + G - (M + F)$$

where S_1 = stock biomass at beginning of time period;
S_2 = stock biomass at end of time period;
R = recruitment of young into fishable stock;
G = addition due to growth;
M = loss due to natural mortality;
F = loss due to fishing mortality.

This expression states that any change in biomass with time will be due to the difference between gains from new young fish entering the fishery plus further growth by existing members of the population and losses due to natural mortality and from those taken by fishing. If one balances the other then there is no net change in biomass and the population is stable although this is a rare event.

In fact, each of these terms is far from simple and all of them are influenced by a variety of environmental and biological factors. Recruitment, for example, will in some fish species depend upon climatic factors such as seasonal temperature, whilst in others it will depend upon the numbers of adult spawning stock.

Each component, therefore, requires its own mathematical description which can then be integrated to form an analytical model for the whole population summarized as a yield equation (Beverton and Holt, 1957). This requires considerable biological information about the fish population and is the basis of most management projections developed for temperate marine fisheries, which have the longest history of fishery management. This approach can be used on fish populations in the tropics if the fish can be aged properly (Section 5.4.3). For one of the *Oreochromis esculentus* populations in Lake Victoria, age and growth (G) could be estimated from biannual rings laid down on the scales. By combining this with the length frequency data and the estimates from catches made by the fishermen, it was possible to assess both natural (M) and fishing mortality (F) (Garrod, 1959, 1963). However, this gives only a part of the information required to predict yields for one species.

Examples of this type of data for tropical fisheries are very sparse. By making a number of simplifying assumptions, Petrere (1983) was able to produce sufficient data on G, M, and F from market samples to calculate optimum yields of *Colossoma macropomum* from the Amazon in the vicinity of Manaus. From this, it was feasible to judge the degree of exploitation of this particular stock and the yield likely to be produced from a given number of trips by the fishermen.

Many inland tropical fisheries are being heavily exploited and answers are needed rapidly. Further complications arise because the mathematical expressions developed in temperate waters to describe patterns of mortality and recruitment may be different in the tropics where mortality rates are generally higher, life spans are shorter, and most of the fish in a population are relatively young (Welcomme, 1979). One additional major point is that, because of the high diversity, tropical fisheries are often based on a large number of species and not just on a single species as are those of temperate regions, which renders the application of models based on single stocks difficult. Much fundamental work, therefore, remains to be carried out but it is possible to adopt more simplified approaches.

Working on catches from rivers with extensive floodplains, Welcomme and Hagborg (1977) deduced that the feature which most significantly influenced the growth of fish in any single year was the area of land flooded, since this was an index of food availability during the prime growing season. During this period, which may last only 3–4 months, the fish can achieve 75% of their annual growth. Mortality, however, was seen to be a function of the area of water left during the dry season when large numbers of fish are concentrated into a progressively smaller area. The major features governing production have, therefore, been interpreted in terms of the characteristics of the hydrological regime rather than fishing effort. Fisheries for juveniles coming down off the floodplain as the waters recede probably do not have an undue effect on the population as a whole since many of these fish would die anyway during the low-water period. Fisheries based on the spawning migrations of the adults, however, are another matter.

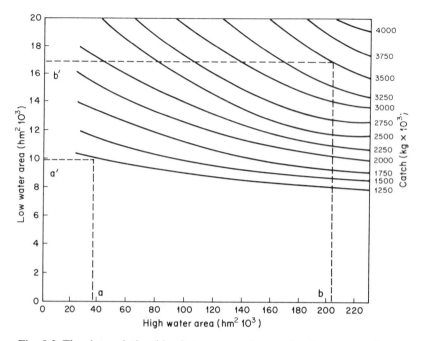

Fig. 8.3 The interrelationship between maximum flood area, minimum low-water area, and fish catches from a model of floodplain fisheries (Welcomme and Hagborg, 1977). Predicted catches following poor and extensive floods are indicated by the intersection of lines a–a′ and b–b′ respectively

In the case of floodplain rivers, it has been possible to produce a preliminary model of the fisheries by integrating the data which are available with what is known about changes in hydrological regime. This gives some indication as to the pattern of fish production and also some predictions as to the order of magnitude of catches which might be expected from the highly variable floodplain system (Fig. 8.3). Basically it shows that when the floods are poor and this is followed by very low water levels during the dry season or inter-flood period, then catches can be expected to be at their lowest (a–a′ in Fig. 8.3). Conversely, when a very extensive flood is followed by a period when the water remains relatively high between the floods, then growth will be high, mortality will be low, and the best catches can be anticipated (b–b′ in Fig. 8.3).

There are a number of assumptions included in this model and the data base is rather restricted owing to the small amount of work carried out on the production and population dynamics of the fishes. It does, however, provide a starting-point and it has also been applied to predict yields from widely fluctuating systems such as man-made lakes with large drawdowns during the dry season, which behave rather like floodplain rivers (Moreau, 1982). Fish production in rivers which do not possess large floodplains may be regulated by other circumstances since there is much less variability in the river area. The huge increase in resources due to flooding of large areas of land does not occur and may not, therefore, be the principal factor determining growth.

Amongst the rivers of Amazonas State, Brazil, where a well-established fishery supplying the urban market of Manaus has developed, Petrere (1983) found that the most significant predictive relationship was between the actual catch and the number of fishermen operating. This was .useful for immediate management purposes although it did not indicate the scope for further expansion of the fishery.

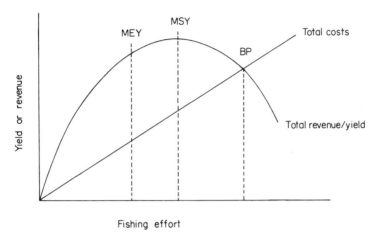

Fig. 8.4 The relationship between fishing effort and yield or economic value of the catch: (MEY) maximum economic yield; (MSY) maximum sustainable yield; (BP) break-even point (open access equilibrium)

In a consideration of the management of most fisheries, particularly where very wide amplitude fluctuations in the environment do not occur, a major factor must be the amount of fishing carried out and the mortality this causes amongst the fish (*F* in the above equation). This additional mortality is the element which can cause problems in the regulation of fish populations and the maintenance of production, whilst at the same time it should be the element most easily controlled. Observations on the relationship of fishing effort to fish yield on a more or less empirical basis can help to predict the future of a fishery. Unexploited fisheries are characterized by having a high proportion of large, old fish. When fishing begins, many of these old fish are caught and their place is taken by recruitment of a large number of smaller, younger fish. These use food more efficiently for growth and therefore the potential yield from the fishery is greater. If fishing effort increases to any further extent, however, there comes a point when too many fish are taken. There are insufficient numbers recruited to replace those taken by fishing and the yield begins to decline (Fig. 8.4). This is a rather simplified approach, but it is evident that the degree of effort has a considerable influence upon the yield from the fishery. There is also the implication that there is a level of effort which will give the maximum return, or what has become known as the maximum sustainable yield (MSY) from a fish stock. In fact, the MSY may not show quite such a sharply defined peak (Gulland, 1968). Nevertheless, generally some intermediate range of effort should produce the highest returns whilst not damaging the stock as a whole. The practical value of the MSY is not without criticism (e.g. Larkin, 1977) but it remains a useful concept.

Fig. 8.5 The most common unit of effort in the tropics, an individual fisherman with his gill net

One common effect of increasing fishing effort on a given fish stock is that the catch per unit effort (c.p.u.e.) goes down. In many tropical inland fisheries the unit of effort is the individual fisherman with a small boat or canoe and perhaps one or two helpers, using gill-nets which are checked daily (Fig. 8.5). A typical pattern of events is illustrated by the history of the gill-net fishery in Lake Victoria (Payne, 1976b). When gill-nets were first introduced in 1905, catches of *Oreochromis esculentus* attained 50–100 individuals per net. In 1928 this had dropped to 6–7 fish per net and by 1954 the average was only 1.6. In an attempt to increase returns the mesh size was dropped from 5 to 4.5 in in order to catch smaller fish and there was a further reduction to 3.5 in by 1966. At this point the mean size of fish retained by the net was very close to the size of first maturity of *O. esculentus* which therefore threatened the reproductive potential. Even switching attention to smaller fish only provided a temporary respite, however, so that by 1968 each fisherman was only catching an average 0.35 rather small fish in each net. In some places where records were available total yields were becoming smaller even though there had been a substantial increase in fishing.

A similar example is that of the fishery in the highly productive Laguna de Bay in the Philippines where the catch declined from 349 748 tonnes in 1963 to 121 880 tonnes in 1973 whilst the number of fishermen had increased from 13 000 to 16 000 (Smith, 1983).

Tropical inland fisheries are commonly unregulated and traditionally have a more or less open access to participants. Some indication of the effect of increasing numbers of fishermen can be gained from the relationship of numbers to catch per unit area for African lakes (Fig. 8.6). The indications are that when there are few fishermen, an increase in their numbers improves the total catch since the water is underexploited, but that beyond the optimum further increases reduce the catch. Social and economic pressures often act to maintain higher densities of fishermen. The reasons for this can be seen if the cost of the effort is superimposed on the relationship between the catch, or its value, and effort (Fig. 8.4). Where the line for total costs cuts the curve for yield revenue lies the point where income just equals expenditure. The effort is unlikely to expand beyond this since costs would exceed income (Christy and Scott, 1965).

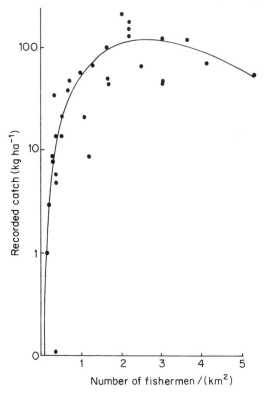

Fig. 8.6 The effect of the numbers of fishermen on catches from African lakes. (*Reproduced by permission of the Food and Agriculture Organization of the United Nations from Henderson and Welcomme, 1974*)

At this break-even point, the fishery ceases to become profitable and fishermen, unable to make a livelihood, would begin to drop out. This, however, reduces the effort and a positive difference between costs and revenue from yield reappears. The fishery would again appear to be marginally profitable, tempting more men to take up fishing again, perhaps as a supplement to their farming activities. The effort will therefore oscillate around this economic break-even point whilst the fishery yields less than its potential maximum and the costs of capture are very much higher than they need be. This is why most fishermen in an open-access fishery are poor. In fact, the greatest difference between costs and the income may appear at a fish yield rather less than the biological maximum of the MSY at a point (See Fig. 8.4) known as the maximum economic yield (MEY). When unit costs are low, however, and a man's time is of no consequence besides his livelihood, then the break-even point may be a considerable way down the declining arm of the catch/effort curve, with the damage which that implies for the fishery. The MEY then becomes a notional or hazy value. Obviously the rational solution is to curtail effort by reducing the numbers of fishermen involved. However, since most fisheries are traditionally free of access and the fishermen themselves are widely dispersed this is extremely difficult to contemplate. Only in new fisheries, such as those planned on man-made lakes, is a limited-entry fishery conceivable.

Since the assemblages can be very diverse (Section 4.8), the exploitation of tropical fisheries often involves the stocks of many different species. It is the community as a whole which is being exploited rather than just a single species as is the case in temperate fisheries. The effect of increasing fishing mortality on a diverse community has only just begun to receive attention. The c.p.u.e. of multispecies fisheries appears to decline with effort in a similar way to those based on a single species, as has recently been demonstrated for the Tanzanian sector of the Lake Victoria gill-net fishery and the Lake Malawi trawl catch (Turner, 1981). It is possible that yield levels close to maximum can be maintained over a wider range of fishing effort by a community, compared to a single species stock, but this will almost certainly be accompanied by a change in species composition as the more vulnerable stocks become exhausted. In Lake Malawi, a rapid growth in the trawl fishery led to a 20% reduction in the number of species in the fishery whilst in both this lake and Lake Victoria, increasing effort led to a decline in the abundance of larger species and an increase in smaller types (Turner, 1981). In general, small species tend to have higher natural mortality and growth rates, both of which will act to maintain productivity. Part of the reason larger fish are eliminated first from the fishery is, of course, due to the fact that one aspect of increasing effort is a reduction in mesh size, as discussed above in relation to the Lake Victoria fishery, which undermines recruitment. However, the larger fish may be the most valuable and so that, although the tonnage of fish may be maintained, its economic value may drop. The tilapia, *Oreochromis esculentus*, used to be the mainstay of the Lake Victoria fishery, for example, and was valued both for domestic consumption and for frozen or processed products. Now the largest single component of the fishery is the small haplochromines which retail at less than half the price of *O. esculentus*. This species now contributes only a few per cent of the total catch and the situation has been further destabilized by the eruption of the large introduced predator *Lates niloticus* (Section 7.6).

The difficulties in maximizing yield whilst conserving particularly valuable species populations is emphasized by some of the Amazon fisheries. Although there are 1300 species of fish described from the Amazon and a further estimated 2000 await description, 75% of the catch on sale at major markets is derived from three characins, the tambaqui (*Colosoma macropomum*), jaraqui (*Semaprochilodus* spp.), and curimata (*Prochilodus nigricans*). The fishery is generally highly selective and tends to prefer large-bodied species which take several years to mature (Junk, 1982). The c.p.u.e. on these types in the vicinity of population centres such as Manaus in central Amazonas State has declined from 80 tonnes per boat annually to 25 tonnes over the period of between 1970 and 1980 as the number of registered boats has increased from around 100 to 700. The total annual catch tends to vary with the height of the floods but, on average, the availability of fish on the market has doubled over the period and, as yet, shows no marked diminishing trend. However, the total recorded catch from the middle and upper Amazon basin is some 85 200 tonnes whilst the theoretical maximum yield could be 300 000 to 350 000 tonnes (Bayley, 1981). These yields will only be approached if the fishery exploits all stocks and switches to utilization of the small, short-life-cycle species. It would be virtually impossible to do this, however, whilst maintaining the stocks of large species intact, since the widespread shift to smaller mesh nets would selectively remove new recruits and premature fish from their populations. The present range of mesh sizes is between 14 and 25 cm but any trend to lower mesh sizes would be difficult to prevent except that, in the Amazon, the extensive damage to small-mesh nets inflicted by the shoals of ferocious piranhas tends to provide its own deterrent (Goulding, 1981).

If we finally sum up our understanding of fisheries in the tropics we can note that in some instances it is possible to predict yields from existing or proposed fisheries but that these predictions are founded on a very narrow data base with considerable room for expansion and refinement. Most are based on the 'black box' synthetic models where relationships between input and output have been observed whilst there is little understanding of the mechanisms between the two. Then there is the generalized observation that too great an increase in effort on a fish community forces the livelihood of the fishermen to become progressively more marginal whilst some stocks, often the more valuable, are eliminated from the fishery. Eventually the total yield can be depressed. However, predictive relationships between catch and effort are in short supply and knowledge of growth, mortality, and recruitment patterns of the type which have been central to many temperate fisheries management programmes is virtually absent (Payne and McCarton, 1986). Finally, there is the considerable difficulty of moderating effort in the unlimited entry, traditional fishery in order to maximize catch and to provide at least some fishermen with a viable income. Even if it is not possible to curtail the effort, an understanding of how the fishery works might indicate how this effort can best be deployed to minimize damage to the stocks and maximize returns within limits.

8.2.3 The Introduction of Exotic Species

Occasionally, particularly when a new environment such as a man-made lake has been created, an identifiable major niche may remain unfilled and the introduction of a species from another locality may be contemplated to make full use of the resources. Often, following the formation of a man-made lake from a river, the riverine fish fauna is sufficiently diverse to colonize most niches, particularly if cichlids are present, since their breeding technique is particularly suitable to non-flowing waters (Section 4.8.2). Numerous cichlids, particularly amongst the tilapias, are also plankton feeders. In Sri Lanka and other parts of South-east Asia, where extensive reservoir construction has been carried out, yields have often been more than doubled by the introduction of some of the tilapias, notably *Oreochromis mossambicus* and *O. niloticus* (Fernando and Furtado, 1975). A particularly dramatic example was an ancient man-made lake, Parakrama Samudra in Sri Lanka, where yields before 1953 were $1.07 \, \text{kg ha}^{-2}$ whilst, following the introduction of the tilapia *O. mossambicus*, they rose to $445 \, \text{kg ha}^{-1}$ by 1978. The indigenous riverine fish faunas of South-east Asia and perhaps also South America may be unable to produce types capable of exploiting the open-water pelagic niche of plankton or fine particle feeders, unlike some of the cichlids and the African freshwater herrings (Fernando and Holcik, 1982). Full utilization of these resources is often the key to obtaining maximum yields from lakes and reservoirs. A particularly good example of this is Lake Kariba on the Zambezi which, until 1972, produced some 2000 tonnes of fish consisting of a mixture of tilapia, *Labeo*, and catfish species. At this time, however, the small freshwater herring, *Limnothrissa miodon*, was introduced from the deep Rift Valley lake, Lake Tanganyika. There was an exponential rise in the catch of this plankton-feeding fish (Section 4.8.3) until in 1981 it totalled 8000 tonnes. It is notable that before the advent of this sardine fishery the relative yields from inshore fishing were of the order of $5.5 \, \text{kg ha}^{-1}$. The MEI of the lake was, however, 2.96 which would have predicted a yield of around $24 \, \text{kg ha}^{-1}$ (Fig. 8.2; note that the actual yield shown for Kariba on this graph was taken from data collected before the sardine fishery had begun); when, however, the sardine yield is incorporated, the relative yield becomes $20.0 \, \text{kg ha}^{-1}$, which is much closer to the predicted catch (Marshall, 1981). Whilst the c.p.u.e. for this fishery has fallen year by year the total yield has risen, reaching 11 131 tonnes in 1981 (Marshall and Mandisodza, 1982), although there are signs that the fishery was nearing its peak. However, the simple but rather specialized dip-net vessels (Fig. 8.7) used for exploiting the fishery have allowed fishing to be licensed and therefore brought under control, which presents a reasonable opportunity for regulating the fishery.

Fig. 8.7 Dip-net vessels built for catching the freshwater sardines, *Limnothrissa miodon*, introduced into Lake Kariba. They are built on low pontoons to give stability in the turbulent weather common on the lake and the small black dinghies which carry carbide lights are moored close to the vessels on the fishing grounds to attract the sardines at night

The introduction of new species therefore offers a tempting prospect, but this may prove to be an illusion if the niche is not really empty. There have been suggestions, for example, for introducing into the natural Lake Malawi the sardine, *Limnothrissa miodon*, which did so well in the recently created Lake Kariba. This lake, however, has its own mature, very diverse fauna which includes a flock of plankton-feeding cichlid species, the utaka, the productivity of which may suffer in competition with the sardine for no net gain. One case in which a high-yielding indigenous fauna has been damaged by introductions concerns the group of cyprinoid species endemic to Lake Lanao in the Philippines. These types alone contribute $27 \, kg \, ha^{-1}$ to the fishery (a relative yield higher than the sardines of Kariba) but some are now facing extinction in competition with a recently arrived and much less valuable goby, *Glossogobius giurus* (Fernando and Holcik, 1982).

8.3 AQUACULTURE

The ultimate aim of the culture of aquatic organisms is to control all aspects of the production process to maximize output and achieve the highest possible efficiency of resource utilization. Aquaculture means the culture of any type of aquatic plant or animal but in fresh waters, fish provide the overwhelming majority of production, although the freshwater prawn, *Macrobrachium rosenbergii*, can be locally important in some tropical areas (Fig. 8.8). Rearing fish is essentially no different to any other intensive livestock production—the animals are kept within a circumscribed area usually at higher than natural densities and therefore require a direct or indirect increase in food supply. The chemical and nutritional conditions need to be manipulated, not only so that the fish will survive but that the total environmental conditions are as close as possible to those optimal for growth to occur (Payne, 1979).

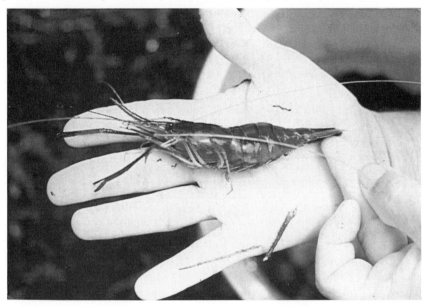

Fig. 8.8 The large freshwater prawn, *Macrobrachium rosenbergii*, which has been successfully cultured in the tropics. (Photo: P. C. Coutin, Coventry Polytechnic)

The current world total aquaculture production is some 6 million tonnes which is less than 10% of the world fisheries catch of 70 million tonnes. Of this 6 million however, 5.2 million is produced in Asia including 0.6 million in tropical and subtropical South-east Asia. This reflects the considerable tradition which has built up over the centuries in this part of the world, whilst in Africa and South America, where interest only extends over the last 30–40 years, the output of fish from culture is only 107 000 and 70 000 tonnes respectively.

More than 90% of the total aquaculture production is in fresh waters. There is very little sustained marine production largely because of the difficulty experienced in rearing the fry of marine fishes up to the present time. Sufficient experience exists to do this on an experimental scale for a variety of species, but no large-scale system has yet been devised to produce young in large enough numbers to sustain marine fish culture on a scale comparable with that of fresh waters. The only marine fish which makes a significant contribution to world production is the milkfish, *Chanos chanos*, in the Philippines, Indonesia, and Taiwan. This fish has a curious life cycle in that the adults are herring-like and thoroughly oceanic, but they release their eggs so that the young drift inshore where they can be captured in large numbers and stocked into brackish-water ponds. Although the adults are naturally marine, the whole culture cycle can be carried out in brackish conditions. In fact, this fish does extremely well in the fish pens of Laguna de Bay (Fig. 8.9) in the Philippines where the salinity rarely exceeds $2–3 \, g \, l^{-1}$ (sea-water is $35 \, g \, l^{-1}$). However, although this is a very successful form of culture the vast majority of the fry for stocking the ponds are still caught wild. The availability or sustained production of sufficient young is, therefore, a crucial factor in determining if a species can provide a reasonable basis for practical culture.

Fig. 8.9 Fish pens for milkfish (*Chanos chanos*) in the Laguna de Bay, Philippines. The fisherman in the foreground emphasizes the conflict that can arise for space between such extensive rearing systems and capture fisheries (see Smith, 1983)

Fig. 8.10 Shallow earth ponds for mullet and tilapia, Egypt

Fig. 8.11 Cropping a small rural fishpond with a seine net in Indonesia. (Photo: P. C. Coutin, Coventry Polytechnic)

Rearing fish which operate at a low trophic level, i.e. herbivores or planktivores, increases the efficiency with which they use natural production in a pond and generally makes them cheaper to feed (Figs. 8.10 and 8.11). In the tropics with high light intensities and temperatures and year-long growing seasons, primary production can be extremely high in shallow fishponds, particularly in those receiving organic and inorganic fertilizers to overcome natural nutrient deficits in the water (Section 5.2.4). For example, Malaysian fishponds achieved daily gross primary productions of $0.12-0.19 \, \text{g C m}^{-2}$ (Prowse, 1972) whilst Indian ponds attained $4.1-9.2 \, \text{g C m}^{-2}$ (Ganapati and Sreenivasan, 1972), which is amongst the most productive rates for tropical waters (Section 5.2.3). The most widely used tropical fishes for culture are the tilapias, which owe part of their suitability to the fact that they feed on leafy vegetation (*Tilapia* proper) or are plankton or algal feeders (*Oreochromis* and *Sarotherodon*). The latter groups are, therefore, particularly suited to exploiting the high levels of primary production which can be achieved (Section 4.8.2). Like many types successfully used for culture, the tilapias are hardy, they are tolerant of high temperatures and low oxygen, and they can readily be handled and transported. In addition, they spawn readily and fry production is rarely a problem; in fact the major difficulty in the use of tilapias is their prolific capacity for reproduction, as discussed further below.

Several other tropical species responsible for high yields are largely herbivorous, including the milkfish and the Indian major carps, *Catla catla*, *Labeo rohita*, and *Cirrhinus mrigala*, although these carps are probably equally detritivorous or omnivorous. Much of the art of culturing the milkfish is directed to producing a carpet of filamentous algae, of the right species composition, on the floor of the ponds upon which the fish can feed. Recently, however, there has been a trend towards the use of deeper ponds and to encourage the fish to feed on the plankton of the water column.

The most productive temperate systems tend to be based upon carps of various kinds including the omnivorous common carp, *Cyprinus carpio*, in Europe and Asia, and a mixture of various species, notably the silver carp, *Hypophthalmichthys molitrix*, which is a plankton feeder, and the grass carp, *Ctenopharyngodon idellus*, in China. The success of these carps can be measured by the annual production of fish from culture in China, which amounts to almost 1.7 million tonnes or a third of the production of Asia. Some of these fishes can be used in the tropics. The common carp, for example, which has an upper lethal temperature of around 35 °C and an optimum for growth at some 28 °C, has been been used in Africa and South-east Asia. It will mature under most circumstances and can be induced to spawn under the appropriate environmental conditions or with hormonal treatment. The Chinese carps, by contrast, always require hormone inducement to spawn, and until this technique became available the use of these valuable fish was confined to China where wild-caught fry were used.

Predatory or carnivorous fishes can be used in culture but because of their requirements for high-grade animal protein, rearing them can be expensive.

Fig. 8.12 Throughflow concrete tanks for the culture of a temperate fish, the rainbow trout (*Salmo gairdneri*), at 2000 m on the slopes of Kilimanjaro, Tanzania

Amongst the commonest in the temperate regions are the various trouts, particularly the rainbow trout, *Salmo gairdneri*, which have been used in the cooler, high altitude areas of the tropics such as the East African highlands (Fig. 8.12) and Lake Titicaca in the Andes. Many of the catfishes such as *Clarias*, although omnivorous, do generally require some animal protein in the diet. If fish need to rely on high-quality diets containing up to 40% protein, as most carnivores do, then the feed becomes a major component of the costs and it is also wasteful from the point of view of food production. The protein content of most commercial pellets of this nature is composed of fish meal and so, in biological terms, fish is being fed to fish to produce fish with a loss of at least 70% due to respiration and other causes in the transformation. Sometimes, however, waste animal products are available such as trash fish or offal which does not compete with other human requirements and is, therefore, less expensive. In some cases predators are intentionally added to mixtures of pond fish to keep down unwanted small fry, but often it is difficult to strike the balance so that the predator confines its activities to the required component and does not turn its attention to the main crop.

Production in fishponds is a function of individual growth and stock density, that is

$$P = D \times G$$

where P = production (kg ha^{-1} unit time^{-1});
D = stock density (number of individuals ha^{-1});
G = mean individual growth rate (kg unit time^{-1}).

Stock density is related to growth rate in that the more closely crowded the fish are, the lower is the individual growth rate (Fig. 8. 13). However, the growth rate is not reduced in direct proportion to the increase in stock density so that the total of smaller increments adds up to a higher overall yield (Fig. 8. 13).

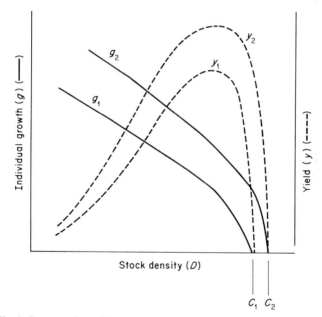

Fig. 8.13 The influence of stocking density and individual growth rate on production from fish ponds

Assume, for example, that a doubling of the stock density only reduces the individual growth to 60% of its former rate, then in terms of the above equation,

$$2D \times 0.6G = 1.2P$$

There has consequently been a net increase in production. As the carrying capacity of the system is reached, that is, the stock density which can just be maintained under the prevailing conditions where $G = O$, the relationship tends to break down, probably due to social factors, and growth and production rapidly approach zero. Fish farming therefore tends to be most effective at producing large numbers of small fish; the production of large fish through rapid growth rates requires a lower density of stocking with the consequent sacrifice of yield. The smaller the size of fish when harvested, the higher will be that total yield, although obviously a compromise must be reached with the minimum acceptable size demanded by the market. Some fish may be less sensitive to this type of relationship than others. For example, tilapia hybrids kept under close to optimal conditions of feeding and oxygen could be kept at up to $80\,000$ ha^{-1} without a reduction in mean growth rate (Rapoport and Sarig, 1975). Fish kept in raceways, that is, very confined conditions with a very rapid throughput of water, may also not be governed by quite the same relationship.

To increase the overall productivity of ponds means shifting the relationship between D and G, shown in Fig. 8. 13, to the right by increasing the resources available or removing some environmental constraint, a symptom of which is an improvement in the carrying capacity (C_1 to C_2 in Fig. 8.13). One of the most obvious ways of doing this is by making more food available to the animals. This can either be done indirectly by improving the general fertility of the ponds or directly by giving a supplementary feed. Indirect increases usually involve the use of fertilizers to enhance primary production. Agricultural fertilizers are commonly used such as superphosphate, NPK mixtures, ammonium sulphate, or even urea. In most temperate situations, phosphate fertilizers are regarded as being the most important but it is evident that, in many tropical waters, nitrogen rather than phosphorus is the principal limiting element (Section 5.2.4). The relative effects of the two in an African fishpond are demonstrated in Fig. 5.8. Agricultural lime may also be of value since it both raises the pH (the most productive waters tend to fall within the range of pH 8–10) and increases the HCO_3 and CO_3 reserves of the water which act as carbon sources for photosynthesis. Organic fertilizers such as animal manures are particularly useful because these not only enrich the water through decomposition but are also often used directly as a food by a variety of fish.

Between 25 and 60 kg ha^{-1} inorganic fertilizer can be added to ponds each week under semi-intensive production conditions. If very dense algal blooms do develop then the diel fluctuation in oxygen conditions may oscillate dramatically and the low oxygen concentrations of the early morning may interfere with growth even if the fish do not actually die (Payne, 1979). Animal manures can also be added to ponds in surprisingly high quantities of up to 400 l day^{-1} per tonne of fish in the ponds (Schroeder and Hepher, 1979) and, with care, intensive manuring can be carried out without damaging the oxygen relations of the pond (Schroeder, 1975). Of the animal manures, poultry waste has consistently proved the most effective.

With similar constraints regarding oxygen concentrations, sewage can be added to promote production. Ponds receiving sewage effluents in tropical and subtropical regions can produce $2\text{--}6\,t\,ha^{-1}\,year^{-1}$ and, in more specialized systems, algal production can reach 110 tonnes ha^{-1} (Payne, 1984). Disposing of effluent through fishponds can actually assist in its purification (Section 8.4).

Most often the introduction of direct supplementary feed in the tropics means the use of agricultural waste such as leaves from horticultural crops, and the waste from milling operations or from processing cotton seed or groundnuts. Usually these are put directly into the water, but formulation into pellets offers advantages of handling and storage so that seasonal peaks in agricultural production can be spread out to provide fish feed over the year. Pellets can also allow a more controlled introduction of food through automatic dispensers or demand feeders which the fish can operate themselves. In this way application can be closely co-ordinated with consumption and wastage can be cut down.

Some form of compromise can be achieved in the use of high-grade, expensive animal protein and cheaper but less efficient plant protein. In Israel the standard pellet for the common carp contains 15% fish meal and 10% plant protein, but even this amount of fish meal would be too expensive for many systems. It is possible to give an almost purely carbohydrate supplement in the form of cereal grains or the waste from milling processes. From this the fish can derive energy for respiration, including activity, and will use all the protein from natural production in the pond for growth. This exploits what is known as the protein-sparing capacity of carbohydrates.

Increasing food availability will naturally promote production within a fish culture system but, for any given set of conditions, the carrying capacity of a pond can be expanded by employing a range of species which occupy different niches, thereby exploiting different food sources within the pond. Some experiments from an Egyptian farm give a good demonstration of this principle (Table 8.1). When carp were stocked alone at a density of 5000 ha^{-1}, a yield of 400 kg ha^{-1} was obtained. A similar yield from carp was also produced when tilapia were added to the system but a further crop was obtained from these to give a total of 11 kg ha^{-1} (El Bolock and Labib, 1967). The carp are botton-feeding omnivores whilst the tilapia introduced were 70% plankton feeders and 30% leaf chewers; consequently little competition would arise in the exploitation of resources within the pond. However, when a third species, the grey mullet *Mugil cephalus*, was present in the mixture the yield from the carp fell although this was more than compensated for by the contribution of the mullet so that, at the end of the production cycle, the total crop from the three species was the highest obtained (Table 8.1). Like the carp, the mullet feeds on the bottom of the pond although it prefers finer particles and detritus. However, there is evidently some competition between the two, hence the fall in carp yield. Whether this is acceptable would depend upon the relative economic value of the two species.

The use of mixtures of species in this way is termed polyculture, as opposed to monoculture of a single species. It is not a new idea, since the highly successful Chinese and Indian traditional systems are both based upon the utilization of

Table 8.1 Experimental polyculture system from Egypt showing the increased yields obtained under a single management system by using mixtures of non-competing species. The carp is a bottom omnivore, the tilapia were 70% planktivorous (*Oreochromis niloticus*) and 30% herbivorous (*Tilapia zillii*), and the grey mullet feed on small particles on the pond bottom such as algae and detritus; the ponds did not receive any fertilizer or supplementary feed (from El Bolock and Labib, 1967).

| | Carp alone | Carp plus tilapia | | carp plus tilapia plus mullet | | |
	Carp	Carp	Tilapia	Carp	Tilapia	Mullet
Number	5000	5000	1250	5000	1250	2500
Weight gain (kg ha^{-1})	377	386	—	311	—	99
Weight harvested (kg ha^{-1})	402	411	300	341	355	104
TOTAL (kg ha^{-1})	402	711		800		

species mixtures. However, the approach is receiving considerable attention and in Israel, for example, successful mixtures of common carp, silver carp, and tilapia have been arrived at through experimentation. Traditional mixtures may also be improved, however. The Indian major carps differ more in where they feed rather than in the nature of the food, with *Catla catla* feeding close to the surface, *Labeo rohita* feeding throughout the water column, and *Cirrhinus mrigala* feeding on the bottom. However, with the addition of a benthic omnivore, the common carp, a phytoplankton feeder, the Chinese silver carp, and the macrophyte-feeding grass carp, it proved possible to produce yields of almost 10 tonnes ha^{-1}, more than twice the average achieved with the more conventional mixtures (Chaudhuri *et al.*, 1975). It is also significant that this yield was obtained without special diets but using large quantities of agricultural waste and manures. Polyculture, therefore, exploits the three-dimensional possibilities of a water body in a way which is rarely possible in agriculture.

The use of polyculture is a way of substantially increasing the stocking density without exceeding the carrying capacity of the system as a whole (Fig. 8.13). To produce optimum results, however, the stocking density of single species or mixtures must be controlled. For most species the problems are usually associated with producing sufficient fry to allow stocking at the optimum density. Where wild-caught fry have to be relied upon, supplies can vary from year to year and there is also the danger that taking too many fry for culture will compete with fisheries which also depend upon the fry for recruitment. This was the case with some of the Chinese carps which were much in demand but would not spawn in confinement, similarly with the milkfish and some of the catfishes. Now, however, it is possible to induce many species to spawn using endocrine treatment with pituitary extracts which contain the equivalent of follicle-stimulating hormone and possibly also luteinizing hormone. It is only because of these techniques that valuable fish such as the silver carp and the grass carp can now be used outside their natural centres of distribution in Asia. In some cases such as the catfish, *Clarias*, for example, it has proved possible to induce the fish to spawn under experimental circumstances (Micha, 1976), but the procedure and early rearing are still not sufficiently reliable to sustain large-scale production for intensive farming. It is probably fry production which is the greatest single constraint upon the availability of a species for culture.

In contrast, the tilapias present just the opposite difficulty. They will mature and spawn readily in ponds to such an extent that numbers increase well beyond the original stocking density. One particular characteristic of tilapias is that, as they become crowded, they begin to mature at a smaller size and also at a younger age. Whereas a species such as *Oreochromis niloticus* might mature in 1–2 years in the wild, it can spawn after 3–4 months in a pond. Such precocious maturation is an unusual feature amongst fishes. Although the individuals are small, a high production rate is maintained as suggested by the model in Fig. 8.5 and, in fact, this has been suggested as a way of maintaining higher production rates from ponds (Iles, 1973). However, most markets require a fish in excess of a few centimetres in length and to maintain a reasonably sized adult fish of over 50 g requires some control over this tendency to overpopulate the pond. This can be achieved either by harvesting the fish just before they become mature (often about 3–5 months), by introducing a predator, or by using only one sex in monosex culture which obviously removes the problem of reproduction. The first method restricts the production period but is the most immediately practicable whilst the second is rather unpredictable. For the third alternative, sexing can be done by hand but this is rather time-consuming. The method received considerable incentive when it was discovered by C. F. Hickling in Malaysia that a cross between two closely related species, *Oreochromis mossambicus* (male) and *O. hornorum* (female), produced 100% male offspring. A number of other crosses have proved to give a similar result (Hickling, 1968) and in Israel, using monosex hybrids of *O. niloticus* and *O. aureus*, a large proportion of tilapia culture is carried out in this fashion. However, the hybrids, although all male, are fertile and will mate with the original stocks thereby complicating the genetic composition of the parental species. This in turn appears to lead to a less predictable outcome of subsequent crosses yet, if the offspring are not all male then overpopulation remains a problem. The maintenance of genetic purity of wild species stocks for hybridization is of some importance, therefore. Similarly, true domestication of most fish species by the types of selective breeding which have been exercised on domestic livestock has yet to be achieved. Selective breeding programmes to produce the most appropriate strains of fish for a given set of conditions will require as broad a genetic base as possible and the availability of true-breeding lines or species will have an important role to play.

Table 8.2 Relative yields from some natural waters and some aquaculture systems (kg/ha)

Fisheries	
open sea	1.4
cont. shelf/inland lakes	14
Culture	
single species + fertilizer	1 000
single/mixed species + fertilizer + feed	3–4 000
mixed species + fertilizer + feed	10 000
(experimental)	
single species + fertilizer + feed + aeration	17 000

The relative yields obtained under any aquaculture system will depend upon the management and the quality of the inputs (Table 8.2). In comparison with the catch from fisheries on an area basis, most aquaculture is more productive but the total areas involved in this are so much smaller than the sea that the world-wide yield, as mentioned earlier, is less than 10% of that of marine fisheries. On this basis it is unlikely that fish farming will ever replace marine fisheries as the principal source of fish although there is considerable scope for development. The relative yields obtained from aquaculture can, nevertheless, be very high. They are not, however, obtained easily. The approach is very similar to that in agriculture in that intensive farming requires constant attention and the appropriate management. The similarity with agriculture may often, in fact, put the two into competition for space, water, fertilizers, and feedstuffs. At the same time, the most successful fish farming systems are those which have been closely integrated with agriculture, as can be seen in China and India. A well-developed agricultural system provides a continuous supply of the large quantities of manure and vegetable waste necessary for intensive pond culture. Without this, reliance must be on put on inorganic fertilizers which are rather less potent at enhancing productivity. The oxygen content of water can vary quite dramatically and generally it is found that, as the oxygen concentration falls in the water, the food consumption of the fish is first affected; as it falls further, below $2–3\,mg\,l^{-1}$, then oxygen availability limits metabolism and the food which is consumed is used less efficiently for growth (Payne, 1979). The use of aeration devices will elevate oxygen concentrations in the water so that oxygen will not limit production. The use of such devices in China have led to an average 14% increase in yields from farm ponds (Tapiador et al., 1977; FAO, 1983), whilst a recent system devised in Taiwan uses a 'red tilapia' hybrid in tanks with high levels of manure and constant aeration so that the fish are virtually being kept in activated sludge and yields in excess of 16 tonnes ha^{-1} have been attained (Liao and Chen, 1984). It is this ability to control and optimize the environment in addition to the possibilities of manipulating the species mixture which gives aquaculture its most important advantage—flexibility.

Providing there is water, even if this is considered unsuitable for agriculture through being too saline or contaminated with sewage, it is generally possible to devise some form of aquaculture. Small-scale units away from major fisheries centres, particularly those in drier areas associated with water conservation or irrigation schemes, may prove particularly valuable.

8.4 POLLUTION

Water pollution is only beginning to become a problem in the tropics as industrialization increases and the human population expands rapidly in this zone. The effects of water pollution are well known from temperate experience and no doubt many will be repeated in the tropics. However, the preceding chapters have indicated differences in the ways aquatic communities from the two areas function and therefore we should look for differences in response and impact.

Owing to enormous increases in population over the last few decades and the progressive concentration of the human population into towns, domestic waste and sewage are probably of the most immediate concern. Most sewage systems open into a watercourse or, if no sewage system is involved, the rains or the nearest piece of water serve the same purpose. In this way large amounts of organic material are released into the water, although some industrial processes such as pulp mills and sugar-processing plants also produce much finely divided organic material as waste products. Organic waste of this type rapidly breaks down through bacterial action which can lead to reduced oxygen levels or even anaerobic conditions in the vicinity of an effluent. The capacity of the microflora to use oxygen in a sample of water over a 5-day period at a standard 20 °C is termed the biochemical oxygen demand (BOD) and is used as an indirect estimate of the amount of oxidizable organic material in the water. In tropical waters, the evidence suggests that decomposition and mineralization proceeds very rapidly and efficiently (Section 5.3). If this is the case then the capacity of such systems to absorb and break down organic waste may be higher than those of temperate areas. Fryer (1972) observed that although the main sewage outflow from the major town of Entebbe, Uganda, flowed into Lake Victoria, there was little evidence of its effects. Nevertheless the capacity of tropical waters is by no means infinite; for example, release of bagasse and molasses waste from sugar-cane plants into the River Kali in India caused severe depletion of oxygen, and fish mortalities on occasions were detected up to 160 km downstream. Only air-breathing fishes such as the catfishes *Heteropneustes fossilis* and *Clarias batrachus* were unaffected (Sharma, 1983).

The recovery from oxygen depletion or from any other form of pollution will always be hastened by the rate of water exchange so that the status of deep, permanently stratified lakes such as Lake Malawi and Lake Tanganyika with exchange times of over 1000 years must be fragile. Large amounts of decomposing organic material falling below the thermocline from, for example, the wood-pulp industry could cause oxygen depletion which would last for considerable periods of time and probably eliminate large sections of the fauna (Fryer, 1972).

In addition to the direct depletion of oxygen, the decomposition of large quantities of organic material in the water produces inorganic nutrients such as ammonia, nitrates, and phosphorus. These enrich the water considerably and give rise to dense algal growth or blooms, which can cause the wide daily fluctuations in oxygen described for fishponds (Section 8.3), and in extreme conditions fish-kills can result. This increased productivity caused by excessive organic loads can cause a decline in water quality and these symptoms of 'overproduction' are generally referred to as 'eutrophication'.

Around the Laguna de Bay in the Philippines live 1.8 million people and the current estimated annual waste input into the water includes 7100 tonnes of nitrogen and 1400 tonnes of phosphorus whilst the natural inflows contribute only 1400 tonnes of nitrogen and 200 tonnes of phosphate (Edra, 1983). Like many tropical waters the Laguna de Bay is nitrogen-limited and dense blooms of algae, dominated by the blue-green *Microcystis*, can produce a biomass up to 145 g m^{-3}. The die-off of these blooms leads to fish-kills which are damaging to the important fish-pen industry on the lake. By the turn of the century it is estimated that 4.2 million people will live around the Laguna and that the nitrogen input will reach 18 600 tonnes and that of phosphorus 3900 tonnes. The situation in Laguna de Bay is further complicated by the outflow through the

Pasig River, which is heavily polluted in its traverse through Manila. During part of the year the inflow into the lake is reduced and the Pasig River flows back into the Laguna, increasing the pollution. A weir system has been constructed to prevent this, but the backflow from the river also allowed in a saline intrusion from the estuary. It is probable that the natural productivity of the lake and its milkfish industry, depend upon this annual inflow of salt water, which may now be prevented. This illustrates the difficulties of controlling pollution when the water is being used for diverse human activities.

Floating plants such as the water-hyacinth, *Eichhornia*, and the water-cabbage, *Pistia* (Section 4.3.9), also tend to flourish under eutrophic conditions and can choke waterways. The early phases of man-made lakes are characterized by high nutrient levels as the flooded terrestrial vegetation decays. This encourages the growth of floating vegetation to such an extent that in Lake Brokopondo in Surinam, South America, within 1 year of its existence a carpet of *Eichhornia* covering 20 km² had grown over the surface and eventually extended over half the area of the lake with a further 65 km² covered by the floating fern, *Ceratopteris*. This surface layer of plants prevents oxygen and light from penetrating the surface and the decomposition of the weeds themselves together with that of the submerged uncleared timber from this lake in the forest zone rendered much of the lake anaerobic and toxic with hydrogen sulphide. A similar experience also occurred in Brazil's first man-made lake, Curua Una, in the Amazon basin. Much of the surface became covered with *Eichhornia* and most of the fish died through the resulting anaerobic conditions. In addition, the acidic waters, which result from the weed cover, caused the steel casings on the turbines to become corroded with a replacement cost of $5 million (Caufield, 1982).

In other parts of the world the story has been the same. In its early years, 22% of the total surface area of Lake Kariba, some 1000 km², became covered with the floating fern *Salvinia molesta* which caused widespread concern at the time (McLachlan, 1974). As the lake matured, nutrient levels declined in the post-filling phase. Conditions suitable for these floating plants may be more short-lived in lakes created in the savanna zone where the flooded vegetation can decompose relatively rapidly, compared to lakes in the forest zone, such as Brokopondo, where the large numbers of sunken trees take a considerable time to break down and continue to release nutrients over a longer period.

The most common cause of eutrophication in rivers is the breakdown of an organic effluent. Initially, close to the outfall some deoxygenation is common, but once mineralization has proceeded and some reoxygenation has taken place, zones of high production can occur. In the River Ganga (Ganges), for example, deoxygenation from organic sludge, even though greatly diluted, extended for several kilometres below an outfall, but further down an abundance of aquatic organisms was found and also a rich fishery for a variety of species including the major carps (Ray and David, 1966).

The distribution of organisms below an outflow of this kind can give some indication as to the presence and severity of organic pollutions. Organic effluents in Indian rivers were shown to change the algal composition; high organic loading reduced numbers of diatoms and green algae, particularly desmids, whilst encouraging the proliferation of blue-greens and euglenoids (Venkateswarlu, 1969). Similar effects can be found in temperate rivers. In South Africa, Chutter (1972) produced a pollution index for the assessment of the severity of organic pollution based on the relative tolerance of macro-invertebrates. Sensitive types, such as most mayflies and caddis-flies, were allocated low scores, whilst tolerant groups, including chironomids and *Simulium*, received high scores. The mean total per individual gave an overall index of pollution on a 0–10 scale. One significant difference from similar temperate biotic indices is the absence of stone-flies from the index, since these are poorly represented in African streams.

Whilst disposal of organic effluent into natural waters can have damaging effects, it has also proved possible to use what can be, in many ways, a valuable resource, by applying partially treated or even untreated sewage into fishponds. The yields from what is effectively an organic fertilizer can be quite considerable (Payne, 1984). Sewage waste is added to 80 000 ha of ponds in West Bengal which can produce some 2.3 tonnes ha^{-1} using traditional methods (Sharma, 1983). The quality of effluents passing through fishponds can show substantial improvements, with significant reductions in organic material as measured by BOD, as well as in nitrogen, phosphate, and also faecal coliforms, many of which may be pathogenic (Payne, 1984).

One other aspect of water pollution which is emerging as a problem in the tropics is the release of toxic agents as a by-product of industrial processes such as mining, paper making, and textile manufacture, and also from agriculture due to the increased use of pesticides and herbicides. In the process of industrialization of developing countries the treatment of factory wastes to maintain water quality has been of low priority compared to the rate of development. The Hooghly, which is a main branch of the Ganga as it flows into the sea near Calcutta, is estimated to receive 1145 million litres of liquid wastes from nearly 161 factories including textile and cotton mills, paper and pulp works, tanneries, and jute mills in addition to the domestic sewage from the towns and cities along its banks (Sharma, 1983). The wastes from paper, distillery, yeast, tanneries, and textile factories have a high organic content with a similar effect upon oxygen and BOD as sewage. In addition, however, there are toxic materials such as bleaches and dyes which compound the effect and may kill off the bacteria responsible for the self-purification process. The total waste discharged daily into the Hooghly consists of 10 611 tonnes BOD load, 23 144 tonnes total solids, 105 753 tonnes suspended solids and 126 111 tonnes dissolved solids (Gosh *et al.*, 1973). The Hooghly is now less productive than the nearby, non-polluted Matlah estuary and, in particular, the important *Hilsa* fishery has declined due to mortality of food organisms, eggs, and fry in the estuary, which are particularly vulnerable at neap tides when the pollutants are least dispersed (Sharma, 1983).

Similar situations are occurring in a number of countries as they shift from an agricultural to an industrial economy. In the Philippines, for example, agriculture now only contributes 29% to the GNP as opposed to 49% in the early 1950s. One result of this is that the Pasig River which passes through Manila now receives a good deal of industrial as well as domestic effluent from the 450 factories along its length. In addition to high levels of organic waste, the river also has elevated heavy metal concentrations of mercury, zinc, and copper (Tamayo-Zafarwalla, 1983). Heavy metals are directly toxic, but in sub-lethal concentrations their uptake from the water or in food occurs more rapidly than their excretion. This leads to bioaccumulation of the compounds within the body which may have detrimental chronic effects on the animal itself but also provides a source for further concentration in consumers as, for example, in the case of fish eaten by man. In the Pasig River, the fishery in this heavily polluted waterway is now confined to the lower reaches and only from October to February after the inflow from the nearby Laguna de Bay flushes out the pollutants. Even this is variable and depends upon the typhoons.

Heavy metal contamination can also be a problem in mining operations, where both acute and chronic effects, including bioaccumulation, are to be looked for. In mining, however, the most immediate problem is the tailings of crushed rocks which are released into the rivers as heavy suspended loads. In the Philippines there have been numerous examples of fishing communities losing their livelihood through mining operations, and also large tracts of land being rendered unfit for agriculture as rivers contaminated with waste have flooded the surrounding area (Tamayo-Zafarwalla, 1983). Fish from a Bolivian lake receiving waste from tin and silver mining showed bioaccumulation of silver, copper, zinc, and cadmium between four and forty times the level normally found (Beveridge *et al.*, 1985). Since these fish form the basis of a large commercial fishery, the metals are most probably passed on to the consumers.

With the development of tropical agriculture and, in particular, with the 'green revolution' which is based upon high-yielding cereal varieties which do, however, require intensive chemical treatment, increasing amounts of pesticides and herbicides find their way into the water system. Many aquatic animals, particularly fish, are very sensitive to these chemicals (Holden, 1963). For example, the concentration of the organochlorine insecticide, Endrin, needed to kill 50% (LC_{50}) of experimental populations of the tilapia *Oreochromis macrochir* was as low as $0.008 \, mg \, l^{-1}$ (Welcomme, 1971). Little information is available on the toxicity of such compounds upon tropical aquatic animals, but work which has been carried out suggests that their susceptibility is similar to that of temperate counterparts. For example, the LC_{50} values and resistance times (time taken for median mortality to occur) of *O. macrochir* to three commonly used organochlorine pesticides, Endrin, Lindane, and Synexa, and to the herbicide TCK were similar to those recorded for other fish in other parts of the world (Welcomme, 1971). With the exception of Endrin, normal patterns of usage of these chemicals on rice or other crops did not exceed concentrations likely to have large-scale acute effects leading to mortality but, particularly with the organochlorine residues which can take several years to break down, there is always the question of accumulation within the bodies of the organism. In some cases, however, the distinctive physiology of some tropical animals does provide a rather different response. When the climbing perch, *Anabas testudineus*, which can respire in air or water, is first exposed to Lindane it quickly shifts to aerial respiration (Bakthavathsalam and Reddy, 1983). Presumably this restricts chemical exchange across the gills and thereby reduces exposure to the compound. Fish which can use air generally appear more resistant to pollution and this is probably one of the reasons.

Widespread application of biocides directly on to water has been carried out in a number of cases either to control insect vectors of human disease, such as the application of DDT to control the mosquito and the more recent spraying of the Volta River catchment area with Abate to control *Simulium* and onchocerciasis, or the use of herbicides to control the spread of floating vegetation. The potential effects of the release of such potent biologically active chemicals are far reaching. Lake Brokopondo in Surinam was sprayed with the herbicide 2,4-D, the active ingredient of the defoliant 'agent orange', to control the large blankets of water-hyacinth on the lake's surface. The river, which had been impounded, and was a source of drinking water, fish, and a route of communication for the local inhabitants became first clogged and eventually poisoned (Caufield, 1982). A certain amount of planning, however, went into the use of the organophosphorus insecticide Abate on the aquatic larvae of the black-fly, *Simulium*, in the waters of the Volta basin. The toxicity of the compound to most animals including fish, was fairly low and, because it breaks down quite rapidly, there was little tendency for bioaccumulation to occur (Lévêque et al., 1978). Use of Abate on fish populations in rivers and dry-season pools showed that, whilst the fish would move downstream to avoid the chemical or show stress movements if avoidance were not possible, actual mortalities in the field were minimal (Abban and Samman, 1982).

Widespread use of such chemicals can be considered if care is taken to minimize damage to the system as a whole, although long-term chronic effects such as interference with reproductive potential or growth are often difficult to predict and require careful investigation to determine such limitations. Generally the nature of the effects of pollution on tropical aquatic systems are similar to those in temperate areas, and there are grounds for thinking that the very efficient decomposer systems of the tropics may enhance the capacity of these waters to deal with pollutants, particularly with those of an organic nature. However, it is also apparent that tropical aquatic communities, whilst complex, are fragile and depend upon a fine equilibrium amongst the components, as has been demonstrated, for example, in one of the best-studied instances, Lake George. Increasing diversity of pollution may rapidly damage some parts of the decomposer chain and thereby reduce the capacity for self-purification. The relative resilience of tropical and temperate waters in this respect is a factor to be resolved.

8.5 HEALTH

Water plays an important role in the dissemination of a large number of important human diseases in the tropics. In some cases, as in many bacterial and viral diseases, it can act as a direct agent of transmission. In the parasitic infections it plays an indirect role by supporting the aquatic phase of the life cycle of the vector or intermediate host. Many of the bacterial diseases are enteric gut infections such as cholera (*Vibrio cholerae*), typhoid and paratyphoid (*Salmonella*), which are spread from faecal-contaminated waters. Drinking water sources are very prone to bacterial contamination in the tropics (Evison and James, 1977), either directly or indirectly by faecal material being washed into water bodies by the rains.

Of these enteric infections, perhaps the one most closely linked with water is cholera, which is endemic in many tropical areas, particularly in Asia and Africa, and is prone to periodic eruptions of epidemic and pandemic proportions. It has particularly rapid effects, symptoms appearing within 5 days of infection, and the presence of one infected person can be sufficient to start an epidemic in a country (Justaz, 1977). In general, the more sluggishly moving the water the more appropriate the conditions for the persistence of the disease; there is therefore a declining prevalence with elevation. Ideal conditions are illustrated by the area around the Hooghly River in West Bengal and the surrounding deltas of the Gangas, the Bramaputra and the Irrawaddy in Burma. These are very flat and frequently flooded, leaving pools of water and swamps, whilst the slightly brackish water can be a particularly suitable environment. Cholera is endemic and a continual problem in these deltas. By contrast, many other estuarine areas in the Far East, even when associated with large ports, do not act as persistent centres for the disease, although they suffer periodic outbreaks. This is due to the open nature of most of these estuaries and the uninterrupted tidal exchange which results in rapid water movements throughout most of the cycle. Water quality also plays a part since the cholera vibrios multiply better and remain infective longer under alkaline conditions. The soil and water of the Gangas Delta are both alkaline; the water storage tanks of this region tend to develop algal blooms which promote alkaline conditions during photosynthesis (Section 5.2.3) and further encourage the prevalence of the disease.

Not all bacteria are discouraged by a low pH and low salt content of the water and substantial numbers of *Salmonella* and *Clostridium* have been detected under these conditions in some West African drinking-water sources (Wright, 1982). In these cases, seasonal effects are also evident with highest bacterial numbers occurring in the dry season. Cholera, too, may show a seasonal cycle coinciding with the rains, although this varies from place to place attaining a peak at the height of the monsoons in the Ganga Valley, for example, whilst the seasonal maximum occurs in the early rains in Calcutta.

Amongst the parasitic diseases associated with water the most widespread are schistosomiasis, malaria, and some of the filarial nematode worms causing debilitating diseases such as river blindness and elephantiasis. Schistosomiasis involves an aquatic animal, a water-snail, as a secondary host whilst in the other two cases water does not play a role in the life cycle of the disease causing organism itself but only in that of the biting fly, the vector, which transmits the disease.

Schistosomiasis or bilharzia affects nearly 200 million people in the warmer areas of the world and is typically a disease of rural areas where people unavoidably come into contact with water regularly in agriculture, fishing, drawing water, travelling by boat, or during domestic activities (Fig. 8.14). *Schistosoma* is a blood fluke, a digenetic trematode flatworm, which lives in the blood and produces eggs which are excreted with urine, in the case of *S. haematobium*, or which are voided with the faeces, in the cases of *S. mansoni* and *S. japonicum*. Only *S. mansoni* occurs in South America and the West Indies, having its widest distribution in Brazil where it affects 14 million people. Africa contains both *S. mansoni* and *S. haematobium* and within the continent are some noted focal points such as the Nile Valley and Lake Victoria where, in some places, up to 100% of the rural populations suffer from the disease. *Schistosoma japonicum* is restricted to relatively small areas of tropical South-east Asia, although it does occur in parts of temperate China and Japan. This species, however, will also infect other hosts apart from man, such as cows or rodents, which makes any attempt at eradication much more difficult.

In all species, when the eggs reach the water, small ciliated forms emerge called miracidia. These ae mobile, positively phototropic and negatively geotropic and therefore tend to swim near the surface where the molluscan hosts are most commonly found. In addition, they do show, in a generalized fashion, a chemotactic response which also assists in finding a host. The miracidium can survive in a free-swimming condition for about 24 hours. The different schistosome species vary in the type of snail they parasitize. The flat, rams-horn planorbid genera *Bulinus* and *Biomphalaria* are used as intermediate hosts by *Schistosoma haematobium* and *S. mansoni* respectively, whilst *S. japonicum* use the snails of the genus *Oncomelania* belonging to the family Hydrobiidae. Once the miracidia make contact with a snail of the appropriate species they burrow through the skin of the host to continue development. Multiplication through sporocysts occurs over a 5–7 week period following which several thousand cercaria larvae emerge, all of which have been derived from the original miracidium since infection by more than one appears to be prevented. Amongst the snails this parasitism, which affects particularly the hepatopancreas and ovotestis, causes a 1.3–3.5 times increase in mortality and eggs laid by infected animals are more commonly sterile. The cercariae can swim for up to 48 hours in search of a human host. Like the miracidia, they swim near the surface but how they locate the host is uncertain. Penetration takes up to 5 minutes and if the skin dries out in this time the cercariae are killed. People in contact with water for longer periods are, therefore, at a higher risk.

Since the schistosomes are species specific with respect to the snails they infect, their distribution is indivisibly linked to that of their intermediate hosts. The absence of *Bulinus* from South America therefore explains the lack of

Fig. 8.14 A child being washed at the edge of Lake Victoria stands a considerable chance of being infected with the blood fluke, *Schistosoma*, which affects 200 million people in the tropics

Schistosoma haematobium in this continent. On a more local scale, the snails are generally not found in fast-flowing waters but prefer pools, swamps, or marginal vegetation. They can be found in large water bodies, such as Lake Victoria, or in small ponds. Even pools which dry out can give seasonal transmission of the disease since the snails can aestivate in the mud until the pool refills. Waters with low ionic concentrations and low pH are not favoured by the snails, particularly if calcium concentration, which is required for shell formation, is poor. However, Lake Victoria, with a calcium content of only $4 \, mg \, l^{-1}$, has an abundance of snails, including those which are hosts of the disease, so the threshold for this element is obviously quite low.

One means of controlling schistosomiasis, as with many of the water-linked diseases, is improved sanitation, but the widespread nature of the problem makes short-term improvements in this way unlikely. The commonest approach has been to attempt to control the snails. Increasing the flow rate of water can be effective although not always possible. Another possibility is the introduction of predators or competitors, such as mollusc-feeding fishes or the snail, *Marisa cornuarietis*, which feeds on the same plants as *Biomphalaria* and also prey upon it. Complete eradication is, however, very difficult to achieve with this approach. A number of chemicals have been used as molluscicides, including copper sulphate and Niclosamide, but there is always the problem of the effects, caused by widespread distribution of these compounds, on other members of the community, particularly those of commercial value such as fish. A particularly effective and relatively specific molluscicide based on saponins is produced by the soapberry bush of Ethiopia, although industrial synthesis has not yet been carried out.

It has not been possible to control schistosomiasis by any of these measures on other than a local scale and the disease is on the increase. Although greater population size is one factor in this, the increased diversity of uses for water together with the lack of control makes the situation worse. The disease has become established in most tropical man-made lakes because the still water and often more alkaline conditions created in this way are conducive to the snails and settlements often spring up on the shores. Moreover, *Schistosoma* is only one of the parasites which have intermediate aquatic hosts. For example, another fluke, *Fasciolopsis buski*, infects the gut and has a similar gastropod, *Planorbis*, as its intermediate host in India, Assam, and China. The cercariae, however, encyst on the roots or fruits of edible plants like the water-chestnut, *Trappa*, to emerge when eaten. Another example is the nematode Guinea worm, *Dracunculus mediensis*, which uses the copepod *Cyclops* as its intermediate host. This is most prevalent in areas of West Africa and the Middle East where water-holes are used extensively and can be controlled by the most basic processing of the water supply since the worm is only transmitted when the copepod is drunk with the water.

There are a large number of diseases relying on insect vectors which spend their larval stages in water. Mosquitoes alone transmit the protozoan malarias, a range of filarial infections and the viral yellow and dengue fevers. The extent and intensity of infection with these diseases are largely determined by the availability of breeding sites for the mosquito. Many mosquitoes are not limited to established pieces of water and will breed in temporary accumulations in tin cans or car tyres for example, although some, such as *Mansonia*, do need established weed beds in more permanent bodies of water. The fact that mosquitoes transmit several diseases can complicate the situation considerably. When malaria was contained in Sri Lanka by the use of DDT against the mosquito there was a fall in infant mortality leading to an increase in population size with a consequent expansion of the suburbs of Colombo. The previously adequate sanitary system became overloaded, producing pools of stagnant polluted water which, along with the increased dumping of car tyres, provided ideal breeding grounds for *Culex fatigans*, the vector of *Wucheria bancrofti*, the filarial nematode which eventually damages and blocks the lymph system over a number of years to produce the symptoms of elephantiasis (Kershaw, 1977). The breakdown of the drainage and sanitary system of the city of Rangoon in the flat delta of the Irrawaddy River following the Second World War had a similar effect and transformed elephantiasis from a curiosity into a real problem.

The control of mosquitoes is often centred upon spraying the adults with insecticide rather than by attacking the larvae. Breeding sites can be many, inconspicuous, and difficult to locate, although in situations where identifiable water bodies are implicated, insecticides have been used in the water or thin films of oil laid on the surface to prevent the larvae from gaining access to the air. Larval control does play its role, therefore, in the eradication of the mosquito.

More susceptible to this approach, however, are insects which have localized breeding sites, such as *Simulium*. The larvae of this fly are to be found where there is rapidly flowing, well-oxygenated, turbulent water. This often means the headwater stream of river basins although rapids and riffles on major rivers also provide suitable habitats (Section 4.7.5). When water flow is seasonal no reproduction will take place in the dry season but, as the rains begin, then the submerged rocks and vegetation at the edge of the river become potential sites, as does the vegetation of flooded marginal areas. In the Niger, the breeding sites tend to be close to the junctions with small tributaries and not in the major part of the river itself which has a muddy or sandy bottom (Kershaw, 1977). Through its bite, *Simulium* transmits the filarial worm which initially occurs near the skin surface but in later stages can cause damage to the retina of the eye and optic nerves. Although *Simulium* occurs in streams and rivers all over the world and can be a serious nuisance because of its irritating bite, only in Africa and central and northern South America does it transmit *Onchocerca*. Unlike most biting flies, the mouthparts of *Simulium* are unsuited for location of individual capillaries and the puncture is more like 'open-cast mining' until capillaries are ruptured which leaves a disproportionally large bite for such a small fly. In at least one species, *S. neavei* from East Africa, the larvae and pupae are to be found not on stones or vegetation as are those of *S. damnosum*, but on the shells of freshwater crabs of the genus *Potomanautes*, although these are used purely as a source of support whilst the larva carries on suspension feeding (Section 4.7.2).

The relatively localized breeding sites of *Simulium* makes it a little more susceptible to control in its aquatic phase than other disease-transmitting organisms. Spraying of headwater streams with insecticides will kill the fly larvae although, if a broad-spectrum insecticide is used, considerable damage to the rest of the community can be expected. Following such treatment, recolonization of upstream sites by natural predators is very slow compared to the ease of downstream reoccupation by surviving pockets of *Simulium* close to the source. Once started therefore, insecticide treatment must continue for some time and, in the case of some of the waterfalls of the upper Nile, it has completely changed the nature of the fauna. Areas in which control is attempted are also liable to reinvasion by adults, which have a flight range of up to 50 km. The most ambitious attempt to clear *Simulium* is the World Health Organization Onchocerciasis Control Programme intended to eliminate the disease in one of the worst-affected endemic areas, the Volta basin in northern Ghana and Upper Volta, where economic blindness was a considerable problem. The sensitive areas were sprayed with the organophosphorus compound Abate (O,O,O-tetramethyl-O,O-thiodi-p-phenylene phosphate) which had proved to be an effective larvicide but whilst effective, concentrations of $0.05-0.1 \, \mathrm{mg \, l^{-1}}$ were considerably lower than the LC_{50} of fish; some invertebrates, however, were susceptible (Samman and Pugh-Thomas, 1978). This project has been continuing for several years but, since the disease is chronic and has its severest effects only after considerable lengths of time, it is difficult to assess the medical repercussions just yet.

GENERAL READING

Caufield, C. 1982. Brazil, energy and the Amazon. *New Scientist*, **96**: 240–243.

Goulding, M. 1981. *Man and Fisheries on an Amazon Frontier*. Junk: The Hague.

Gribbin, J. 1982. Menace of the giant dams. *New Scientist*, **96**: 493–495.

Gulland, J. 1968. *The Concept of the Maximum Sustainable Yield and Fishery Management*. FAO Fisheries Circular.

Howe, G. M. 1977. *A World Geography of Human Diseases*. Academic Press: London.

Payne, A. I. 1976. *The Exploitation of African Fisheries. Oikos*, **27**, 356–366.

Payne, A. I. 1979. Physiological and ecological factors in the development of fish culture. In 'Fish Phenology'. *Symp. Zool. Soc. Lond.*, **44**: 383–415.

Pruginin, Y., and Hepher, B. 1981. *Commercial Fish Farming*. Wiley: New York.

Stanley, N. F., and Alpes, M. P. 1975. *Man-made Lakes and Human Health*. Academic Press: London.

Welcomme, R. H. 1979. *The Fisheries Ecology of Floodplain Rivers*. Longman: London.

BIBLIOGRAPHY

Abban, E. K., and Samman, J. 1982. Further observations on the effect of Abate on fish catches. *Environ. Pollut.*, Ser. A, **27**: 245–254.

Adeniji, H. A. 1978. Diurnal vertical distribution of zooplankton during stratification in Kainji Lake, Nigeria. *Verh. Internat. Verein Limnol.*, **20**: 1677–1683.

Adeniji, H. A. 1981. Circadial vertical migration of zooplankton during stratification and its significance to fish distribution and abundance in Kainji Lake, Nigeria. *Verh. Internat. Verein Limnol.*, **21**: 1021–1024.

Angermeir, P. L., and Karr, J. R. 1983. Fish communities along environmental gradients in a system of tropical streams. *Environ. Biol. Fish.*, **9**: 1.

Aruga, Y., and Monsi, M. 1963. Chlorophyll amount as an indicator of organic matter productivity in bio-communities. *Plant & Cell Physiol.*, **4**: 29–39.

Arumugam, P. T., and Furtado, J. I. 1980. Physico-chemistry, destratification and nutrient budget of a lowland eutrophicated Malaysian reservoir and its limnological implications. *Hydrobiologia*, **70**: 11–24.

Bagenal, T. 1978. *Methods for Assessment of Fish Production in Freshwaters*, IBP Handbook 3. Oxford: Blackwell.

Bagenal, T. B., and Tesch, F. W. 1978. Age and growth. In *Methods for Assessment of Fish Production in Freshwaters*, Bagenal, T. (ed.), IBP Handbook 3, pp. 101–136. Oxford: Blackwell.

Baker, J. H., and Bradnam, L. A. 1976. The role of bacteria in the nutrition of aquatic detritivores. *Oecologia (Berl.)*, **24**: 95–104.

Bakthavathsalam, R., and Reddy, Y. S. 1983. Changes in bimodal oxygen uptake of an obligate air breather, *Anabas testudineus* (Block), exposed to Lindane. *Water Research*, **17**: 1221–1226.

Balon, E. K. 1973. Results of fish population size assessments in Lake Kariba coves (Zambia) a decade after their creation. In *Manmade Lakes: Their Problems and Environmental Effects*, Ackerman, W. C., White, G. F., and Worthington, E. B. (eds.), *Geophysical Monogr.*, **17**, American Geophysical Union, Washington DC, pp. 149–158.

Bannister, K. E., and Clarke, M. A. 1980. A revision of the large *Barbus* (Pisces, Cyprinidae) of Lake Malawi with a reconstruction of the history of the Southern African Rift Valley Lakes. *J. Nat. Hist.*, **14**: 483–542.

Barbosa, F. A. R., and Tundisi, J. G. 1980. Primary production of phytoplankton and environmental characteristics of a shallow Quaternary lake at Eastern Brazil. *Arch. Hydrobiol.* **90**: 139–161.

271

Barbour, C. D., and Brown, J. H. 1974. Fish species diversity in lakes. *Am. Nat.*, **108**: 473–489.

Barel, C. D. N., Van Oijen, M. J. P., Witte, ¨., and Witte-Maas, E. L. M. 1977. An introduction to the taxonomy and morphology of the haplochromine Cichlidae from Lake Victoria. *Neth. J. Zool.*, **27**: 333–389.

Bayley, P. B. 1981. Fish yields from Amazon in Brazil: comparison with African river yields and management possibilities. *Trans. Am. Fish. Soc.*, **110**: 351–359.

Bazigos, G. P. 1974. The design of fisheries statistical surveys—inland waters. *FAO Fish. Tech. Pap.*, **133**, 122 pp.

Beadle, L. C. 1981. *The Inland Waters of Tropical Africa.* Longman: London, 475 pp.

Begg, G. W. 1976. The relationship between the diurnal movements of some of the zooplankton and the sardine *Limnothrissa miodon* in Lake Kariba, Rhodesia. *Limnol. Oceanogr.*, **21**: 529–539.

Beveridge, M. C. M., Stafford, E., and Coutts, R. 1985. Metal concentrations in the commercially exploited fishes of an endorehic saline lake in the tin–silver province of Bolivia. *Aquaculture and Fish. Man.*, **1**: 41–53.

Beverton, R. J., and Holt, S. J. 1957. On the dynamics of exploited fish populations. Fish Invest. Minist. Agric. Fish Food UK (Ser. 2), No. 19: 533 pp.

Bishop, J. E. 1973. Limnology of a small Malayan river Sungai Gombak. *Monographiae Biologicae,* Vol. 22. The Hague: Junk.

Biswas, S. 1975. Phytoplankton in Volta Lake, Ghana during 1964–73. *Verein. Internat. Verein Limnol.*, **19**: 1928–1934.

Bonetto, A. A. 1972. *Report on IBP/PF Projects.* Instituto Nacional de Limnologia, Santo Tome, Argentine Republic, 1971.

Boulanger, G. A. 1915. *Catalogue of the Freshwater Fishes of Africa in the British Museum (Nat. Hist.).* London: British Museum (Nat. Hist.), 4 Vols.

Bourrelly, P. G., and Couté, A. S., 1982. Quelques algues d'eau douce de la Guyanne Française. *Amazoniana,* **7**: 221–292.

Bowen, S. H. 1976. Mechanism for digestion of detrital bacteria by the cichlid fish *Sarotherodon mossambicus* (Petrs). *Nature,* **260**: 137.

Brinkhurst, R. O. 1974. *The Benthos of Lakes.* London: Macmillan, 190 pp.

Brinson, M. M., and Nordlie, F. G. 1975. Lake Izabal, Guatemala. *Verh. Internat. Verein. Limnol.*, **19**: 1468–1479.

Brylinsky, M. 1980. Estimating the productivity of lakes and reservoirs. In *The Functioning of Freshwater Ecosystems,* Le Cren, E. D., and Lowe-McConnell, R. H. (eds.), pp. 411–454. Oxford: Blackwell.

Brylinsky, M., and Mann, K. M. 1978. An analysis of factors governing productivity in lakes and reservoirs. *Limnol. Oceanogr.*, **18**: 1–14.

Burgis, M. T. 1971. The ecology and production of copepods, particularly *Thermocyclops hyalinus*, in tropical Lake George, Uganda. *Freshwat. Biol.*, **1**: 169–192.

Burgis, M. J. 1974. Revised estimates for the biomass and production of zooplankton in Lake George, Uganda. *Freshwat. Biol.*, **4**: 535–541.

Burgis, M. J., Darlington, J. P. E. C., Dunn, I. G., Ganf, G. G., Gwahaba, J. J., and McGowan, L. M. 1973. The biomass and distribution of organisms in Lake George, Uganda. *Proc. Roy. Soc. Lond.,* **B184**: 271–298.

Burgis, M. J., and Dunn, U. G. 1978. Production in three contrasting ecosystems. In *Ecology of Freshwater Fish Production*, Gerking, S. D. (ed.), pp. 137–158. Oxford: Blackwell.

Carlander, K. D. 1955. The standing crop of fish in lakes. *J. Fish. Res. Bd. Canada,* **12**: 543–570.

Carmouze, J. P. 1976. La régulation hydrogéochimique du lac Tchad. *Trav. Doc. ORSTOM,* **58**, 418 pp.

Carmouze, J. P., Dejoux, C., Durand, J. R., Gras, R., Iltis, A., Lauzanne, L., Lemoalle, J., Lévêque, C., Loubens, G., and Saint-Jean, L. 1972. Contribution à la connaissance du Bassin Tchadien. Sommaire. Grandes zones écologiques du Lac Tchad. *Cah. ORSTOM Ser. Hydrobiol.*, **6**: 103–169.

Carmouze, J. P., Durand, J. R., and Lévêque, C. (eds.) 1983a. *Lake Chad, Ecology and Productivity of a Shallow Tropical Ecosystem*. The Hague: Junk. 518 pp.

Carmouze, J. P., Durand, J. R., and Lévêque, C. 1983b. The lacustrine phase during the 'Normal Chad' period and the drying phase. In *Lake Chad, Ecology and Productivity of a Shallow Tropical Ecosystem*, Carmouze, J. P., Durand, J. R., and Lévêque, C. (eds.), pp. 527–560. The Hague: Junk.

Carter, G. S. 1955. *The Papyrus Swamps of Uganda*. Cambridge: Heffer, 15 pp.

Caufield, C. 1982. Brazil, energy and the Amazon. *New Scientist*, **96**: 240–243.

Chaudhuri, H., Chakrabarty, R. B., Sen. P. R., Rao, M. G. S., and Jena, S. 1975. A new high in fish production in India with record yields by composite fish culture in freshwater ponds. *Aquaculture*, **6**: 343–356.

Chilvers, L. M., and Gee, J. M. 1974. The food of *Bagrus docmac* and its relationship with *Haplochromis* in Lake Victoria. *J. Fish. Biol.* **6**: 483–505.

Chouret, A. 1978. La persistance des effects de la sécheresse sur le Lac Tchad. In *Symposium on River and Flood-plain Fisheries in Africa*, Welcomme, R. L. (ed.), pp. 74–91. Rome: FAO.

Christy, F. T., and Scott, A. 1965. *The Commonwealth in Ocean Fisheries; Some Problems of Growth and Economic Allocation*. Baltimore: Johns Hopkins, 281 pp.

Chutter, F. M. 1972. An empirical biotic index of the quality of water in South African streams and rivers. *Water Res.*, **6**: 19–30.

Cole, G. A. 1968. Desert limnology. In *Desert Biology*, Brown, G. W. (ed.) Vol. 1, New York: Academic Press.

Corbet, P. S. 1961. The food of non-cichlid fishes in the Lake Victoria basin, with remarks on their evolution and adaptation to lacustrine conditions. *Proc. Zool. Soc. Lond.*, **136**: 1–101.

Daget, J. 1952. Mémoire sur la biologie des poissons du Niger Moyen. I. Biologie et croissance des espèces du genre *Alestes*. *Bull. Inst. fr. Afr. noire*, **14**: 191–225.

Daget, J. 1957. Données récentes sur la biologie des poissons dans la delta central du Niger. *Hydrobiologia, 9*: 321–347.

Daget, J. 1962. Les poissons du Fouta Diallon et de la Basse Guinée. *Mem. Inst. fr. Afr. noire*, **65**: 1–61.

Darlington, J. P. E. C. 1976. Temporal and spatial variation in the benthic invertebrate fauna of Lake George, Uganda. *J. Zool. Lond.*, **180**: 95–112.

De Silva, S. S. 1985. Status of the introduced cichlid *Sarotherodon mossambicus* (Peters) in the reservoir fishery of Sri Lanka: a management strategy and ecological implications. *Aquaculture and Fisheries Management*, **1**: 91–102.

De Silva, S. S., Cumaranatunga, P. R. T., and De Silva, C. D., 1980. Food feeding ecology and morphological features associated with feeding of four co-occuring cyprinids (Pisces: Cyprinidae). *Neth. J. Zool.*, **30**: 54–73.

Dokulil, M., Bauer, K., and Silva, I. 1983. An assessment of the phytoplankton biomass and primary productivity of Parakrama Samudra, a shallow man-made lake in Sri Lanka. In *Limnology of Parakrama Samudra—Sri Lanka*, Schiemer, F. (ed.), pp. 49–76. The Hague: Junk.

Dudgeon, D. 1983. An investigation of the drift of aquatic insects in Tanpo Kau forest stream New Territories, Hong Kong. *Arch. Hydrobiol.*, **96**: 434–447.

Dudgeon, D. 1984. Seasonal and long-term changes in the hydrobiology of the Lam Tsuan River, New Territories, Hong Kong with special reference to benthic macro--invertebrate distribution and abundance. *Arch. Hydrobiol.*, **69**: 55–129.

Duncan, A. and Gucati, R. D. 1981. Parakrama Samudra (Sri Lanka) Project, a study of a tropical lake ecosystem. III. Composition, density and distribution of the zooplankton in 1979. *Verh. Internat. Verein Limnol.* **21**: 1001–1008.

Eccles, D. H. 1974. An outline of the physical limnology of Lake Malawi (Lake Nyasa). *Limnol. Oceanogr.*, **19**: 730–742.

Edmondson, W. T., and Winberg, G. G. 1971. *A Manual on Methods for the Assessment of Secondary Production in Fresh Waters*, IBP Handbook 17. Oxford: Blackwell.

274

Edra, R. B. 1983. Laguna de Bay—an example of a fresh and brackishwater fishery under the stress of the multiple-use of a river basin. *FAO Fish. Rep.*, **288**: 119–124.

Egborge, A. B. M. 1981. The phosphate content of the bottom sediments of a small West African impoundment. *Verh. Internat. Verein Limnol.*, **21**: 1025–1070.

Eisma, D., Kalf, J., and Van der Gaast, S. J. 1978. Suspended matter in the Zaire estuary and the adjacent atlantic coast. *Neth. J. Sea Res.*, **12**: 382–406.

El Bolock, A. R., and Labib, W. 1968. Carp culture in the U.A.R. *Proc. World Symp. Warm-water Fish Culture*, Pillay, T. V. R. (ed.), FAO Fish Rep., **44**, FR II/R-1, pp. 165–174.

Elliott, J. M. 1977. Some methods for the statistical analysis of samples of benthic invertebrates, 2nd edn. *Sci Pub. Freshwat. Biol. Assoc.*, **25**, 160 pp.

Elliott, J. M., and Drake, C. M. 1981. A comparative study of seven grabs used for sampling macroinvertebrates in rivers. *Freshwat. Biol.* **11**: 99–120.

Emiliani, F. 1984. Oligotrophic bacteria: seasonal fluctuations and correlations with environmental variables (Middle Parana River, Argentina). *Hydrobiologia*, **111**: 31–36.

Entz, B. 1976. Lake Nasser and Nubia. In *The Nile. Biology of an Ancient River*, Rzoska, J. (ed.), pp. 271–297. The Hague: Junk.

Estèves, A. F. 1983. Levels of phosphate calcium, magnesium and organic matter in the sediments of some Brazilian reservoirs and implications for the metabolism of ecosystems. *Arch. Hydrobiol.*, **96**: 129–138.

Evison, L. M., and James, A. 1977. Microbiological criteria for tropical water quality. In *Water, Wastes and Health in Hot Climates*, Feacham, R. G., McGarry, M. G., and Mara, D. D. (eds.), Ch. 3. Chichester: Wiley.

FAO 1983. Freshwater aquaculture development in China. *FAO Fish Techn. Pap.*, **215**: 1–124.

Fee, E. J. 1969. A numerical model for the estimation of photosynthetic production, integrated over time and depth, in natural waters. *Limnol. Oceanogr.*, **14**: 906–911.

Ferguson, A. J. D. 1975. Invertebrate production in Lake Turkana (Rudolf). Symposium on the hydrobiology and fisheries of Lake Turkana, 25–29 May 1975. 28 pp. mimeo.

Fernando, C. H. 1980a. The species and size composition of tropical freshwater zooplankton with special reference to the oriental region (south East Asia). *Int. Rev. ges. Hydrobiol.*, **65**: 411–426.

Fernando, C. H. 1980b. The freshwater zooplankton of Sri Lanka, with a discussion of tropical freshwater zooplankton composition. *Int. Rev. ges. Hydrobiol.*, **65**: 85–125.

Fernando, C. H., and Furtado, J. I. 1975. Reservoir fishery resources of South-East Asia. *Bull. Fish. Res. Sta. Sri Lanka*, **26**: 83–95.

Fernando, H., and Holcik, J. 1982. The nature of fish communities: a factor influencing the fishery potential and yields of tropical lakes. *Hydrobiologia*, **97**: 127–140.

Finlayson, C, M., Farrell, T. P., and Griffiths, D. J. 1980. Studies on the hydrobiology of a tropical lake in North-Western Queensland. II. Seasonal changes in thermal and dissolved oxygen characteristics. *Aust. J. Mar. Freshwater Res.*, **31**: 589–596.

Fisher, T. R., and Parsley P. E. 1979. Amazon lakes: Water storage and nutrient stripping by algae. *Limnol. Oceanogr.*, **24**: 547–553.

Fittkau, E. J. 1964. Remarks on the limnology of Central Amazon rain-forest streams. *Verh. Intern. Verein. Limnol.*, **15**: 1092–1096.

Fittkau, E. J. 1970. Role of caimans in the nutrient regime of mouth lakes of Amazon effluents. *Biotropea*, **2**: 138–142.

Fittkau, E. J., Irmler, U., Junk, W. J., Reiss, F., and Schmidt, G. W. 1975. Productivity, biomass and population dynamics in Amazonian water bodies. In *Tropical Ecological Systems, Trends in Terrestrial and Aquatic Research*, Golley, F. B., and Medina, E. (eds.), pp. 289–312. Berlin: Springer-Verlag.

Fogg, G. E. 1955. *Algal cultures and phytoplankton ecology.* London: Athlone Press pp. 112.

Fogg, G. E. 1972. *Photosynthesis.* London: EUP.

Frey, D. G. 1969. A limnological reconnaissance of Lake Lanao. *Verh. Internat. Limnol.,* **17**: 1090–1102.

Fryer, G. 1959. The trophic inter-relationships and ecology of some littoral communities of Lake Nyasa with especial reference to the fishes, and a discussion of the evolution of a group of rock-frequenting Cichlidae. *Proc. Zool. Soc. Lond.,* **132**: 153–281.

Fryer, G. 1965. Predation and its effects on migration and speciation in African fishes: a comment. *Proc. Zool. Soc. Lond.,* **144**: 301–322.

Fryer, G. 1969. Speciation and adaptive radiation in African lakes. *Verh. Internat. Verein Limnol.,* **17**: 303–322.

Fryer, G. 1972. Conservation of the Great Lakes of East Africa: a lesson and a warning. *Biological Conservation,* **4**: 256–262.

Fryer, G. 1977a. The atyid prawns of Dominica. *Freshwat. Biol. Assoc. Annual Rep. 1976–77:* 48–54.

Fryer, G. 1977b. Studies on the functional morphology and ecology of the atyid prawns of Dominica. *Phil. Trans. R. Soc.* **B277**, 8–128.

Fryer, G. 1977c. Evolution of species flocks of cichlid fishes in African lakes. *Z. zool. syst. Evolut.-forsch.,* **15**: 141–165.

Fryer, G., and Iles, T. D. 1972. *The Cichlid Fishes of the Great Lakes of Africa: Their Biology and Evolution.* Oliver and Boyd: Edinburgh.

Ganapati, S. V., and Sreenivasan, A. 1972. Energy flow in aquatic ecosystems in India. In *Problems of Production in Freshwaters,* Kajak, Z., and Hillmecht-Ilkowska, A. (eds.), 457–476. Krakow, Polish Scientific Publishers.

Ganf, G. G. 1974. Phytoplankton biomass and distribution in a shallow eutrophic lake (Lake George, Uganda). *Oecologia (Berl.),* **16**: 9–29.

Ganf, G. G. 1975. Photosynthetic production and irradiance–phytosynthesis relationships of the Phytoplankton from a shallow equatorial lake (Lake George, Uganda). *Oecologia (Berl.),* **18**: 165–183.

Ganf, G. G., and Horne, A. J. 1975. Diurnal stratification, photosynthesis and nitrogen-fixation in a shallow, equatorial Lake (Lake George, Uganda). *Freshwat. Biol.,* **5**: 13–39.

Ganf, G. G., and Viner, A. B. 1973. Ecological stability in a shallow equatorial lake (Lake George, Uganda). *Proc. R. Soc. Lond.,* **B184**: 321–346.

Garrod, D. J. 1959. The growth of *Tilapia esculenta* Graham in Lake Victoria. *Hydrobiologia,* **12**: 268–298.

Garrod, D. J. 1963. An estimation of mortality rates in a population of *Tilapia esculenta* Graham (Pisces: Cichlidae) in Lake Victoria, East Africa. *J. Fish. Res. Bd. Canada,* **20**: 195–227.

Garrod, D. J., and Newell, B. S. 1958. Ring formation in *Tilapia esculenta. Nature,* **181**: 1411–1412.

Gaudet, J. J., and Melack, J. M. (1981). Major ion chemistry in a tropical African lake basin. *Freshwat. Biol.,* **11**: 309–333.

Gaudet, G. J., and Mutmuri, F. M. 1981. Nutrient regeneration in a shallow tropical lake water. *Verh. Internat. Verein Limnol.,* **21**: 725–729.

Geisler, R., Schmidt, G. W., and Sookvibul, S. 1979. Diversity and biomass of fishes in three typical streams of Thailand. *Int. Rev. ges. Hydrobiol.,* **64**: 673–697.

Gessner, F., and Hammer, L. 1967. Limnologische Untersuchungen an Seen der Venezolanischen Hochanden. *Int. Rev. ges. Hydrobiol.,* **52**: 301–320.

Ghosh, B. B., Ray, P., and Gopolakrishnan, V. 1973. Survey and characterisation of waste waters discharged into the Hoogly estuary. *J. Inland Fish. Soc. India,* **5**; 82–101.

Gibbs, R. J. 1967a. The geochemistry of the Amazon river system: Part 1. The factors that control the salinity and composition and concentration of the suspended solids. *Bull. geol. Soc. Am.,* **78**: 1203–1232.

Gibbs, R. J. 1967b. Amazon river: environmental factors that control its dissolved and suspended load. *Science N.Y.,* **156**: 1734–1737.

Gibbs, R. J. 1970. Mechanisms controlling world water chemistry. *Science N.Y.,* **170**: 1088–1090.

Gibbs, R. J. 1972. Water chemistry of the Amazon River. *Geochim. Cosmochim. Acta,* **36**: 1061–1066.

Gibson, F. M., and Statzner, B. 1985. Longitudinal zonation of lotic insects in the Bandama River System (Ivory Coast). *Hydrobiologia,* **122**: 61–64.

Gliwicz, Z. M. 1975. Effect of zooplankton grazing on photosynthetic activity and composition of phytoplankton. *Verh. Internat. Verein Limnol.,* **19**: 1490–1497.

Golterman, H. L. 1973. Natural phosphate sources in relation to phosphate budgets: a contribution to the study of eutrophication. *Water Res.,* **7**: 3–17.

Golterman, H. L. 1975. Chemistry. In *River Ecology,* Whitton, B. A. (ed.), pp. 39–80. Oxford: Blackwell.

Golterman, H. L. 1976. Sediments as a source of phosphate for algal growth. In *Interactions Between Sediments and Fresh Water,* Golterman, H. L. (ed.), pp. 286–293. The Hague: Junk.

Golterman, H. L., Clymo, R. S., and Ohnstad, M. A. M. 1978. *Methods for Chemical and Physical Analysis of Fresh Waters,* IBP Handbook 8. Oxford: Blackwell.

Gosse, J. P. 1963. Le milieu aquatique et l'écologie des poissons dans la région de Yangambi. *Annls Mus. r. Afr. Cent. Zool.,* Ser. 8v, Sci. Zool., **116**: 113–270.

Goulding, M. 1981. *Man and Fisheries on an Amazon Frontier.* 137 pp. The Hague: Junk, 137 pp.

Green, J. 1972. Latitudinal variation in associations of planktonic Rotifera. *J. Zool. Lond.,* **167**: 31–39.

Green, J., Corbet, S. A., Watts, E., and Lan, O. B. 1976. Ecological studies on Indonesian lakes. Overturn and restratification of Ranu Lamongan. *J. Zool. Lond.,* **180**: 315–354.

Greenwood, P. H. 1953. Feeding mechanism of the cichlid fish, *Tilapia esculenta* Graham. *Nature,* **172**: 207–208.

Greenwood, P. H. 1965. The cichlid fishes of Lake Nabugabo, Uganda. *Bull. Brit. Mus. Nat. Hist. (Zool.),* **12**: 315–357.

Greenwood, P. H. 1974. Cichlid fishes of Lake Victoria, East Africa: the biology and evolution of a species flock. *Bull. Brit. Mus. Nat. Hist. (Zool.),* suppl. **26**: 1–134.

Greenwood, P. H. 1976. Lake George, Uganda. *Phil. Trans. R. Soc. Lond.,* **B274**: 375–391.

Greenwood, P. H. 1979. Macroevolution—myth or reality? *Biol. J. Linn. Soc.,* **12**: 293–304.

Greenwood, P. H. 1984a. What is a species flock? In *Evolution of Fish Species Flocks,* Echelle, A. A., and Kornfield, I. (eds.), pp. 13–19. Orono: University of Maine Press.

Greenwood, P. H. 1984b. African cichlids and evolutionary theories. In *Evolution of Fish Species Flocks,* Echelle, A. A., and Kornfield, I. (eds.), pp. 141–154. Orono: University of Maine Press.

Greenwood, P. H., and Gee, J. M. 1969. A revision of the Lake Victoria *Haplochromis* species, pt. vii, *Bull. Brit. Mus. Nat. Hist. (Zool.),* **18**: 1–65.

Greenwood, P. H., and Lund, J. 1973. A discussion on the biology of an equatorial lake: Lake George, Uganda. *Proc. Roy. Soc. Lond.,* **B184**: 299–319.

Grobbelaar, J. U. 1983. Availability to algae of nitrogen and phosphorus absorbed on suspended solids in turbid waters of the Amazon river. *Arch. Hydrobiol.,* **96**: 302–316.

Grove, A. T. 1972. The dissolved and solid load carried by some West African rivers: Senegal, Niger, Benue and Shari. *J. Hydrol.,* **16**: 277–300.

Grubert, M. W. 1974. Podostomaceen-studien. Teil I. Zur Okologie einiger venezolanischer Podostomaceen. *Beitr. Biol. Pflanzen*, **50**: 321–391.

Grubert, M. W. 1975. Ökologie extrem adaptierter Blutenpflanzen tropischer Wasserfalle. *Biologie in unserer Zeit*, **5**: 19–24.

Gulland, J. A. 1968. *The Concept of the Maximum Sustainable Yield and Fishery Management*, FAO Fisheries Circular FRs/T70, 13 pp.

Gunatilaka, A. 1983. Phosphorus and phosphatase dynamics in Parakrama Samudra based on diurnal observations. In *Limnology of Parakrama Samudra—Sri Lanka*, Scheimer, F. (ed.), pp. 35–47. The Hague: Junk.

Gwahaba, J. J. 1975. The distribution, population density and biomass of fish in an equatorial lake, Lake George, Uganda. *Proc. Roy. Soc. Lond.*, **B190**: 393–414.

Harbott, B. J. 1976. Preliminary observations on the feeding of *Tilapia nilotica* Linn. in Lake Rudolf. *Afr. J. Trop. Hydrobiol. Fish*, **4**: 27–37.

Harrison, A. D. (1965). River zonation in Southern African streams. *Arch. Hydrobiol.* **61**: 380–386.

Harrison, A. B., and Rankin, J. J. 1975. Forest litter and stream fauna on a tropical island, St. Vincent, West Indies. *Verh. Internat. Verein Limnol.*, **19**: 1480–1483.

Hart, R. C. 1981. Population dynamics and production of tropical freshwater shrimp *Caridina nilotica* (Decapoda: Atyidae) in the littoral of Lake Sibaya. *Freshwat. Biol.*, **11**: 531–547.

Hart, R. C., and Allanson, B. R. 1975. Preliminary estimates of production by a calanoid copepod in subtropical Lake Sibaya. *Verh. Internat. Verein Limnol.*, **19**: 1434–1441.

Hart, R. C., and Allanson, B. R. 1981. Energy requirements of the tropical freshwater shrimp *Caridina nilotica* (Decapoda: Atyidae). *Verh. Internat. Verein Limnol.*, **21**: 1597–1602.

Hawkes, H. A. 1975. River zonation and classification. In *River Ecology*, Whitton, B. A. (ed.), pp. 312–374. Oxford: Blackwell.

Hecky, R. E., and Fee, E. J. 1981. Primary production and rates of algal growth in Lake Tanganyika. *Limnol. Oceanogr.*, **26**: 532–547.

Hegewald, E., Aldave, A., and Hakuli, T. 1976. Investigations on the lakes of Peru and their phytoplankton. 1. Review of literature, description of the investigated waters and chemical data. *Arch. Hydrobiol.*, **78**: 494–506.

Henderson, H. F., and Welcomme, R. L. 1974. The relationship of yields to Morpho-Edaphic Index and numbers of fishermen in African inland fisheries. *CIFA Occas. Pap. 1*, 19 pp.

Henderson, H. F., Ryder, R. A., and Kudhongania, A. W. 1973. Assessing fishery potentials of lakes and reservoirs. *J. Fish Res. Bd. Can.* **30**: 2000–2009.

Henry, R., and Tundisi, J. G. 1982. Evidence of limitation by molybdenum and nitrogen on the growth of phytoplankton community of the Lobo Reservoir (São Paulo, Brazil). *Rev. Hydrobiol. trop.*, **15**: 201–208.

Henry, R., and Tundisi, J. G. 1983. Responses of the phytoplankton community of a tropical reservoir (São Paulo, Brazil) to the enrichment with, nitrate, phosphate and EDTA. *Int. Rev. ges. Hydrobiol.*, **68**: 853–862.

Hickling, C. F. 1968. Fish hybridisation. *Proc. World Symp. Warm-water Fish Culture*, Pillay, T. V. R. (ed.), FAO Fish Rep IV, **44**r-1, 1–11. Rome: FAO.

Holcik, H. 1970. Standing crop, abundance, production and some ecological aspects of fish populations in some inland waters of Cuba. *Vestnik. Cs spol. zool. (Acta soc. zool. Bohemoslov.)*, **34**: 184–201.

Holden, M. J. 1963. The populations of fish in the dry season pools of the River Sokoto. *Colon Off. Fish Publ.*, **19**: 1–58, HMSO.

Holden, M. J., and Green, J. 1960. The hydrology and plankton of the River Sokoto. *J. Anim. Ecol.*, **29**: 65–84.

Hopkins, C. D. 1972. Sex differences in electric signalling in an electric fish. *Science*, **176**: 1035–1037.

Hopkins, C. D. 1973. Lighting as background noise for communication among electric fish. *Nature,* **242**: 268–270.

Howard-Williams, C., and Junk, W. J. 1977. The chemical composition of Central Amazonian macrophytes with special reference to their role in the ecosystem. *Arch. Hydrobiol.* **79**: 446–464.

Howard-Williams, C., and Lenton, G. M. 1975. The role of the littoral zone in the functioning of a shallow tropical ecosystem. *Freshwat. Biol.,* **5**: 445–459.

Hulbert, E. M., Ryther, J. H., and Guillard, R. R. L. 1960. The phytoplankton of the Sargasso Sea off Bermuda. *J. Cons. int. explor. Mer.,* **25**: 115–128.

Hutchinson, G. E. 1957a. *A Treatise on Limnology:* Vol. 1, *Geography, Physics and Chemistry.* New York: Wiley, 1016 pp.

Hutchinson, G. E. 1957b. Concluding remarks. *Cold Spring Harbor Symp. Quant. Biol.,* **22**: 415–427.

Hutchinson, G. E. 1959. Homage to Santa Rosalia or why there are so many kinds of animals. *Am. Nat.,* **93**: 145–159.

Hutchinson, G. E. 1965. *The Ecological Theater and the Evolutionary Play* New Haven, Conn.: Yale Univ. Press.

Hutchinson G. E. 1967. *A Treatise on Limnology:* Vol. 2, *Introduction to Lake Biology and the Limnoplankton.* New York: Wiley, 115 pp.

Hyder, M. 1970. Gonadal and reproductive patterns in *Tilapia leucosticta* (Teleostei: Cichlidae) in an equatorial lake, Lake Naivasha (Kenya). *J. Zool. Lond.,* **162**: 179–195.

Hynes, H. B. N. 1971. Zonation of the invertebrate fauna in a West Indian stream. *Hydrobiologia,* **38**: 1–8.

Hynes, J. D. 1975. Annual cycles of macro-invertebrates in a river in southern Ghana, *Freshwat. Biol.,* **5**: 71–83.

Ikusima, I, 1978. Primary production and population ecology of the aquatic sedge, *Lepironia articulata,* in a tropical swamp, Tasek Bera, Malaysia. *Aquatic Botany* **4**: 269–280.

Ikusima, I. 1982. Primary production–growth of *Lepironia* stands. In *Tasek Bera, The Ecology of a Freshwater Swamp,* Furtado, J. I. and Mori, S. (eds.), pp. 254–256. The Hague: Junk.

Ikusima, I., Kumano, S., and Mizuno, T. 1982. Population density and standing crop—emergent macrophytes. In *Tasek Bera, The Ecology of a Freshwater Swamp,* Furtado, J. I. and Mori, S. (eds.), pp. 244–248. The Hague: Junk.

Iles, T. D. 1973. Dwarfing or stunting in the genus *Tilapia* (Cichlidae) a possibly unique recruitment mechanism. In *Fish Stocks and Recruitment,* Parish B. P. (ed.), pp. 247–254. Rapp. Proc. Verb. Re-unions Cons. Int. Explor. Mer., Charlottenlund.

Illies, J. 1961. Gebirgsbäche in Europa und in Sudamerika—ein limologischer Vergleich. *Verh. Internat. Verein theor. angew Limnol.,* **14**: 517–523.

Illies, J. 1964. The invertebrate fauna of the Huallaga, a Peruvian tributary of the Amazon River, from the sources down to Tingo Maria. *Verh. Internat. Verein theor. angew. Limnol.,* **15**: 1077–1083.

Illies, J., and Botosaneanu, L. 1963. Problèmes et méthodes de la classification et de la zonation écologigue des eaux courantes, considerées surtout du point de vue faunistique. *Mitt. Internat, Verein theor. angew. Limnol.,* **12**: 1–57.

Iltis, A. 1982a. Peuplements algaux des rivières de Côte d'Ivoire. *Rev. Hydrobiol. trop.,* **15**: 231–239.

Iltis, A. 1982b. Peuplements alguax des rivières de Côte d'Ivoire. III. Etude du périphyton. *Rev. Hydrobiol. trop.* **15**: 303–312.

Imevbore, A. M. A. 1970. The chemistry of the River Niger in the Kainji Reservoir area. *Arch. Hydrobiol.,* **67**: 412–431.

Infante, A., and Reihl, W. 1984. The effect of cyanophyta upon zooplankton in a eutrophic tropical lake (Lake Valencia, Venezuela). *Hydrobiologia,* **113**: 293–298.

Jackson, P. B. N. 1961. The impact of predation especially tiger fish (*Hydrocynus vittatus* Cast.) in African freshwater fishes. *Proc. Zool. Soc. Lond.,* **136**: 603–622.

Jana, B. B., and De, U. K. 1983. Primary production of phytoplankton, environmental covariates and fish growth in West Bengal ponds under polyculture systems. *Int. Rev. ges. Hydrobiol.,* **68**: 45–58.

Junk, W. J. 1970. Investigations on the ecology and production-biology of the 'floating meadows' (Paspalo-Echinochloetum) on the middle Amazon. I. The floating vegetation and its ecology. *Amazoniana,* **2**: 449–495.

Junk, W. J. 1975. The bottom fauna and its distribution in Bung Borapet, a reservoir in Central Thailand. *Verh. Internat. Verein Limnol.,* **19**: 1935–1946.

Junk, W. J. 1982. Amazonian floodplains: their ecology, present and potential use. *Rev. Hydrobiol. trop.,* **15**: 285–301.

Justaz, M. 1977. Cholera. In *A World Geography of Human Diseases,* Howe, M. E. (ed.), pp. 131–144. London: Academic Press.

Kapetsky, J. M. 1974. Growth, mortality and production of five fish species of the Kafue river floodplain, Zambia. Ph.D. dissertation, University of Michigan, 194 pp.

Kershaw, W. E. 1977. Filariasis. In *A World Geography of Human Diseases,* Howe, M. E. (ed.), pp. 33–60. London: Academic Press.

Kira, T., Ogawa, H., Yoda, K., and Ogino, K. 1967. Comparative ecological studies in three main types of forest vegetation in Thailand. IV. Dry matter production, with special reference to the Khao Chong rain forest. *Nature and Life in S.E. Asia,* **5**: 149–174.

Knippling, E. B., West, S. H., and Haller, W. T. 1970. Growth characteristics, yield potential and nutritive content of water hyacinths. *Soil and Crop Sci. Soc. Fla. Proc.,* **30**: 51–63.

Knoppel, H. A. 1972. Food of Central Amazon fishes. *Amazoniana,* **2**: 257–352.

Kornfield, I., Smith, D. C., Gagnon, P. S., and Taylor, J. M. 1982. The Cichlid fish of Cuatro Ciengas, Mexico: direct evidence of conspecificity among trophic morphs. *Evolution,* **36**: 658–664.

Kortmulder, K. 1982a. Ecology and behaviour in tropical freshwater fish communities. Symposium on sustainable clean waters, Kuala Lumpur, mimeo.

Kortmulder, K. 1982b. Etho-ecology of seventeen *Barbus* species (Pisces; Cyprinidae). *Neth. J. Zool.,* **32**: 144–168.

Kosswig, C., and Villwock, W. 1965. Das Problem der intralakustrischen Speciation im Titicaca und im Lanosee. *Verh. Deut. Zool. Ges. Keil 1964 Zool. Anz. Suppl.,* **28**: 95–102.

Kramer, D. C. 1978. Reproductive seasonality in the fishes of a tropical stream. *Ecology,* **59**: 976–985.

Kramer, D. C. 1983. Aquatic surface respiration in the fishes of Panama: distribution in relation to the risk of hypoxia. *Env. Biol. Fish,* **8**: 49–54.

Ladle, M. 1972. Larval Simulidae as detritus feeders in chalk streams. *Mem. Ist. Ital. Idrobiol.,* **29**: 429–439.

Ladle, M., and Hansford, R. G. 1981. The feeding of the larvae of *Simulium austeni* Edwards and *Simulium* (Wilhelmia) spp. *Hydrobiologia,* **78**: 17–24.

Lagler, F. E., Kapetsky, J. M., and Stewart, D. J. 1971. *The Fisheries of the Kafue Flats, Zambia, in Relation to the Kafue George Dam.* Univ. Michigan. Tech. Rept., FAO, Rome, FI:SF/ZAM 11 Tech. Rept. **1**: 1–161.

Langbein, W. B., and Dawdy, D. R. 1964. Occurrence of dissolved solids in surface waters in the United States. *Prof. Pap. US geol. Surv.,* **501-D**, D115–D117.

Larkin, P. A. 1977. An epitaph for the concept of maximum sustainable yield. *Trans. Am. Fish. Soc.,* **106**: 1–11.

Lehman, J. T., Borkin, D. B., and Likens, G. E. 1975. The assumptions and rationales of a computer model of phytoplankton population dynamics. *Limnol. Oceanogr.*, **20**: 343–364.

Lemasson, L., and Pages, J. 1982. Apports de phosphore et d'azote par la pluie en zone tropical (Côte d'Ivoire). *Rev. Hydrobiol. trop.*, **15**: 9–14.

Lemoalle, J. 1979. Biomasse et production phytoplanktoniques du lac Tchad (1968–1976). Relations avec les conditions de milieu. Thèse Doct. es sciences, Univ. Paris 6, 287 pp.

Lemoalle, J. 1979–80. Application des données landsat à l'estimation de la production du phytoplankton dans le Lac Tchad. *Cah. ORSTOM ser. Hydrobiol.*, **8**: 35–46.

Lemoalle, J. 1981. Photosynthetic production and phytoplankton in the euphotic zone of some African and temperate lakes. *Rev. Hydrobiol. trop.*, **14**: 31–37.

Lemoalle, J. 1983. Phytoplankton production. In *Lake Chad, Ecology and Productivity of a Shallow Tropical Ecosystem*, Carmouze, J. P., Durand, J. R., and Lévêque, C. (eds.), pp. 357–384. The Hague: Junk.

Lévêque, C., Odei, H. A., and Pugh-Thomas, M. 1978. The Onchocerciasis control programme and the monitoring of its effect on the riverine biology of the Volta River Basin. In *Ecological Effects of Pesticides*, Perring, F. H., and Mellanby, K. (eds.), pp. 133–143. London: Academic Press.

Lévêque, C., and Saint-Jean, L. 1983. Secondary production (zooplankton and benthos). In *Lake Chad, Ecology and Productivity of a Shallow Tropical Ecosystem*, Carmouze, J. P., Durand, J. R., and Lévêque, C. (eds.), pp. 357–384. The Hague: Junk.

Lewis, W. M. 1970. Morphological adaptations of cyprinodonts for inhibiting oxygen deficient waters. *Copeia*, 1970, 319–326.

Lewis, W. M. 1974. Primary production in the plankton community of a tropical lake. *Ecological Monographs*, **44**: 377–409.

Lewis, W. M. 1975. Distribution and feeding habits of a typical *Chaoborus* population. *Verh. Internat. Verein Limnol.*, **19**: 3106–3119.

Lewis, W. M. 1977. Feeding selectivity of a tropical *Chaoborus* population. *Freshwat. Biol.*, 7: 311–325.

Lewis, W. M. 1978a. A compositional, phytogeographical and elementary structural analysis of the phytoplankton in a tropical lake: Lake Lanao, Philippines. *J. Ecol.*, **66**: 213–226.

Lewis, W. M. 1978b. Dynamics and succession of the phytoplankton in a tropical lake: Lake Lanao, Philippines. *J. Ecol.*, **66**: 849–880.

Lewis, W. M. 1984a. A five year record of temperature, mixing and stability for a tropical lake (Lake Valencia, Venezuela). *Arch. Hydrobiol.*, **99**: 340–346.

Lewis, W. M. 1984b. Temperature, heat and mixing in Lake Valencia, Venezuela. *Limnol. Oceanogr.*, **28**: 273–286.

Lewis, W. M., and Weibezahn, F. H. 1976. Chemistry, energy flow, and community structure in some Venezuelan freshwaters. *Arch. Hydrobiol. Suppl.*, **50**: 145–207.

Liao, I. C., and Chen. T. P. 1984. Status and prospects of tilapia culture in Taiwan. In *Proceedings of the International Symposium on Tilapia in Aquaculture*, Fishelson, L., and Yaron, A. (eds.), pp. 588–598. Israel: Tel Aviv University Press.

Liem, K. F. 1980. Adaptive significance of intra and interspecific differences in the feeding repertoires of cichlid fishes. *Am. Zool.*, **20**: 295–314.

Liem, K. F., and Kaufman, L. 1984. Intraspecific macroevolution: functional biology of the polymorphic cichlid species *Cichlasoma minckleyi*. In *Evolution of Fish Species Flocks*, Echelle, A. A., and Kornfield, I. (eds.), pp. 203–216. Orono: University of Maine Press.

Livingstone, D. A. 1963. Chemical composition of rivers and lakes. In *Data of Geochemistry*, Fleischer, M. (ed.). Washington, DC, Government Printing Office.

Löffler, H. 1964. The limnology of tropical high-mountain lakes. *Verh. Internat. Verein Limnol.*, **15**: 176–193.

Lotrich, V. A. 1973. Growth, production and community composition of fishes inhabiting a first, second and third order stream of eastern Kentucky. *Ecological Monographs,* **43**: 377–397.

Lowe-McConnell, R. H. 1956. Observations on the biology of *Tilapia* (Pisces-Cichlidae) in Lake Victoria, East Africa. *E. Afr. Fish. Res. Org. Suppl. Publ.,* No. 1: 1–72.

Lowe-McConnell, R. H. 1964. The fishes of the Rupununi district of British Guiana. Pt. 1. Groupings of fish species and effects of the seasonal cycles of the fish. *J. Linn. Soc. (Zool.),* **45**: 103–144.

Lowe-McConnell, R. H. 1969. Speciation in tropical freshwater fishes. *Biol. J. Linn. Soc.,* **1**: 51–75.

Lowe-McConnell, R. H. 1975. *Fish Communities in Tropical Freshwaters.* 337 pp. London: Longman.

Lowe-McConnell, R. H. 1977. On environmental stability and its effects on fish populations in tropical freshwaters. *Actas del IV Simposium Internacional de Ecologia Tropical, Panama,* Vol. II, pp. 695–710. Panama: Organizaciones Patrocinadones.

Lowe-McConnell, R. H. 1979. Ecological aspects of seasonality in fishes of tropical waters. *Symp. Zool. Soc. Lond.,* **44**: 219–241.

MacArthur, R. H. 1957. On the relative abundance of bird species. *Proc. Nat. Acad. Sci.,* **43**: 293–295.

MacArthur, R. H. 1960. On the relative abundance of species. *Amer. Nat.,* **98**: 387–397.

McGowan, L. M. 1974. Ecological studies on *Chaoborus* (Diptera, Chaoboridae) in Lake George, Uganda. *Freshwat. Biol.,* **4**: 483–505.

Maciolek, J. A. 1969. Freshwater lakes in Hawaii. *Verh. Internat. Verein Limnol.,* **17**: 386–391.

McKaye, K. R. 1977. Competition for breeding sites between the cichlid fishes of Lake Jiloa, Nicaragua. *Ecology,* **58**: 291–302.

McKaye, K. R., and Marsh, A. 1983. Food switching by two specialised algae-scraping cichlid fishes in Lake Malawi, Africa. *Oecologia,* **56**: 245–248.

McLachlan, A. J. 1974. Development of some lake ecosystems in tropical Africa with special reference to the invertebrates. *Biol. Rev.,* **49**: 365–397.

Mclaren, I. A. 1963. Effects of temperature on growth of zooplankton and the adaptive value of vertical migration. *J. Fish, Res. Bd. Canada,* **20**: 685–727.

Mann, K. H. 1975. Patterns of energy flow. In *River Ecology,* Whitton, B. A. (ed.), pp. 248–263. Oxford: Blackwell.

Margalef, R. 1968. *Perspectives in Ecological Theory.* University of Chicago Press, pp. 111.

Marker, A. F., and Jinks, S. 1982. The spectrophotometric analysis of chlorophyll *a* and phaeopigments in acetone, ethanol and methanol. *Arch. Hydrobiol. Beih.,* **16**: 3–17.

Marlier, G. 1967. Etudes sur les lacs de l'Amazonie centrale. *Cardernos de Amazonia, Manaus, INPA,* **5**: 1–51.

Marsh, A. C. and Ribbink, A. J. 1981. A comparison of the abilities of three sympatric species of *Petrotilapia* (Cichlidae, Lake Malawi) to penetrate deep water. *Env. Biol. Fish.,* **6**: 367–369.

Marshall, B. E. 1981. A review of the sardine fishery in Lake Kariba, 1973–1980. *Lake Kariba Fish. Res. Inst. Proj. Rep.,* **40**, 15 pp.

Marshall, B. E. 1984. Predicting ecology and fish yields in African reservoirs from preimpoundment physico-chemical data. *FAO/CIFA Tech. Pap./Doc. Tech. CPCA 12,* 36 pp.

Marshall, B. E., and Mandisodza, J. W. 1982. 1981 Fisheries statistics, Lake Kariba (Zimbabwe waters). *Lake Kariba Fish. Res. Inst. Proj. Rep.,* **44**, 31 pp.

Matthes, H. 1964. Les poissons du lac Tumba et de la région d'Ikela. Etude systématique et écologique. *Bull. aquatic. Biol.,* **3**: 1–15.

May, R. H. 1975. Stability in ecosystems: some comments. In *Unifying Concepts in Ecology,* van Doblen, W. H., and Lowe-McConnell, R. H. (eds.), pp. 161–168. The Hague: Junk.

Melack, J. M. 1976. Primary productivity and fish yields in tropical lakes. *Trans. Am. Fish Soc.*, **105**: 575–580.

Melack, J. M. 1979. Temporal variability of phytoplankton in tropical lakes. *Oecologia (Berl.)* **44**: 1–7.

Melack, J. M. 1980. An initial measurement of photosynthetic productivity in Lake Tanganyika. *Hydrobiologia*, **72**: 243–247.

Melack, J. M. 1981. Photosynthetic activity of phytoplankton in tropical African soda lakes. *Hydrobiologia*, **81**: 71–85.

Melack, J. M. 1982. Photosynthetic activity and respiration in an equatorial African soda lake. *Freshwat. Biol.*, **12**: 381–400.

Melack, J. M., and Fisher, T. R. 1983. Diel oxygen variations and their ecological implications in Amazon flood-plain lakes. *Arch. Hydrobiol.*, **98**: 422–442.

Melack, J. M., and Kilham, P. 1974. Photosynthetic rates of phytoplankton in East African alkaline, saline lakes. *Limnol. Oceanogr.*, **19**: 743–755.

Micha, J. C. 1976. Synthèse des essais de réproduction, d'alevinage et de production chez un silure Africain: *Clarias lazera. Symposium of Aquaculture in Africa*, pp. 450–473, FAO/CIFA Tech. Paper, **4** Suppl. 1. Rome: FAO.

Mizuno, T., Gose, K., Lim, R. P., and Furtado, J. I. 1982. Secondary production— benthos and attached animals. In *Tasek Bera, the Ecology of a Freshwater Swamp*, Furtado, J. I., and Mori, S. (eds.), pp. 286–320. The Hague: Junk.

Moreau, J. 1982. Le Lac Ilotry. lac plat hypersale (Madagascar). *Rev. Hydrobiol. trop.*, **15**: 71–80.

Moriarty, D. J. W. 1973. The physiology of digestion of blue-green algae in the cichlid fish. *Tilapia nilotica. J. Zool. Lond.*, **171**: 25–39.

Moriarty, D. J. W., Darlington, J. P. E. C., Dunn, I. G., Moriarty, C. M., and Tevlin, M. P. 1973. Feeding and grazing in Lake George, Uganda. *Proc. R. Soc. Lond.*, **B184**: 299–320.

Mortimer, C. H. 1941. The exchange of dissolved substances between mud and water in lakes: I and II. *J. Ecol.*, **29**: 280–329.

Mortimer, C. H. 1942. The exchange of dissolved substances between mud and water in lakes: III and IV. *J. Ecol.*, **30**: 147–201.

Moss, B. 1969. Limitation of algal growth in some central African waters. *Limnol. Oceanogr.*, **14**: 591–601.

Moss. B. 1973. Diversity in freshwater phytoplankton. *Am. Midl. Nat.*, **90**: 341–355.

Motwani, M. P., and Kanwai, Y. 1970. Fish and fisheries of the coffer-dammed right channel of the R. Niger at Kainji. In *Kainji Lake Studies*, Vol. 1, *Ecology*, Visser, S. A. (ed.), pp. 27–48. Nigerian Inst. Soc. Econ. Res., Ibadan University Press.

Moyle, P. B., and Senanayake, F. R. 1984. Resource partitioning among fishes of rainforest streams in Sri Lanka. *J. Zool. Lond.*, **202**: 195–224.

Munro, J. L. 1967. The food of a community of East African freshwater fishes. *J. Zool. Lond.*, **151**: 389–415.

Myers, G. S. 1951. Freshwater fishes and East Indian zoogeography. *Stanford Ichthyol. Bull.*, **4**: 11–21.

Myers, G. S. 1960. The endemic fish fauna of Lake Lanao, and the evolution of higher taxonomic categories. *Evolution*, **14**: 323–333.

Nogrady, T. 1983. Succession of planktonic rotifer populations in some lakes of the East African Rift Valley, Kenya. *Hydrobiologia*, **98**: 45–54.

Odum, M. T. 1956. Primary production in flowing waters. *Limnol. Oceanogr.*, **1**: 103–117.

Oglesby, R. T. 1977. Relationships of fish yield to lake phytoplankton standing crop, production and morphoedaphic factors. *J. Fish. Res. Bd. Canada*, **34**: 2271–2279.

Okemwa, E. N. 1984. Potential of Nile perch, *Lates niloticus* Linne (Pisces; Centropomidae) in Nyanza Gulf of Lake Victoria, East Africa. *Hydrobiologia*, **108**: 121–126.

Owens, M. 1969. Measurements of non-isolated natural communities—in running waters. In *A Manual on Methods for Measuring Primary Production in Aquatic Communities*, IBP Handbook 12, pp. 92–96. Oxford: Blackwell.

Pandian, T. J. 1967. Intake, digestion, absorption and conversion of food in the fishes *Megalops cyprinoides* and *Ophiocephalus striatus*. *Mar. Biol.*, **1**: 16–32.

Patrick, R. 1964. A discussion on natural and abnormal diatom communities. In, *Algae and Man*, Jackson, D. F. (ed.), pp. 185–204. New York, Plenum.

Pauly, D., and David, N. 1981. ELEFAN 1, a BASIC programme for the objective extraction of growth parameters from length frequency data. *Meeres forsch.*, **28**: 205–211.

Payne, A. I. 1971. An experiment on the culture of *Tilapia esculenta* Graham and *Tilapia zillii* (Gervais) in fish ponds. *J. Fish. Biol.*, **3**: 325–340.

Payne, A. I. 1975. The reproductive cycle, condition and feeding in *Barbus liberiensis*, a tropical stream-dwelling cyprinid. *J. Zool. Lond.*, **176**: 247–269.

Payne, A. I. 1976a. The determination of age and growth in *Barbus liberiensis* (Pisces, Cyprinidae). *J. Zool. Lond.*, **180**: 455–465.

Payne, A. I. 1976b. The exploitation of African fisheries. *Oikos*, **27**: 356–366.

Payne, A. I. 1978. Gut pH and digestive strategies in estuarine grey mullet (Mugilidae) and tilapia (Cichlidae). *J. Fish. Biol.*, **13**: 627–630.

Payne, A. I. 1979. Physiological and ecological factors in the development of fish culture. *Symp. Zool. Soc. Lond.*, **44**: 383–415.

Payne, A. I. 1984. Use of sewage waste in warm water aquaculture. *Symposium on the Reuse of Sewage Effluent*, pp. 117–131. Institute of Civil Engineers, London, Thomas Telford Press.

Payne A. I., and McCarton, B. 1985. Estimation of population parameters and their application to the development of fishery management models in two African rivers. *J. Fish. Biol.*, **27**: (suppl. A) in press.

People, W., and Rogoyska, M. 1969. The effect of the Volta River hydroelectric project on the salinity of the lower Volta River. *Ghana J. Science*, **8**: 9–20.

Peters, R. H., and MacIntyre, S. 1976. Orthophosphate turnover in East African lakes. *Oecologia (Berl.)*, **25**: 313–319.

Petr, T. 1969. The development of the bottom fauna in the man-made Volta Lake in Ghana. *Verh. Internat. Verein Limnol.*, **17**: 273–287.

Petr, T. 1970. Macro-invertebrates of flooded trees in the man-made Volta Lake (Ghana) with special reference to the burrowing mayfly *Povilla adusta* Navas. *Hydrobiologia*, **36**: 399–418.

Petr, T. 1976. Some chemical features of two Papuan freshwaters (Papua New Guinea). *Aust. J. Mar. Freshwater Res.*, **27**: 467–474.

Petrere, M. 1983. Yield per recruit of the Tambaqui, *Colossoma macropomum* Cuvier, in the Amazonas State, Brazil. *J. Fish. Biol.*, **22**: 133–144.

Pimm, S. L. 1984. The complexity and stability of ecosystems. *Nature*, **307**: 321–326.

Poll, M. 1959. Recherches sur la faune ichthyologique de la région du Stanley Pool. *Annls Mus. r. Congo Belge, Sci. Zool.*, **71**: 75–174.

Power, M. E. 1983. Grazing responses of tropical freshwater fishes to different scales of variation in their food. *Env. Biol. Fish*, **9**: 10.

Prowse, G. A. 1972. Some observations on primary and fish production in experimental fish ponds in Malacca, Malaysia. In *Productivity Problems of Freshwaters*, Kajack, Z., and Hillbricht-Ilkowska, A. (eds.), pp. 55–561. Warsaw and Krakow: Polish Scientific Publishers.

Prowse, G. A., and Talling, J. F. 1958. The seasonal growth and succession of plankton algae in the White Nile. *Limnol. Oceanogr.*, **3**: 222–238.

Rai, H. 1978. Distribution of carbon, chlorophyll-*a* and pheo-pigments in the black water lake ecosystem of Central Amazon Region. *Arch. Hydrobiol.* **82**: 74–87.

Rai, H. 1979. Microbiology of Central Amazon lakes. *Amazoniana*, **6**: 583–599.

Rai, H., and Hill, G. 1978. Bacteriological studies on Amazonas, Mississippi and Nile waters. *Arch. Hydrobiol.* **81**: 445–461.

Rai, H., and Hill, G. 1980. Classification of central Amazon lakes on the basis of their microbiological and physico-chemical characteristics. *Hydrobiologia,* **72**: 85–99.

Rai, H., and Hill, G. 1981. Observations on heterotrophic activity in Lago Janauari: a ria/varzea lake of central Amazonia. *Verh. Internat. Verein Limnol.,* **21**: 715–720.

Ramadan, F. M. 1982. Characterisation of Nile waters prior to the High Dam. *Wasser- und Abwasser-Forschung,* **1**: 21–24.

Rapoport, U., and Sarig, S. 1975. The results of tests in intensive growth of fish at the Genosar (Israel) station ponds in 1974. *Bamidgeh,* **27**: 75–82.

Ray, P., and David, A. 1966. Effects of industrial wastes and sewage upon the chemical and biological composition and fisheries of the River Ganga at Kampur (U.P.). *Environ. Health* **8**: 61–69.

Raymont, J. E. G. 1980. *Plankton and Productivity of the Oceans,* Vol. 1, *Phytoplankton.* 489 pp. Oxford: Pergamon.

Regier, H. A., and Wright, R. 1972. *Evaluation of Fisheries Resources in African Freshwaters Sympos. Eval. Fish Res.,* Fort Lamy, Chad, 1972. FAO, Rome CIFA/72/S.3 1–14.

Reiss, F. 1977. Qualitative and quantitative investigations on the macrobenthic fauna of central Amazon lakes. I. Lago Tupé, a blackwater lake on the lower Rio Negro. *Amazoniana,* **6**: 203–235.

Reynolds, C. S. 1982. Phytoplankton periodicity: its motivation, mechanisms and manipulation. *Freshwat. Biol. Assoc. Annual Report 1981–82,* pp. 60–75.

Reynolds, C. S., 1984. Phytoplankton periodicity: the interactions of form, function and environmental variability. *Freshwat. Biol.,* **14**: 111–142.

Reynolds, C. S., Tundisi, J. G., and Hino, K. 1983. Observations on a metalimnetic *Lyngbya* population in a stably stratified tropical lake (Lago Carioca, eastern Brazil). *Arch. Hydrobiol.,* **97**: 7–17.

Reynolds, C. S., and Walsby, E. A. 1975. Water-blooms. *Biol. Rev.,* **50**: 437–481.

Ribbink, A. J., and Lewis, D. S. C. 1982. *Melanochromis crabro* sp. nov.: a cichlid fish from Lake Malawi which feeds on ectoparasites and catfish eggs. *Neth. J. Zool.* **32**: 72–87.

Ribbink, A. J., Marsh, B. A., Marsh, A. C., Ribbink, A. C., and Sharp, B. J. 1983. A preliminary survey of the cichlid fishes of rocky habitats in Lake Malawi. *S. Afr. J. Zool.* **18**: 149–310.

Richardson, J. L., and Jin, L. T. 1975. Algal productivity of natural and artificially enriched freshwaters in Malaya. *Verh. Internat. Verein Limnol.,* **19**: 1383–1389.

Richey, J. E. 1981. Particulate and dissolved carbon in the Amazon river: a preliminary annual budget. *Verh. Internat. Verein Limnol.,* **21**: 914–917.

Rickleffs, R. E. 1973. *Ecology,* London: Nelson, 861 pp.

Ridder, M. de 1981. Some considerations on the geographical distribution of rotifers. *Hydrobiologia,* **85**: 209–225.

Robarts, R. D., and Southall, G. C. 1977. Nutrient limitation of phytoplankton growth in seven tropical man-made lakes, with special reference to lake McIlwain, Rhodesia. *Arch. Hydrobiol.,* **79**: 1–35.

Roberts, T. R. 1972. Ecology of fishes in the Amazon and Congo Basins. *Bull. Mus. Comp. Zool. Harv.,* **143**: 117–147.

Rogers, A. 1980. The consequences of river impoundment and dam construction on the wildlife resources of the Rafiji Basin. Department of Zoology, University of Dar-es-Salaam, 84 pp.

Room, P. M. 1983. 'Falling apart' as a lifestyle : the rhizome architecture and population growth of *Salvinia molesta. J. Ecol.,* **71**: 349–365.

Ruttner, F. 1931. Hydrographische und hydrochemische Beobachtungen auf Java, Sumatra und Bali. *Arch. Hydrobiol. Suppl.,* **8**: 197–454.

Ruttner, F. 1940. *Grundriss der Limnologie.*, Berlin: Walter de Gruyter, 167 pp.

Ruttner, F. 1952. Planktonstundien der Deutschen Limnologischen Sunda-Expedition. *Arch. Hydrobiol. Suppl.,* **21**: 1–274.

Ryder, R. A. 1965. A method for estimating the potential fish production of north temperate lakes. *Trans. Am. Fish. Soc.,* **94**: 214–218.

Ryder, R. A., and Henderson, H. F. 1975. Estimates of potential fish yield for the Nasser Reservoir, Arab Rep. of Egypt. *J. Fish. Res. Bd. Can.* **32**: 2137–2157.

Ryder, R. A., Kerr, S. R., Loftus, K. M., and Regier, H. A. 1974. The morphoedaphic index, a fish yield estimator—review and evaluation. *J. Fish. Res. Bd. Canada,* **31**: 663–688.

Sage, R. D., and Selander, R. K. 1975. Trophic radiation through polymorphism in cichlid fishes. *Proc. Nat. Acad. Sci.,* **72**: 4669–4673.

Sale, P. F. 1977. Maintenance of high diversity in coral reef fish communities. *Am. Nat.,* **111**: 337–359.

Sale, P. F. 1978. Co-existence of coral reef fishes—a lottery for living space. *Environ. Biol. Fish.,* **3**: 85–102.

Sale, P. F. 1982. Stock-recruitment relationships and regional co-existence in a lottery competitive system: a simulation study. *Am. Nat.,* **120**: 139–159.

Samman, J., and Pugh-Thomas, M. 1978. Effect of an organophosphorus insecticide, Abate, used in control of *Simulium damnosum* on non-target benthic fauna. *Int. J. Environ. Stud.,* **12**: 141–144.

Sato, O., Tezuka Y., and Joyama, T. 1982. Decomposition: origin of black water. In *Tasek Bera, the Ecology of a Freshwater Swamp,* Furtado, J. I., and Mori, S. (eds.), Monographia Biologicae Vol. 47, pp. 170–190. The Hague: Junk.

Schlosser, I. J. 1982. Fish community structure and function along two habitat gradients in a headwater stream. *Ecol. monogr.,* **52**: 395–414.

Schmidt, G. W. 1973a. Primary production of phytoplankton in the three types of Amazonian waters. II. The limnology of a tropical flood-plain lake in Central Amazonia (Lago do Castanho). *Amazoniana,* **4**: 135–138.

Schmidt, G. W. 1973b. Primary production of phytoplankton in the three types of Amazonian waters. III. Primary productivity of phytoplankton in a tropical flood-plain lake of Central Amazonia, Lago do Castanho, Amazonas, Brazil. *Amazoniana,* **4**: 379–404.

Schmidt, G. W. 1976. Primary production of phytoplankton in three types of Amazonian waters. IV. On the primary productivity of phytoplankton in a bay of the Rio Negro (Amazonas, Brazil). *Amazoniana,* **5**: 517–528.

Schmidt, G. W. 1982. Primary production of phytoplankton in three types of Amzon waters. V. Some investigations on the phytoplankton and its primary productivity in the clear water of the lower Rio Tapajoz (Para, Brazil). *Amazoniana,* **7**: 335–348.

Schroeder, G. 1975. Some effects of stocking fish in waste treatment ponds. *Water Res.,* **9**: 591–593.

Schroeder, G., and Hepher, B. 1979. Use of agricultural and urban wastes in fish culture. In *Advances in Aquaculture,* Pillay, T. V. R. and Dill, W. A. (eds.), pp. 478–486. Farnham: Fishing News Books.

Serruya, C., and Pollingher, U. 1983. Lakes of the Warm Belt. Cambridge: CUP.

Shannon, C. E., and Weaver, W. 1949. *The Mathematical Theory of Communication.* Urbana: Univ. of Illinois Press.

Sharma, K. P. 1983. Multipurpose use of water resources in relation to inland fisheries in India. *FAO Fish. Rep.,* **288**: 150–166.

Sharma, K. P., and Pradnan, V. N. 1983. Study on growth and biomass of underground organs of *Typha angustata*. Bory & Chaub. *Hydrobiologia,* **99**: 89–94.

Siddiqui, A. Q. 1977. Reproductive biology, length/weight relationship and relative condition of *Tilapia leucosticta* (Trewavas) in Lake Naivasha, Kenya. *J. Fish. Biol.* **10**: 251–260.

286

Sinada, F., and Kerin, A. G. A. 1984. Primary production and respiration of the phytoplankton in the Blue and White Niles at Khartoum. *Hydrobiologia*, **110**: 57–59.
Sioli, H. 1964. General features of the limnology of Amazonia. *Verh. Internat. Verein theor. angew. Limnol.*, **15**: 1053–1058.
Sioli, H. 1968. Principal biotopes of primary production in the waters of Amazonia. *Proc. symp. Recent Adv. Trop. Ecol., Varanasi, India*, pp. 591–600.
Sioli, H. 1975. Tropical river: the Amazon. In *River Ecology*, Whitton, B. A. (ed.), pp. 461–488. Oxford: Blackwell.
Sioli, H., Shwabe, G. H., and Klinge, H. 1969. Limnological outlooks on landscape-ecology in Latin America. *Trop. Ecol.*, **10**: 72–82.
Slobodkin, L. B., and Sanders, H. L. 1969. On the contribution of environmental predictability to species diversity. In *Brookhaven Symposium* **12**: 82–95.
Smith, I. R. 1983. Management of inland fisheries and some corrective measures. *FAO Fish. Rep.*, **288**: 88–100.
Soeder, C. J., and Talling, J. F. 1969. The enclosure of phytoplankton communities. In *A Manual on Methods for Measuring Primary Production in Aquatic Communities*, IBP Handbook 12, pp. 62–69. Oxford: Blackwell.
Starmuhlner, F., and Therezien, Y. 1982. Résultats de la mission hydrobiologique austro-française de 1979 aux iles de la Guadeloupe de la Dominique et de la Martinique (Petites Antilles). II. Etude général de la Dominique et de la Martinique. *Rev. Hydrobiol. trop.*, **15**: 325–346.
Steinitz-Kannan, M., Colinvaux, P. A., and Kannan, R. 1983. Limnological studies in Ecuador: 1. A survey of chemical and physical properties of Ecuadorian lakes. *Arch. Hydrobiol.* **65** (suppl.): 61–105.
Strickland, J. D. H. 1960. Measuring the production of marine phytoplankton. *Bull. Fish. Res. Bd. Canada*, **122**: 1–172.
Sydenham, D. H. J. 1975. Observations on the fish populations of a Nigerian forest stream. *Rev. Zool. afr.*, **89**: 257–272.
Talling, J. F. 1957. Diurnal changes of stratification and photosynthesis in some tropical African waters. *Proc. R. Soc. Lond.*, **B147**: 57–83.
Talling, J. F. 1965. The photosynthetic activity of phytoplankton in East African lakes. *Int. Rev. ges. Hydrobiol.*, **50**: 1–32.
Talling, J. F. 1966. The annual cycle of stratification and phytoplankton growth in Lake Victoria (East Africa). *Int. Rev. ges. Hydrobiol.*, **51**: 545–621.
Talling, J. F. 1969. The incidence of vertical mixing, and some biological and chemical consequences, in tropical African lakes. *Verh. Internat. Verein Limnol.*, **17**: 998–1012.
Talling, J. F. 1976a. Phytoplankton: composition, development and productivity. In *The Nile, Biology of an Ancient River*, Rzoska, J. (ed.), pp. 385–402. The Hague: Junk.
Talling, J. F. 1976b. Water characteristics. In *The Nile, Biology of an Ancient River*. Rzoska, J., (ed.), pp. 357–384. The Hague: Junk.
Talling, J. F. 1982. Utilization of solar radiation by phytoplankton. In *Trends in Photobiology*, Helene, C., Charlier, M., Montenay-Garestier, T., and Lanstriat, G. (eds.), pp. 619–631. New York: Plenum.
Talling, J. F., and Driver, D. 1963. Some problems in the estimation of chlorophyll-*a* in phytoplankton. *Proceedings, Conference of Primary Productivity Measurement, Marine and Freshwater, Hawaii 1961*. US Atomic Energy Comm., TID-7633, pp. 142–146.
Talling, J. F., and Rzoska, J. 1967. The development of plankton in relation to hydrological regime in the Blue Nile. *J. Ecol.*, **55**: 637–662.
Talling, J. F., and Talling, I. B. 1965. The chemical composition of African lake waters. *Int. Rev. ges. Hydrobiol.*, **50**: 421–463.
Talling, J. F., Wood, R. B., Prosser, M. V., and Baxter, R. M. 1973. The upper limit of photosynthetic productivity by phytoplankton; evidence from Ethiopian soda lakes. *Freshwat. Biol.*, **3**: 57–76.
Tamayo-Zafarwalla, M. 1983. Conflicts between fisheries and mining, industrialization and pollution of inland waters in the Philippines. *FAO Fish. Rep.*, **288**: 150–166.

Tapiador, D. D., Henderson, H. F., Delmondo, M. N., and Tsusui, H. 1977. Freshwater fisheries and aquaculture in China. *FAO Fish techn. pap.*, No. 168: 1–84.

Tesch, F. W. 1968. Age and growth. In *Methods for Assessment of Fish Production in Freshwaters*, IBP Handbook 3, pp. 93–123, Oxford: Blackwell.

Thompson, K. 1976. The primary productivity of African wetlands with particular reference to the Okavango delta. In *Proc. Symp. Okavango Delta and its Future Utilisation*, pp. 67–79. Gaborone, Botswana.

Thompson, K., Shewry, P. R., and Woolhouse, H. W. 1979. Papyrus swamp development in the Upemba Basin, Zaire: studies of population structure in *Cyperus papyrus* stands. *Bot. J. Linn. Soc.*, **78**: 299–316.

Thorson, G. 1957. Bottom communities. In *Treatise in Marine Ecology and Paleoecology*, Vol. 1, Geol. Soc. Amer. Memoir., **67**: 461–534.

Toews, D. R., and Griffith, J. S. 1979. Empirical estimates of potential yield for the Lake Bangweulu system, Zambia, Central Africa. *Trans. Am Fish. Soc.*, **108**: 241–252.

Trewavas, E. 1983. *Tilapiine Fishes of the Genera Sarotherodon, Oreochromis and Danakilia*. London: British Museum (Nat. Hist.), 583 pp.

Trewavas, E., Green, J., and Corbet, S. A. 1972. Ecological studies on crater lakes in West Cameroon. Fishes of Barombi Mbo. *J. Zool. Lond.*, **166**: 15–30.

Tundisi, J. G., Forsberg, B. R., Devol, A. M., Zaret, T. M., Tundisi, T. M., Dos Santos, A., Zibeiro, J. S., and Hardy, E. R. 1984. Mixing patterns in Amazon lakes. *Hydrobiologia*, **108**: 3–15.

Turcotte, P., and Harper, P. P. 1982a. The macro-invertebrate fauna of a small Andean stream. *Freshwat. Biol.*, **12**: 411–419.

Turcotte, P., and Harper, P. ·P. 1982b. Drift pattern in a high Andean stream. *Hydrobiologia*, **89**: 141–152.

Turner, J. L. 1981. Changes in multispecies fisheries when many species are caught at the same time. *FAO/CIFA Tech. Paper*, **8**: 202–211.

Uherkovich, G. 1976. Algen aus den Flussen Rio Negro und Rio Tapajoz. *Amazoniana*, **5**: 465–515.

Uherkovich, G. 1981. Algen aus einigen Gewassern Amazoniens. *Amazoniana* **7**: 191–219.

Ulfstrand, S. 1967. Microdistribution of benthic species (Ephemeroptera, Plecoptera, Trichoptera, Diptera: Simulidae) in Lapland streams. *Oikos*, **18**: 293–310.

Vaas, K. F., Sachlan, M., and Wiraatmadja, G. 1953. On the ecology and fisheries of some inland waters along the Rivers Ogan and Komering in S.E. Sumatra. *Contr. Inl. Fish. Res. Sta. Bogor, Indonesia*, **3**: 1–32.

Van Oijen, M. J. P., Witte, F. and Witte-Maas, E. L. M. 1977. Interim report from the Haplochromis Ecology Survey Team (HEST). Morph. Dept. Zool. Lab., State Univ. Leiden, Netherlands, pp. 1–51.

Vareschi, E. 1979. The ecology of Lake Nakuru (Kenya). II. Biomass and spatial distribution of fish. *Oecologia*, **37**: 321–335.

Vareschi, E. 1982. The ecology of Lake Nakuru (Kenya). III. Abiotic factors and primary production. *Oecologia*, **55**: 81–101.

Vareschi, E., and Jacobs, J. 1984, The ecology of Lake Nakuru (Kenya). V. Production and consumption of consumer organisms. *Oecologia*, **61**: 83–98.

Vareschi, E., and Vareschi, A. 1984. The ecology of Lake Nakuru (Kenya). VI. Biomass and distribution of consumer organisms. *Oecologia*, **61**: 70–82.

Venkateswarlu, V. 1969. An ecological study of the algae of the River Mosi, Hyderbad (India) with special reference to water pollution. II. Factors influencing the distribution of algae. *Hydrobiologia*, **33**: 352–363.

Verbeke, J. 1957. Recherches ichtyologiques sur la faune des grands lacs de l'est du Congo Belge. *Result. Sci. Explor. Hydrobiol. Lacs Kivu, Eduard Albert (1952–54)*, **3**: 182–201.

288

Viner, A. B. 1973. Response of a tropical mixed phytoplankton population to nutrient enrichments of ammonia and phosphate, and some ecological implications. *Proc. R. Soc. Lond.,* **B183**: 351–370.

Viner, A. B. 1975a. The supply of minerals to tropical rivers and lakes (Uganda). In *An Introduction to Land–Water Relationships,* Olson, G. (ed.), pp. 227–261. New York: Springer-Verlag.

Viner, A. B. 1975b. Sediments of Lake George (Uganda). II: Release of ammonia and phosphate from an undisturbed mud surface. *Arch. Hydrobiol.,* **76**: 368–378.

Viner, A. B. 1982. A quantitative assessment of the nutrient phosphate transported by particles in a tropical river. *Rev. Hydrobiol. trop.,* **15**: 3–8.

Viner, A. B., and Smith, I. R. 1973. Geographical, historical and physical aspects of Lake George. *Proc. R. Soc. Lond.,* **B184**: 235–270.

Visser, S. A. 1961. Chemical composition of rainwater in Kampala, Uganda and its relation to meteorology and topographical conditions. *J. Geophys. Res.,* **66**: 3759–3765.

Visser, S. A. 1974. Composition of waters of lakes and rivers in East and West Africa. *Afr. J. Trop. Hydrobiol. Fish.,* **3**: 43–60.

Vollenweider, R. A. 1969. *A Manual on Methods for Measuring Primary Production in Aquatic Environments,* IBP Handbook 12. Oxford: Blackwell.

Walsby, A. E., and Reynolds, C. S. 1980. Sinking and floating. In *The Physiological Ecology of Phytoplankton,* Morris, I. (ed.), pp. 371–412. Oxford: Blackwell.

Watanabe, J., Kumano, S., and Mizuno, T. 1982. Population density and standing crop—epiphytic and benthic algae. In *Tasek Bera, the Ecology of a Freshwater Swamp,* Furtado, J. I., and Mori, S. (eds.), pp. 238–240. The Hague: Junk.

Watt, K. E. F. 1972. *Principles of Resource Management.* New York: McGraw-Hill.

Weers, E. T., and Zaret, T. M. 1975. Grazing effects on nannoplankton in Gatun Lake, Panama. *Verh. Internat. Verein Limnol.,* **19**: 1480–1483.

Welcomme, R. L. 1967. Observations on the biology of the intròduced species of *Tilapia* in Lake Victoria. *Revue Zool. Bot. afr.,* **76**: 249–279.

Welcomme, R. L. 1971. The toxicity of four insecticides and a herbicide under tropical conditions. *Afr. J. Trop. Hydrobiol. Fish,* **1**: 107–114.·

Welcomme, R. L. 1976. Some general and theoretical considerations on the fish yield of African rivers. *J. Fish Biol.,* **8**: 351–364.

Welcomme, R. L. 1979. *Fisheries Ecology of Floodplain Rivers.* London: Longman, 317 pp.

Welcomme, R. L., and Hagborg, D. 1977. Towards a model of a floodplain fish population and its fishery. *Environ. Biol. Fish.,* **2**: 7–24.

Westlake, D. F. 1969. Sampling techniques and methods for estimating quantity and quality of biomass—macrophytes. In *A Manual on Methods for Measuring Primary Production in Aquatic Communities,* IBP Handbook 12, pp. 25–32. Oxford: Blackwell.

Westlake, D. F. 1981. The development and structure of aquatic weed populations. *Proceedings, Aquatic Weeds and their Control,* pp. 33–47.

Westlake, D. F. 1982. The primary production of water plants. In *Studies on Aquatic Vascular Plants,* Symoens, J. J., Hooper, S. S., and Compere, P. (eds.), Brussels: Royal Botanical Society of Belgium.

Whitehead, P. J. 1959. The anadromous fishes of Lake Victoria. *Rev. Zool. Bot. afr.* **59**: 329–363.

Whyte, S. A. 1975. Distribution, trophic relationships and breeding habits of the fish populations in a tropical lake basin (Lake Bosumtwi, Ghana). *J. Zool. Lond.,* **177**: 25–56.

Widmer, C., Kittel, T., and Richerson, P. J. 1975. *Verh. Internat. Verein Limnol.,* **19**: 1504–1510.

Wiegert, R. G., and Owen, D. F. 1971. Trophic structure, available resources and population density in terrestrial vs. aquatic ecosystmes. *J. Theor. Biol.,* **30**: 69–87.

Williams, P. J. B., Heinemann, K. R., Marra, J., and Purdic, D. A. 1983. Comparison of ^{14}C and O_2 measurements of phytoplankton production in oligotrophic waters. *Nature*, **305**: 49–50.

Winberg, G. G. 1956. *Rate of Metabolism and Food Requirements of Fishes*. Translated from the Russian by J. Fish. Bd. Canada trans. ser. (1960), 253 pp.

Witte, F. 1981. Initial results of the ecological survey of the haplochromine cichlid fishes from the Mwanza Gulf of Lake Victoria (Tanzania): breeding patterns, trophic and species distribution. *Neth. J. Zool.*, **31**: 175–202.

Worthington, E. B. 1984. *The Ecological Century*. Oxford: OUP. 206 pp.

Worthington, E. B., and Worthington, S. 1933. *Inland Waters of Africa*. London: Macmillan. 259 pp.

Wright, R. 1982. Seasonal variation in water quality of a West African river (R. Jong in Sierra Leone). *Rev. Hydrobiol. trop.*, **15**: 193–199.

Wurtsbaugh, W. A., Vincent, W. F., Alfaro Tapia, R., Vincent C. L., and Richerson, P. J. 1985. Nutrient limitation of algal growth and N-fixation in a tropical alpine lake, Lake Titicaca (Peru/Bolivia). *Freshwat. Biol.*, **15**: 185–195.

Wyngaard, G. A. 1983. 'In situ' life table of sub tropical copepods. *Freshwat. Biol.*, **13**: 275–282.

Zaret, T. M. 1975. Strategies for existence of zooplankton prey in homogenous environments. *Verh. Internat. Verein Limnol.*, **19**: 1480–1483.

Zaret, T. M., Devol, A. H., and Dos Santos, A. 1981. Nutrient addition experiments in Lago Jacartinga, central Amazonia, Brazil. *Verh. Internat. Verein Limnol.*, **21**: 721–724.

Zaret, T. M., and Paine, R. T. 1973. Species introduction in a tropical lake. *Science*, **182**: 449–455.

Zaret, T. M., and Rand, A. J. 1971. Competition in tropical stream fishes: support for the competitive exclusion principle. *Ecology*, **52**: 336–342.

Zaret, T. M., and Suffern, J. S. 1976. Vertical migration of zooplankton as a predator avoidance mechanism. *Limnol. Oceanogr.*, **21**: 804–813.

Index

Pages in italics show halftones

292

294